Eternity

OXFORD **PHILOSOPHICAL** CONCEPTS

OXFORD PHILOSOPHICAL CONCEPTS

Christia Mercer, Columbia University

Series Editor

PUBLISHED

Efficient Causation
Edited by Tad Schmaltz

Sympathy
Edited by Eric Schliesser

The Faculties
Edited by Dominik Perler

Memory
Edited by Dmitri Nikulin

Eternity
Edited by Yitzhak Y. Melamed

FORTHCOMING

Health
Edited by Peter Adamson

Evil
Edited by Andrew Chignell

Dignity
Edited by Remy Debes

Animals
Edited by G. Fay Edwards
and Peter Adamson

Space
Edited by Andrew Janiak

Self-Knowledge
Edited by Ursula Renz

Pleasure
Edited by Lisa Shapiro

Consciousness
Edited by Alison Simmons

Moral Motivation
Edited by Iakovos Vasiliou

OXFORD PHILOSOPHICAL CONCEPTS

Eternity
A HISTORY

Edited by Yitzhak Y. Melamed

Oxford University Press is a department of the University of Oxford. It furthers the University's objective of excellence in research, scholarship, and education by publishing worldwide. Oxford is a registered trade mark of Oxford University Press in the UK and certain other countries.

Published in the United States of America by Oxford University Press
198 Madison Avenue, New York, NY 10016, United States of America.

© Oxford University Press 2016

All rights reserved. No part of this publication may be reproduced, stored in a retrieval system, or transmitted, in any form or by any means, without the prior permission in writing of Oxford University Press, or as expressly permitted by law, by license, or under terms agreed with the appropriate reproduction rights organization. Inquiries concerning reproduction outside the scope of the above should be sent to the Rights Department, Oxford University Press, at the address above.

You must not circulate this work in any other form
and you must impose this same condition on any acquirer.

Library of congress cataloging-in-publication data
Names: Melamed, Yitzhak Y., 1968– editor.
Title: Eternity: a history / edited by Yitzhak Y. Melamed.
Description: New York, NY: Oxford University Press, 2016. |
Series: Oxford philosophical concepts series |
Includes bibliographical references and index.
Identifiers: LCCN 2015037326 | ISBN 978-0-19-978186-7 (pbk.: alk. paper) |
ISBN 978-0-19-978187-4 (hardcover: alk. paper)
Subjects: LCSH: Infinite. | Eternity.
Classification: LCC BD411.E84 2016 |
DDC 115—dc23
LC record available at http://lccn.loc.gov/2015037326

"The idea of eternity being the greatest, the most terrible and most frightening of all those that astonish the mind and strike the imagination, it is necessarily accompanied by a large following of ancillary ideas, which all have a considerable effect upon the mind because of their relation to this great and terrible idea of eternity."
—NICOLAS MALEBRANCHE, *Search after Truth* 327 (translated by Thomas M. Lennon and Paul J. Olscamp)

"When the mid-point of eternity comes, it can be said of God that half of his life passed."
—G. W. LEIBNIZ, *De Summa Rerum* 9 (translated by G.H.R. Parkinson)

"אדון עולם אשר מלך בטרם כל יציר נברא
לעת נעשה בחפצו כל אזי מלך שמו נקרא
ואחרי ככלות הכל לבדו ימלוך נורא"

—A medieval liturgical poem, commonly attributed to Salomon Ibn-Gabirol

For Alfa in eternal love

Contents

ACKNOWLEDGMENTS X

CONTRIBUTORS XI

SERIES EDITOR'S FOREWORD XV

Introduction 1
YITZHAK Y. MELAMED

1 Eternity in Ancient Philosophy 14
JAMES G. WILBERDING

Reflection: Eternity and Astrology in the Work of Vettius Valens 56
DORIAN GIESELER GREENBAUM

Reflection: Eternity and World-Cycles 64
GODEFROID DE CALLATAŸ

Reflection: The Eternality of Language in India 70
ANDREW OLLETT

2 Eternity in Medieval Philosophy 75
PETER ADAMSON

Reflection: "Eternalists" and "Materialists" in Islam: A Note on the *Dahriyya* 117
HINRICH BIESTERFELDT

Reflection: Eternity and the Trinity 124
CHRISTOPHE ERISMANN

3 Eternity in Early Modern Philosophy 129
YITZHAK Y. MELAMED

Reflection: Out of Time: Dante and Eternity 168
AKASH KUMAR

Reflection: Perpetuum Mobiles and Eternity 173
MARIUS STAN

4 Eternity in Kant and Post-Kantian European Thought 179
ALISTAIR WELCHMAN

Reflection: Eternity in Early German Romanticism 226
JUDITH NORMAN

Reflection: Eternity in Hasidism: Time and Presence 231
ARIEL EVAN MAYSE

Reflection: On White Eternity in the Poetry of Mahmoud Darwish 239
ABED AZZAM

5 Eternity in Twentieth-Century Analytic Philosophy 245
KRIS MCDANIEL

Reflection: Borges on Eternity 277
WILLIAM EGGINTON

Reflection: Music and Eternity 283
WALTER FRISCH

Reflection: The Kaddish 290
ABRAHAM P. SOCHER

GLOSSARY 297

BIBLIOGRAPHY 301

INDEX 323

Acknowledgments

I would like to thank Christia Mercer, Peter Ohlin, Lucy Randall, and Stephen Menn whose advise and support contributed tremendously to the design of this book.

Yitzhak Y. Melamed
Pikesville, February 2016

Contributors

PETER ADAMSON is professor of late ancient and Arabic philosophy at the Ludwig-Maximilians-Universität in Munich. He has authored monographs on the Arabic Plotinus and al-Kindi, whose philosophical works he has translated (with P. E. Pormann). He has edited or coedited numerous books, including *The Cambridge Companion to Arabic Philosophy* and *Interpreting Avicenna: Critical Essays*. Two collections of his articles are now available from Ashgate's Variorum series.

ABED AZZAM teaches philosophy at the University of Potsdam and the University of Marburg. He is the author of *Nietzsche versus Paul* (Columbia University Press, 2015).

HINRICH BIESTERFELDT is a retired professor of Arabic and Islamic studies at the Ruhr Universität, Bochum. He has published on classical Arabic literature and the history of medicine and philosophy in Islam.

GODEFROID DE CALLATAŸ is professor of Arabic and Islamic studies at the Oriental Institute of the University of Louvain. He has specialized in the history of Arabic sciences and philosophy and the role played by Islam in the transmission of knowledge from Greek antiquity to the Latin west during the Middle Ages. Among other subjects, he has published extensively on the encyclopedic corpus known as *Rasā'il Ikhwān al-Ṣafā'* (*Epistles of the Brethren of Purity*). Since 2012, he has directed Speculum Arabicum, a project on comparative medieval encyclopedism at the University of Louvain.

CHRISTOPHE ERISMANN held positions at the Universities of Cambridge, Helsinki, and Lausanne, where he taught medieval philosophy for several years as Swiss National Science Foundation Professor. Since autumn 2015, he leads the European Research Council project (CoG 648298) "Reassessing Ninth Century Philosophy. A Syncronic Approach to the Logical Tradition" at the Institute for Byzantine Studies, University of Vienna. His research focuses on the reception of Greek logic (mainly Aristotle's *Categories* and Porphyry's *Isagoge*) in late ancient, Patristic, and early medieval philosophy. He has published on the problem of universals, individuality, causality, and relation. He is the author of *L'homme commun: la genèse du réalisme ontologique durant le haut Moyen Âge* (Paris 2011).

WALTER FRISCH is H. Harold Gumm/Harry and Albert von Tilzer Professor of Music at Columbia University, where he has taught since 1982. He has published on music of the Austro-German sphere in the nineteenth and twentieth centuries, especially Brahms and Schoenberg. He is the author of *Music in the Nineteenth Century* (Norton, 2012).

WILLIAM EGGINTON is Andrew W. Mellon Professor in the Humanities at Johns Hopkins University. His most recent book is *The Man Who Invented Fiction: How Cervantes Ushered in the Modern World*.

DORIAN GIESELER Greenbaum is a tutor at the University of Wales Trinity St David. She received her Ph.D. from the Warburg Institute in 2009. Her areas of interest are the history of astrology and the concept of the daimon in antiquity. Her book *The Daimon in Hellenistic Astrology: Origins and Influence* has just been published by Brill (November 2015).

AKASH KUMAR is a Core Lecturer in Italian at Columbia University. His research focuses on the importing of science and philosophy in the early Italian lyric and on the broad implications of intercultural mingling in the Italian Middle Ages. He is currently working on his first book, *Dante's Elements: Translation and Natural Philosophy from Giacomo da Lentini to the Commedia*.

ARIEL EVAN MAYSE holds a Ph.D. in Jewish studies from Harvard University. In addition to having published a number of popular and scholarly articles on Kabbalah and Hasidism, he is a coeditor of the two-volume collection *Speaking Torah: Spiritual Teachings from Around the Maggid's Table* (Jewish Lights, 2013) and editor of the recent *From the Depth of the Well: An Anthology of Jewish*

Mysticism (Paulist Press, 2014). He is currently a research fellow at the Frankel Institute for Advanced Judaic Studies at the University of Michigan, Ann Arbor.

KRIS MCDANIEL is a professor at Syracuse University. He works in metaphysics, history of philosophy, and ethics.

YITZHAK Y. MELAMED is a professor in the Department of Philosophy at Johns Hopkins University. He is the author of *Spinoza's Metaphysics: Substance and Thought* (Oxford UP 2013) and of numerous studies in Modern Philosophy and German Idealism. He has been awarded the Fulbright, Mellon, and American Academy for Jewish Research Fellowships. Recently, he has also won the ACLS Burkhardt (2011), NEH (2010), and Humboldt (2011) fellowships for his forthcoming book on Spinoza and German Idealism.

JUDITH NORMAN is a professor of philosophy at Trinity University in San Antonio, Texas. She publishes on nineteenth-century philosophy and has translated works by Schopenhauer and Nietzsche. She is currently coediting (with Elizabeth Millán) *A Companion to German Romantic Philosophy*.

ANDREW OLLETT is a junior fellow at the Harvard Society of Fellows. He works on the literary and intellectual traditions of premodern India, with a focus on theories, concepts, and practices of language.

ABRAHAM SOCHER is a professor of religion and director of Jewish studies at Oberlin College. He is the author of *The Radical Enlightenment of Solomon Maimon: Judaism, Heresy and Philosophy* (Stanford, 2006) and the editor of the *Jewish Review of Books*.

MARIUS STAN is an assistant professor of philosophy at Boston College. His research is on the history of modern natural philosophy with a special interest in Leibniz, Newton, and Kant.

ALISTAIR WELCHMAN is associate professor of philosophy at the University of Texas at San Antonio. He has written widely on nineteenth-century German philosophy and contemporary French thought and is currently working on a book about Deleuze and German idealism.

JAMES G. WILBERDING is Professor of Ancient and Medieval Philosophy at the Ruhr Universität, Bochum. He has published widely on ancient philosophy, especially on Plato and the Platonic tradition.

Series Editor's Foreword

Oxford Philosophical Concepts (OPC) offers an innovative approach to philosophy's past and its relation to other disciplines. As a series, it is unique in exploring the transformations of central philosophical concepts from their ancient sources to their modern use.

OPC has several goals: to make it easier for historians to contextualize key concepts in the history of philosophy, to render that history accessible to a wide audience and to enliven contemporary discussions by displaying the rich and varied sources of philosophical concepts still in use today. The means to these goals are simple enough: eminent scholars come together to rethink a central concept in philosophy's past. The point of this rethinking is not to offer a broad over-view, but to identify problems the concept was originally supposed to solve and investigate how approaches to them shifted over time, sometimes radically.

Recent scholarship has made evident the benefits of reexamining the standard narratives about western philosophy. OPC's editors look beyond the canon and explore their concepts over a wide philosophical landscape. Each volume traces a notion from its inception as a solution to specific problems through its historical transformations to its modern use, all the while acknowledging its historical context. Each OPC volume is *a history* of its concept in that it tells a story about changing solutions to specific problems. Many editors have found it appropriate to include long-ignored writings drawn from the Islamic and Jewish traditions and the philosophical contributions of women. Volumes also explore ideas drawn from Buddhist, Chinese, Indian, and other philosophical cultures when doing so adds an especially helpful new perspective. By combining scholarly innovation with focused and astute analysis, OPC encourages a deeper understanding of our philosophical past and present.

One of the most innovative features of *Oxford Philosophical Concepts* is its recognition that philosophy bears a rich relation to art, music, literature, religion, science, and other cultural practices. The series speaks to the need for informed interdisciplinary exchanges. Its editors assume that the most difficult and profound philosophical ideas can be made comprehensible to a large audience and that materials not strictly philosophical often bear a significant relevance to philosophy. To this end, each OPC volume includes Reflections. These are short [no comma here] stand-alone essays written by specialists in art, music, literature, theology, science, or cultural studies that *reflect on* the concept from other disciplinary perspectives. The goal of these essays is to enliven, enrich, and exemplify the volume's concept and reconsider the boundary between philosophical and extra-philosophical materials. OPC's Reflections display the benefits of using philosophical concepts and distinctions in areas that are not strictly philosophical, and encourage philosophers to move beyond the borders of their discipline as presently conceived.

The volumes of OPC arrive at an auspicious moment. Many philosophers are keen to invigorate the discipline. OPC aims to provoke philosophical imaginations by uncovering the brilliant twists and unforeseen turns of philosophy's past.

Christia Mercer
Gustave M. Berne, Professor of Philosophy
Columbia University in the City of New York
June 2015

Eternity

OXFORD PHILOSOPHICAL CONCEPTS

Introduction

Yitzhak Y. Melamed

1 ETERNITY

Eternity is a unique kind of *existence* that is supposed to belong to the most real being or beings. It is an existence that is not shaken by the common wear and tear of time. Over the two-and-a-half-millennia history of western philosophy we find various conceptions of eternity, yet one sharp distinction between two notions of eternity seems to run throughout this long history: eternity as *timeless* existence, as opposed to eternity as existence in *all times*. Both kinds of existence stand in sharp contrast to the coming in and out of existence of ordinary beings, like hippos, humans, and toothbrushes: were these eternally timeless, for example, a hippo could not eat, a human could not think or laugh, and a toothbrush would be of no use. Were a hippo an eternal-everlasting creature, it would not have to bother itself with nutrition in order to extend its existence. Everlasting human beings might appear similar to us, but their mental life and patterns of behavior would most likely be very different from ours.

The distinction between eternity as timelessness and eternity as everlastingness goes back to ancient philosophy, to the works of Plato and Aristotle, and even to the fragments of Parmenides's philosophical poem. In the twentieth century the concept of eternity seemed to go out of favor, though one could consider to be eternalists those proponents of realism in philosophy of mathematics, and those of timeless propositions in philosophy of language (i.e., propositions that are said to exist independently of the uttered sentences that convey their thought-content). However, recent developments in contemporary physics and its philosophy have provided an impetus to revive notions of eternity, due to the view that time and duration might have no place in the most fundamental ontology.

The importance of eternity is not limited to strictly philosophical discussions. It is a notion that also has an important role in traditional biblical interpretation. The Tetragrammaton, the Hebrew name of God considered to be most sacred, is derived from the Hebrew verb for being, and as a result has been traditionally interpreted as denoting eternal existence (in either one of the two senses of eternity). Hence, Calvin translates the Tetragrammaton as *l'Eternel*, and Mendelssohn as *das ewige Wesen* or *der Ewige*. Eternity also plays a central role in contemporary South American fiction, especially in the works of J. L. Borges. The representation of eternity poses a major challenge to both literature and arts (just think about the difficulty of representing eternity in music, a thoroughly temporal art). This book aims at providing a history of the philosophy of eternity surrounded by a series of short essays, or reflections, on the role of eternity and its representation in literature, religion, language, liturgy, science, and music. Thus, our aim is to provide a history of philosophy as a discipline that is in constant commerce with various other domains of human inquisition and exploration. Finally, we would like to stress our commitment to expanding the horizons of the philosophical curriculum as taught in Anglo-American universities. Against the still widespread attitude that identifies the history of (especially medieval) philosophy with the history of Christian philosophy, we see no place for such an attitude, which is not only immoral but also erroneous and deeply misleading.

2 A History of Eternity

The five chapters of this book attempt to trace the development of the concept of eternity, explore the variety of philosophical problems leading to the development of the concept(s) of eternity, and investigate the variety of philosophical problems resulting from it.

Chapter 1, by James Wilberding, studies the emergence of the concept of eternity in ancient Greek philosophy and its close scrutiny in late antiquity. The early history of the concept of eternity turns out to be as slippery as the concept itself. It is generally agreed on that by the end of late antiquity the concept of eternity had emerged, but when exactly it developed and who developed it remains a matter of controversy. Added to these problems are those concerned with the content of the concept of eternity itself. In this chapter, Wilberding investigates the evidence on the notion of eternity in antiquity. He approaches the evidence by looking to see what philosophical problems the introduction of (some notion of) eternity is meant to solve.

The chapter is divided into four sections. In the brief first section, Wilberding introduces and discusses the vocabulary the ancients used to discuss eternity. Here he pays particular attention to the Greek term *aiôn*, which by the end of late antiquity comes to refer to eternity but which was originally used to denote "life," with varying connotations. These connections to life become important in subsequent sections. The second section is devoted to Parmenides, who is taken by some to be the first thinker to have articulated some notion of timeless eternity. The third section is devoted to Plato and Aristotle, both of whom grapple with one of the central problems of eternity in antiquity: determining how the sensible world, which is changing and in time, can be caused by an eternal principle. Although both Plato and Aristotle see eternity (*aiôn*) as an alternative to time and thus as timeless, Wilberding argues that there is no pressing reason to assume that they are working with a concept of durationless eternity. Here an attempt is also made to unravel Plato's enigmatic characterization of time as a "moving image of eternity [*aiôn*]" in the *Timaeus*. The final and longest section is devoted to later Platonic theories of eternity. This section begins with an in-depth examination of Plotinus's understanding of eternity. After introducing some necessary background on Plotinus's metaphysics, Wilberding argues that on Plotinus's view eternity is the durationless manner of presence of the plurality of forms as they are contemplated by the intellect. This discussion will show that emanationist metaphysics provides a metaphysical background that makes the notion of eternity as metaphysical *life* more comprehensible by allowing for timeless activities that have a bearing on the temporal goings-on in the sensible world. The chapter concludes with a brief look at a selection of post-Plotinian Christian Platonists, Augustine, Philoponus, and Boethius, who explore various puzzles that God's eternity poses for creation and divine omniscience.

The chapter dedicated to medieval discussions of eternity is by Peter Adamson. The topic of this book as a whole provides an unusually good opportunity for tracing a philosophical concept from the late ancient period through the Islamic philosophical tradition and on into the Scholastic Latin west. This chapter looks at three distinct but frequently interacting conceptions of eternity in the Islamic world before moving on to look at the impact of this tradition on Christian medieval philosophy. The look at eternity in the Islamic world will include Jewish authors as well as Muslims: both Saʿādia Gaon and Maimonides are prominent contributors to the debate over the eternity of the world in Arabic philosophy.

Of the three conceptions of eternity considered in this chapter, two derive from the ancient Greek tradition and will already have been explored in the chapter by Wilberding (chapter 1). These conceptions are (1) eternity as timelessness, and (2) eternity as an infinite duration supervening on an infinite motion. The former comes into the Islamic world in late Platonic texts, especially translations of Plotinus and Proclus produced in al-Kindi's circle in the ninth century. Accordingly, it is unsurprising to find al-Kindi saying: "God is above time, since He is the cause of time." Al-Kindi favors this idea of eternity over the more Aristotelian idea of infinite duration supervening on motion: he is unusual among Islamic philosophers in rejecting the eternity of the created world. On the other hand al-Kindi is also influenced by a third concept indigenous to the Islamic world (though certainly resonating with the late Platonic conception): (3) eternity as a near-synonym of divinity. Al-Kindi, like other authors of his period more associated with *kalam* (Islamic theology), assumes that "eternal" means "on an ontological par with God." This identification of eternity and divinity was not broken decisively until the work of the Ash'arites in theology and Avicenna in philosophy in the tenth to eleventh centuries.

Another figure who draws on these conceptions, negatively and positively, is the ninth- to tenth-century philosopher al-Razi. He is another unusual case among Islamic philosophers in that he had Plato's *Timaeus* as his primary influence rather than the late Platonists or Aristotle. Under this influence, he set forward a strikingly original theory that departs from both the late Platonist and the Aristotelian conceptions. For al-Razi, eternity is an infinite extension that does *not* supervene on motion. He calls this not only "eternity" but also "absolute time." Here eternity is not atemporal but is rather the "empty" temporal extension within which God creates the world. (Thus al-Razi treats it as similar to void, which he calls "absolute place.") Al-Razi is arguably the first philosopher to put forward a conception of time as both infinitely extended and independent from motion (though he draws here on Galen's critique of Aristotle in the lost *On Demonstration*).

These discussions provide the background for what may be the most prominent single philosophical issue in Islamic philosophy (or at least in western perceptions of that tradition): the eternity of the world. I have already mentioned that al-Kindi rejected the world's eternity, apparently in order to avoid putting the world on a par with God. Saʿādia Gaon does the same, using some of the same arguments drawn from the late Platonist John Philoponus. More famous, though, is the clash over this issue in al-Ghazali and Averroes. The question receives a more aporetic answer in both Maimonides and (following him) Aquinas, both of whom try to show that because neither side has compelling arguments, the issue can only be decided by recourse to revealed truth. Thus, the debate over the eternity of the world becomes largely a methodological one—what sorts of argument could in principle settle the question? This issue is closely linked to the different conceptions of eternity canvassed above. The idea that the eternity of the world can be proven *physically* is linked to the Aristotelian conception of eternity as infinite duration supervening on time, whereas the idea that revelation or metaphysics has the last word on the subject goes hand-in-hand with the late Platonic and Islamic theological conceptions of eternity as timelessness and/or divinity.

Modernity seemed to be the autumn of eternity. The secularization of European culture provided little sustenance to the concept of eternity with its heavy theological baggage. Yet our hero would not leave the stage without an outstanding performance of its power and temptation. Indeed, in the first three centuries of the modern period—the subject of the third chapter, by me—the concept of eternity will play a crucial role in the great philosophical systems of the period. The first part of this chapter concentrates on the debate about the temporality of God. While most of the great metaphysicians of the seventeenth century—Suárez, Spinoza, Malebranche, and Leibniz—ascribed to God eternal, nontemporal existence, a growing number of philosophers conceived God as existing in time. For Newton, God's eternity was simply the fact that "He was, he is, and is to come." A similar view of God as being essentially in time was endorsed by Pierre Gassendi, Henry More, Samuel Clarke, Isaac Barrow, John Locke, and most probably Descartes as well. In the second part of the chapter I examine the concept of eternal truth and its relation to the emerging notion of Laws of Nature. The third part of the chapter explicates Spinoza's original understanding of *eternity* as a *modal* concept. For Spinoza, eternity is a unique kind of necessary existence: it is existence that is *self-necessitated* (unlike the existence of other things whose necessity derives from external causes). Eternity is the existence of God or the one substance. Yet Spinoza claims that if we conceive finite things adequately—"sub specie aeternitatis"—as nothing but modes flowing from the essence of God, even finite things (like our minds) can take part in God's eternity. The fourth and final part of the chapter is mostly focused on the reception of Spinoza's original conception of eternity by Leibniz and other eighteenth-century philosophers.

Kant's philosophy decisively reorients the understanding of the eternal for European thought of the next two centuries, claims Alistair Welchman in the fourth chapter, whose subject is the conceptions of eternity in nineteenth- and twentieth-century continental philosophy. At one level, Kant's mature thought clearly involves a critique of the metaphysical speculation characteristic of the early modern era, including speculation about the eternal. At the same time, one of the core claims of Kant's positive doctrine of transcendental idealism is that space and time are subjective forms of appearance. A well-known—though controversial—corollary of this doctrine is that things in themselves are nontemporal and hence eternal. To the extent that the post-Kantian tradition of German idealism emphasizes such issues, it is sometimes regarded as regressing to a precritical metaphysical position. But there is, Welchman argues, a marked change after Kant, even if metaphysical and theological issues do not altogether disappear. That change is the result of Kant's famous dictum that he has denied knowledge of things in themselves in order to make way for faith, specifically the faith that we, as human persons understood as things in ourselves, might still be free even though we, as appearances, are rigorously causally determined. As a result of this change, Kant inaugurates an association of the eternity of things in themselves with the subjective experience of human freedom that was to be taken up by a number of thinkers in the nineteenth century, including Schelling, Schopenhauer, Kierkegaard, and Nietzsche. A brief first part of the chapter paints an appropriate picture of this Kantian background.

Metaphysical and theological issues are by no means absent from many of these nineteenth-century discussions, especially from Schelling and Kierkegaard. But we can begin to note a change, due in large part to the intervening philosophical influence of Kant, in the way the concept of eternity is deployed. The concept comes to be not simply associated with metaphysical descriptions of God but also increasingly connected with questions of subjectivity, of the concrete manner in which we experience the world and especially our freedom within it. This of course is true in different ways and to different degrees in the thought of Schelling and Kierkegaard. However, this concept further introduces the possibility of a secular, sometimes even psychological, dimension, which one sees continuing to influence subsequent thinkers and, for example, remaining in force right up to Nietzsche's conception of eternal return.

Even where Schelling and Kierkegaard do engage in a more conventional metaphysics of eternity, they draw on literary methods to sustain some ironic distance from the content of their discussions: Schelling by means of a mythological register and Kierkegaard by masking himself under his pseudonyms. Eternity, for Schelling, describes both what is "prior to" temporality as well as the perspective that transcends temporality. It describes the nature of the act that constitutes temporality itself, that is, the act of divine creation; but this act is also one that is recapitulated in the free act of deciding in which a human being constitutes himself or herself as an authentic individual. Schelling's conception of eternity as a moment of radical and unmotivated choice of character is a piece of anti-Hegelianism (it does not stand in a dialectical relation to temporality) that links him to Kierkegaard. Kierkegaard is concerned with eternity as a lived experience of freedom—encapsulated in his theory of the moment—as the condition for the possibility both of an authentic grounding in temporality and of religious redemption. Of particular importance for all these thinkers—starting with the Kant of the *Religion* text—is the idea that human freedom (and responsibility) is linked to an act that takes place in eternity, "outside" time. This paradoxical thought will form the link between the thinkers discussed in the longer second part of the chapter.

In the—again shorter—third part, Welchman investigates the concept of eternity's contemporary fate in European thought. Until very recently, this fate has been, it seems, ignominious. The development of the (at least potentially or problematically) secular understanding of eternity in the nineteenth century, one centered on the experience of human freedom, has been stymied by developments in the twentieth century. There is still some talk of the eternal in a number of significant twentieth-century thinkers such as Martin Heidegger, Henri Bergson, and Gilles Deleuze. But Welchman argues that it is mainly diagnostic in intent. These thinkers are primarily interested in the concept of *time*. They want to develop a deeper thought of temporality (often in relation to lived human experience) than that of empirical ("clock") temporality, and they see the metaphysical and theological notions of eternity as conceptually confused ways of doing this. Correlatively, there is some discussion of the concept in avowedly religious thinkers like Emmanuel Lévinas and, more recently, Giorgio Agamben. As a result, one might argue that the fate of the concept of eternity in the twentieth century suggests that the continued interest of nineteenth-century thinkers in the eternal was a result of an incomplete movement toward secularization. But this result is contradicted by what appears to be a genuine revival in recent years of a full-fledged and unapologetic conception of eternity in the work of Alain Badiou, a radical Maoist and philosopher of mathematics. His interest in mathematics suggests a rapprochement with Anglophone interests in eternity as descriptive of the status of numbers and propositions. But Badiou's *political* interests also point to a revival of nineteenth-century notions of the relation of human freedom with the eternal.

The fifth and final chapter, on the conception of eternity in the twentieth century and contemporary analytic philosophy, is written by Kristopher McDaniel. At the beginning of the twentieth century, claims McDaniel, many attempts were made to demonstrate by means of speculative metaphysics that time is a mere appearance. The most famous of these purported demonstrations is McTaggart's argument for the unreality of time. Although this argument is not widely viewed as successful, it did set the agenda for analytic philosophers pursuing the philosophy of time in the second half of the twentieth century.

Complementing the arguments of speculative metaphysics are the arguments of speculative philosophy of physics. The theory of special relativity seemed to some to show that temporality per se was not metaphysically fundamental but should instead be seen as an aspect of spatiotemporality. Kurt Gödel attempted to argue from the unreality of time by appeal to considerations stemming from the theory of general relativity. More recently, some physicists and philosophers of physics have entertained the hypothesis that spatiotemporality is itself a derivative feature that emerges from a more fundamental non-spatiotemporal framework.

In the first part of this chapter, McDaniel discusses in more detail arguments for the eternality of some entities, specifically focusing first on the case for ideal meanings, including propositions, and then turning to questions concerning the purported eternity of God. In the second part of the chapter, McDaniel begins by critically discussing some of the arguments of speculative metaphysics for the unreality of time and then tracing some of the highlights from the twentieth-century philosophy of time. The third part of the chapter turns to a discussion of the hypothesis in speculative philosophy of physics that space-time derives from a more fundamental basis. This hypothesis has received comparatively little attention from metaphysicians, despite the tempting prospects for speculation it offers. Accordingly, McDaniel explores how the truth of this hypothesis would impact various other disputes in metaphysics, including disputes about what it is to be an abstract rather than a concrete object, the nature of material composition, and the relationship between necessity and eternity.

3 Problems with Eternity (and with the Writing of Its History)

The issue of eternity is replete with problems. If you are a thinker who enjoys reflecting on a good question: welcome aboard! The five chapters of the book will unfold and investigate in a chronological order the study of these problems throughout the history of philosophy. But just as a starter (wait for the meal!), here is a cluster of problems and questions surrounding the issue of eternity: Why is eternal existence considered by many to be more real than temporal existence?—What is the relation between timeless eternity and the present tense (both have no temporal measure)? What kinds of relations, if any, obtain between timeless and temporal entities? (Are these causal relations? If so, what kind of causality is operative? Explanatory relations? If so, must the explanatory relation be asymmetric?)—Is the notion of *timeless action* consistent? If God is supposed to be "living," in what sense can he be eternal?—Can nonexisting things be eternal?—Can material things be eternal?—Are there eternal truths, and if so, how can temporal minds access them?—Are numbers eternal?—Can there be more than one eternity? (If so, how are they to be distinguished?)—Can we make sense of the predicate "eternally eternal" and, if so, how is it to be distinguished from just "eternal"?

Next to these philosophical questions, studying the history of the philosophy of eternity raises basic and crucial questions regarding the methodology of the history of philosophy. I have already mentioned that the English term "eternity" and its Latin original, *aeternitas*, have the two distinct senses of (1) being in all times, and (2) timeless existence. However, this is nothing but a historical accident, and we can easily conceive of languages in which each of these senses of "eternity" has its own distinct term. In fact, we can easily conceive of situations in which one and the same term in the same language loses or gains a new sense. (This is one of the most natural processes of living languages.) For this reason we attempted to fix our attention on the concept(s) of eternity and its development, rather than the terms. Still, in order to complement the five chapters that study the development of the concept of eternity, we have added a brief glossary at the end of the book that aims at providing an overview of the terminology of eternity in some of the major languages that have been employed in the two and half millennia of western philosophy.

4 A Future for Eternity?

A quick survey of various recent companions, handbooks, and guides to metaphysics would hardly uncover any discussion of eternity. In many, eternity does not even appear as a marginal term in the index. Usually, we relegate eternity to the somewhat less rigorous and less prestigious field of "philosophy of religion." This need not be the case. Metaphysics as an independent discipline has a surprisingly short history. Till the early eighteenth century, many, perhaps even most, of the writers on "metaphysics" had primarily the outstanding work of Aristotle in mind. In the writings of the early eighteenth-century German rationalists—Christian Wolff and Alexander Baumgarten— we find a conception of metaphysics that is no longer tied to Aristotle's great work. But metaphysics as a discipline was not blessed with longevity, as a dozen years or so before Louis XVI it was condemned to the guillotine by Kant's first critique. The fate of metaphysics after the Kantian revolution is a story that still needs to be told, but it would be fair to say, I think, that for the past two centuries engagement with heavy metaphysical concepts, such as eternity, has been taken to be either a form of backwardness (religious or otherwise) or a kind of eccentricity.

Luckily, things seem to have changed over the past twenty years. Suddenly, for example, we are seeing debates about monism appear in the most mainstream journals. (This could hardly be imagined in the 1970s or 1980s.) The emergence of interest in metaphysical monism seems to open a window of opportunity through which eternity might again take her rightful seat as a fundamental notion of metaphysics.

CHAPTER ONE

Eternity in Ancient Philosophy

James G. Wilberding

Background: Eternity and Life

Although it is common today to use the word "eternity" to refer to an endless or even simply a very long period of time, this word actually derives from a Latin term, *aeternitas*, that was used in philosophical circles of late antiquity to refer to a mode of being that was removed from time altogether (and in the case of some philosophers, as I will show, from duration as well). Thus, Boethius (480–525/6 CE) distinguished the timeless eternity (*aeternitas*) of God from the everlasting temporal duration (*sempiternitas*) of the heavens,[1] and a similar distinction between αἰών (eternity) and ἀιδιότης (everlastingness) can be found in the Greek Platonists of late antiquity.[2] This chapter aims

---

[1] Boethius *Trin.* 4 (Claudio Moreschini (ed.) *Boethius De consolatione philosophiae. Opuscula theologica* [Munich and Leipzig: K.G. Saur, 2005]) ll. 244–245.
[2] See, for example, Olympiodorus *In Meteor.* 146.15–23. This distinction, however, was in some sense a work in progress. Although all Platonists of late antiquity shared some distinction between timeless eternity and everlastingness in time, the terminology employed to capture this distinction varied. Proclus, for example, sometimes uses αἰών and ἀιδιότης for eternity and everlastingness,

to present an account of the development of this concept of timeless eternity.

One puzzle that will be addressed here is rooted in the very etymology of the Greek word αἰών, which was originally used above all in poetry and tragedy to denote "life," with varying connotations, emphasizing either the force or principle of life, the temporal span of one's life, or the content and quality of one's life (and thus one's fated life).[3] Even as this term comes to take on the sense of "eternity,"[4] this sense of "life" appears never to be entirely abandoned. Consider, for example, the following selection of statements on eternity:

> For the life of God is not time, but the archetype and paradigm of time—eternity. (Philo, c. 15 BCE—45 CE)[5]

> Seeing all this, one sees eternity in seeing a life that abides in the same and always has the all present to it (Plotinus 205–270/1 CE)[6]

---

respectively (e.g., *Theol. Plat.* 3.55.1:–14). but he is also comfortable speaking of two senses of ἀιδιότης: one eternal and one in time (*Elem. Theol.* 55). Plotinus explicitly raises the question whether these terms are synonymous (*Enn.* 3.7.3.3), and although he decides that they are not, for Plotinus they refer to two different aspects of timeless eternity (*Enn.* 3.7.5.15–18). To be clear, however, there was broad agreement among Platonists of late antiquity that αἰών denoted timeless eternity.

3 For a comprehensive study of the development of the sense of αἰών, see C. Lackeit, *Aion: Zeit und Ewigkeit in Sprache und Religion der Griechen* (Königsberg: Hartungsche, 1916). Similar conclusions have been drawn by E. Benveniste, "Expression indo-européenne de l'Éternité," *Bulletin de la société linguistique de Paris* 38 (1937): 103–112, and R. B. Onians, *The Origins of European Thought about the Body, the Mind, the Soul, the World, Time and Fate* (Cambridge: Cambridge University Press, 1951), 200–216. There appears even to be some evidence that αἰών at some point came to possess the meaning "marrow," which was taken to be the physical principle of life (Onians, *Origins*, 205–206; Benveniste, "Expression," 109; Lackeit, *Aion*, 10–11 and 13).

4 An attempt has been made (see Lackeit, *Aion*, 23–37 and 53–56 to explain the transition from the sense of "life" to "eternity" by pointing to an intermediate evolution in the sense of αἰών in which it goes from the sense of one's individual life span to increasingly larger periods of time (consider, by way of comparison, the English word "age"), and this might be right, though the passages below may serve as an initial point of evidence that the connotations of life never entirely disappear.

5 Philo *Quod deus sit immutabilis* 6.32: καὶ γὰρ οὐ χρόνος, ἀλλὰ τὸ ἀρχέτυπον τοῦ χρόνου καὶ παράδειγμα αἰὼν ὁ βίος ἐστὶν αὐτοῦ. Unless otherwise noted, all translations are my own.

6 Plotinus *Enn.* 3.7.3.16–17 (*Opera. Porphyrii vita Plotini. Enneades I-VI*, 3 vols., ed. P. Henry and H.-R. Schwyzer [Oxford: Clarendon Press, 1964–1982], translation from *Enneads*, 7 vols., ed. and trans. A. H. Armstrong [Cambridge, MA: Harvard University Press, 1966–88]): ταῦτα πάντα ἰδὼν αἰῶνα εἶδεν ἰδὼν ζωὴν μένουσαν ἐν τῷ αὐτῷ ἀεὶ παρὸν τὸ πᾶν ἔχουσαν.

But whereas eternity is a measure of the life of the intelligible living thing, time is the measure of the life of this sensible cosmos. (Proclus, 410–485 CE)[7]

Eternity therefore is the whole and perfect possession all together of a life which cannot end. (Boethius)[8]

As these passages illustrate, there was a central connection between the concepts of "eternity" and "life." Yet this connection has struck many modern philosophers as problematic or even incoherent on the grounds that life necessarily involves activity, and activity must involve change and time, which is incompatible with the purported timelessness of eternity.[9] To be sure, this criticism raises an important problem, but as I will show, the Platonists of late antiquity, such as Plotinus, put themselves in a position to address this problem effectively by developing a very sophisticated metaphysics of the intelligible world and, thus, of intelligible life.

Parmenides

Parmenides of Elea, a Presocratic philosopher of the early fifth century BCE, has frequently been hailed as the first philosopher to give expression to a notion of eternity, although he never uses any of the terms that later came to be used to denote everlastingness (ἡ ἀϊδιότης, ἀΐδιος) or eternity (ὁ αἰών, αἰώνιος), but there has been a significant amount of debate in recent decades regarding the accuracy of this attribution.[10] Clear expressions of it can be traced back as far as the Platonic

---

7 Proclus *In Remp.* 2.11.22–24.
8 Boethius *Consol.* 5.6.4 (Moreschini ed., ll. 9–10): "Aeternitatas igitur est interminabilis vitae tota simul et perfecta possessio"; translation from R. Sorabji, *Time, Creation and the Continuum* (London: Duckworth, 1983), 119.
9 E.g., L. Tarán, "Perpetual Duration and Atemporal Eternity in Parmenides and Plato," *Monist* 62 (1979): 47: "Even the unmoved mover is alive, and cannot therefore be eternal." See R. Sorabji, *Time, Creation and the Continuum*, 110, on the problem of the vitality of the Paradigm in Plato.
10 For a survey of some of the views that have been defended, see Sorabji, *Time, Creation and the Continuum*, 99–108. L. Tarán, *Parmenides: A Text with Translation, Commentary and Critical Essays* (Princeton, NJ: Princeton University Press, 1965), 175 n. 1, and M. Schofield, "Did Parmenides Discover Eternity?," *Archiv für Geschichte der Philosophie* 52 (1970): 113 n. 2, provide a list of earlier scholars who attribute some notion of eternity to Parmenides. The interpretation offered here, namely that being for

interpretations of late antiquity,[11] but in this matter we cannot accord too much weight to their views. For the Platonists of late antiquity were reading Parmenides nearly a millennium after his death, and their exegetical objectivity may be questioned not only because they were working from the privileged historical perspective of those already in possession of a sophisticated notion of eternity—as I will show, it was the Platonists of late antiquity who were chiefly responsible for articulating and developing the notion of eternity—but also because they viewed Parmenides himself as a major forerunner of the Platonic tradition and were keen to emphasize points of continuity between his thought and their own. While a rigorous examination of Parmenides would well exceed the boundaries of this survey, the following should give the reader an adequate idea both of the reasons why Parmenides has been credited with the first articulation of a concept of eternity and of some of the serious problems this attribution encounters.

The focus of the controversy revolves around two of the approximately 150 lines that have survived of his sole work, a hexameter poem traditionally (though probably not authentically) entitled "On Nature." In this poem Parmenides's central concern is to show that the impossibility and unintelligibility of non-being leads to certain consequences for the nature of being, which he programmatically sets out in lines 3–4 of fragment 8:

ὡς ἀγένητον ἐὸν καὶ ἀνώλεθρόν ἐστιν,
οὖλον μουνογενές τε καὶ ἀτρεμὲς ἠδὲ τέλειον.

Being is ungenerated and indestructible,
whole and of a single kind and unshaken and complete.[12]

Parmenides is merely everlasting, is argued by H. Fränkel, *Wege und Formen frühgriechischen Denkens* (Munich: Beck, 1970); Tarán, *Parmenides*, 175–188; Schofield, "Did Parmenides Discover Eternity?"; and D. O'Brien, "Temps et intemporalité chez Parménide," *Les études philosophiques* 3 (1980), 257–72, and "L'être et l'éternité," in *Études sur Parménide*, vol. 2, ed. P. Aubenque (Paris: J. Vrin, 1987), 135–162.

11 See, for example, Proclus *In Parm.* 639.20–21 (Proclus, *In Platonis Parmenidem commentaria*, 3 vols., ed. C. Steel [Oxford: Clarendon Press, 2007–9], and Ammonius (c. 440–after 517) *In de Int.* 136.22–25.

12 Accepting Simplicius's οὖλον μουνογενες over Plutarch's ἐστι γὰρ οὐλομελές and the emendation of ἠδὲ τέλειον for Simplicius's ἠδ' ἀτελεστον. For a defense of this text, see G. E. L. Owen, "Eleatic Questions," *Classical Quarterly* 10 (1960): 101–102.

What follows is a series of arguments aimed at establishing these consequences: being is ungenerated and indestructible (8.5–21), whole and continuous (8.22–25), not subject to change, that is, "unshaken" (8.26–31) and complete (8.32–49). None of these features is necessarily incompatible with a conception of being that endures through time. The question then is whether Parmenides takes the additional step of envisioning being as transcending temporal extension altogether. In the critical lines 5–6 of fragment 8, he has often been taken to be doing precisely this:

οὐδέ ποτ' ἦν οὐδ' ἔσται, ἐπεὶ νῦν ἔστιν ὁμοῦ πᾶν,
ἕν, συνεχές·[13]

The most straightforward translation of these lines would suggest that it is incorrect to speak of being in the past and future tenses:

Neither was it ever nor will it be, since it is now all together, one, continuous.

If this translation gets Parmenides's meaning right, then he would indeed seem to have put his finger on some notion of eternity as timelessness. For on this line of interpretation his idea would seem to be that being's unity would be undermined by temporal extension, since then all of it would not be together in the present. If this is his idea, however, it hardly receives the development one would expect. For one thing, Parmenides's willingness in these lines to characterize being as existing "now" (νῦν) would already seem to speak against the strict timelessness of being, though perhaps this can be either excused as an infelicitous slip by someone struggling to articulate a radically original

---

[13] This is the text as Simplicius (and in part Proclus) has preserved it. A variation on this text that eliminates the temporal terms "ever" and "now" has been preserved by other commentators of roughly the same period as Simplicius. For a defense of Simplicius's text over the others, see O'Brien, "Temps et intemporalité," 263–264.

idea or accepted as a refinement, in which case he would be proposing not timelessness per se but an eternal present. More importantly, if Parmenides means to place being outside time, it would have been reasonable for him to have mentioned and discussed the notion of time at some point, but the preserved fragments—and the arguments for the nature of being appear to be completely preserved in fragment 8—contain no reference to the concept of time (χρόνος). Moreover, one would expect Parmenides to return to this notion of timelessness (or tenselessness) in each of the subsequent arguments for completeness, changelessness, and continuity, but when one examines these arguments, a concern for being's timelessness is absent at best and, at worst, even contradicted by Parmenides's own words.

The argument for the completeness of being (8.32–49), for example, is far from affirming that temporal extension would compromise being's completeness. In fact, Parmenides uses tensed language in connection with being only once in this stretch of argument (8.36–37), and here he appears to accept as unproblematic that being *will* exist in the future:

οὐδὲν γὰρ <ἢ> ἔστιν ἢ ἔσται
ἄλλο πάρεξ τοῦ ἐόντος

For nothing else is or will be
except for being.

Moreover, if Parmenides's aim in the completeness argument was to rule out temporal extension, one would expect Parmenides to be at least equally concerned about spatial extension, but spatial extension appears to be taken for granted in these lines, so much so that Parmenides notoriously even allocates a shape to being—"a sphere well-rounded on every side" (8.43). His real concern is rather to show that being's spatial extension is not perforated with gaps of non-being, which would rob it of its completeness: "non-being would stop it from reaching [ἱκνεῖσθαι] its like" (8.46). Likewise, timelessness does not appear to be an issue in the argument for changelessness (8.26–31), since otherwise Parmenides would have presumably resisted describing being as "persisting" (μένον [8.29], μένει [8.30]) in the same state and place. Finally, as has been often noted, there is no indication at all that the argument at 8.22–25 is directed at temporal as opposed to spatial continuity, and this is particularly significant, since this argument appears to establish the very thing that in 8.5–6 was supposed to rule out being's temporal extension, namely that being is "all continuous" (ξυνεχὲς πᾶν, 8.25). So here more than anywhere else one might have expected Parmenides to make his concern with past and future tenses clear. Moreover, even if we assume that this argument is about temporal duration, it seems to show the opposite of what it is supposed to show, namely that being *is* extended in time. For his conclusion states that being "is all continuous because being approaches being" (ξυνεχὲς πᾶν ἐστιν· ἐὸν γὰρ ἐόντι πελάζει, 8.25), which on this assumption would mean that being at one time is continuous with being at the next time.

If, then, Parmenides is not advancing the view that being is eternal, then how should we understand the critical lines 5–6 of fragment 8? An alternative translation of these lines appears to capture his meaning better: "It is not the case that it ever *was* in existence [but is no longer] nor that it ever will be [but is not yet], since it is now all together, one, continuous."[14] According to this translation, Parmenides is denying not that being existed in the past and will exist in the future, but only that its existence is wholly contained in the past or in the future. This interpretation better accounts not only for the structure of the three arguments discussed above but also for the immediate context of these lines themselves. For what immediately follows lines 5–6 is an argument (8.6–21) that aims to establish not that being is timeless but only that it is ungenerated and indestructible. The beginning of

---

14 This translation has been offered and defended by Schofield, "Did Parmenides Discover Eternity?," 127. Translations along similar lines have been offered and defended by Tarán, *Parmenides*, and "Perpetual Duration"; Fränkel, *Wege und Formen frühgriechischen Denkens*; O'Brien, "Temps et intemporalité" and "L'être et l'éternité."

this argument makes clear that it is intended to support the claim in lines 5–6,[15] and its conclusion is that "coming-to-be is extinguished and destruction unheard of" (8.21). It also offers a picture of a more continuous development of Eleatic philosophy, insofar as Melissus, a fifth-century philosopher who adopts and explores the consequences of Parmenides's thought, explicitly describes being in the tensed language at issue here: "It always was whatever it was and always will be" (ἀεὶ ἦν ὅ τι ἦν καὶ ἀεὶ ἔσται, fragment DK 30B1.1),[16] which would be a strange step backward had Parmenides intended being to be timeless. All in all, then, it appears less contentious to understand Parmenides as envisioning being merely as beginningless and everlasting as opposed to eternal.

Plato and Aristotle

Even if one were to credit Parmenides with a conception of timeless eternity, one striking difference would remain between the framework of his engagement with this "eternal" being and that of his successors. For Parmenides gives no indication that he views this "eternal" being as a cause of the temporal sensible world. Indeed, if we set aside the familiar difficulties of reconciling the two halves of his poem, his interest lies rather in showing that this "eternal" being has no relation at all to the sensible world because there can be no sensible world After Parmenides, however, what is eternal is also seen to be a cause of the temporal sensible world, which raises a new problem that will exercise all subsequent philosophers who grapple with the concept of eternity: if there is an eternal cause of the sensible world, and if the sensible world came to be at some point, then it would seem to follow that the

---

15 The argument begins with the question τίνα γὰρ γένναν διζήσεαι αὐτοῦ; The γάρ makes clear that this is supposed to support the thought articulated in the preceding lines.
16 And see fragment DK 30B7.9–10: "Now if it should become different by one hair in ten thousand years, it will all perish in all of time"; translation from P. Curd and R. D. McKirahan, Jr., *A Presocratics Reader: Selected Fragments and Testimonia* (Indianapolis: Hackett, 1996).

cause in question has undergone some change, from not creating the sensible world to creating it, which would seem incompatible with its eternity, as eternity is universally understood to involve changelessness. Plato and Aristotle, both of whom introduce the term αἰών to denote a conception of eternity that is clearly meant as an alternative to time, offer glimpses of two fundamentally different strategies for dealing with this problem.

Aristotle's solution lies primarily in denying that the sensible cosmos ever came to be. Rather, for him the cosmos is bidirectionally everlasting, which is to say that it always has existed and always will exist.[17] Thus, Aristotle has more freedom to develop a conception of an eternal divine cause of the cosmos in the form of his God, the Prime Mover. Although Aristotle does not frequently address the issue of eternity, he does allow us in one passage from his *On the Heavens* to infer that he sees eternity as an alternative to time: "Wherefore neither are the things there [i.e., outside the cosmos] born in place, nor does time cause them to age, nor does change work in any way upon any of the beings whose allotted place is beyond the outermost motion: changeless and impassive, they have uninterrupted enjoyment of the best and most independent life for all of eternity [αἰών]."[18] Yet both the sequel to this passage and his account of the Prime Mover in the *Metaphysics* make clear that eternity is not extradurational. In the *Metaphysics*, for example, Aristotle twice contrasts human happiness with divine happiness by saying that the former can only obtain for brief periods, while the latter is *always* obtaining,[19] and in the sequel

---

17 See *De Caelo* 1.10–12.
18 *De Caelo* 279a18–22.
19 See *Metaphysics* 1072b14–16: διαγωγὴ δ' ἐστὶν οἵα ἡ ἀρίστη μικρὸν χρόνον ἡμῖν (οὕτω γὰρ ἀεὶ ἐκεῖνο· ἡμῖν μὲν γὰρ ἀδύνατον); *Metaphysics* 1072b24–30: εἰ οὖν οὕτως εὖ ἔχει, ὡς ἡμεῖς ποτέ, ὁ θεὸς ἀεί, θαυμαστόν· εἰ δὲ μᾶλλον, ἔτι θαυμασιώτερον. ἔχει δὲ ὧδε. καὶ ζωὴ δέ γε ὑπάρχει· ἡ γὰρ νοῦ ἐνέργεια ζωή, ἐκεῖνος δὲ ἡ ἐνέργεια· ἐνέργεια δὲ ἡ καθ' αὐτὴν ἐκείνου ζωὴ ἀρίστη καὶ ἀΐδιος. φαμὲν δὴ τὸν θεὸν εἶναι ζῷον ἀΐδιον ἄριστον, ὥστε ζωὴ καὶ αἰὼν συνεχὴς καὶ ἀΐδιος ὑπάρχει τῷ θεῷ· τοῦτο γὰρ ὁ θεός. See 1073a7, where the Prime Mover is said to κινεῖν τὸν ἄπειρον χρόνον. J. Whittaker, "The 'Eternity' of the Platonic Forms," *Phronesis* 13 (1968): 141–142, and D. O'Brien, "Temps et éternité dans la philosophie grecque," in *Mythes et représentations du temps*, ed. D. Tiffenau (Paris: CNRS, 1985), 65–66, adequately show that the Prime Mover is not exempt from duration.

to our passage, in which Aristotle spells out his understanding of eternity, duration seems to be presupposed: "Indeed, our forefathers were inspired when they made this word αἰών. The completion (τέλος) which circumscribes the time-span (χρόνος) of life of every living thing, and which cannot naturally be exceeded, they named the αἰών of each."[20] The idea appears to be that although for Aristotle time and eternity both involve duration, temporal duration differs from eternal duration insofar as what is in time is subject to change whereas the eternal remains always the same. Thus, Aristotle's eternal Prime Mover is to be construed as an intellect that is everlastingly and immutably engaged in a single activity, self-contemplation, and as such as an everlastingly enduring cause of the everlastingly enduring cosmos.

When we turn to Plato, there is a well-known disagreement, dating back to Plato's earliest exegetes and persisting today, that gives way to two divergent renderings of Plato's response to this problem. For in the *Timaeus* Plato famously presents a sequential cosmogony in which the cosmos initially does not exist but then is brought into existence by the Demiurge. The disagreement concerns whether this sequential account is best understood literally or metaphorically. On the metaphorical reading, Plato is presenting a cosmological account in the form of a sequential cosmogony for didactic reasons.[21] In this way one may maintain that Plato in fact holds the cosmos to be bidirectionally everlasting, and that the "cosmogony" is simply a tool to elucidate the causal relations that bear upon the everlasting cosmos. It follows that on this reading Plato's solution to the reconciliation problem follows the same general lines as Aristotle's. On the literal

---

20 *De Caelo* 279a23–25.
21 Ancient defenders of the metaphorical interpretation may be said to include the majority of subsequent Platonists, including Xenocrates, Speusippus, Crantor (on all of whom see J. Dillon, *The Heirs of Plato* [Oxford: Oxford University Press, 2003]), and nearly all Platonists after and including Plotinus. Plotinus's metaphorical interpretation will be considered below. M. Baltes, "Τέγονεν (Platon, *Tim.* 28b7). Ist die Welt real entstanden oder nicht?," in *ΔΙΑΝΟΗΜΑΤΑ Kleine Schriften zu Platon und zum Platonismus*, ed. M Baltes (Stuttgart: Teubner, 1999), 304, provides a list of modern metaphorical interpretations, to which I believe T. K. Johansen, *Plato's Natural Philosophy* (Cambridge: Cambridge University Press, 2004), may now be added (see 91). Aristotle might well also belong to this group (*De Caelo* 280a1; but see Dillon, *Heirs of Plato*, 25 n. 49).

reading, by contrast, according to which the existence of the cosmos really does have a datable starting point, Plato would appear to pursue a radically different strategy for dealing with this problem, which rests on the line that he draws between two distinct causes of the cosmos: the Demiurge and the Forms. Whereas Plato consistently, both in the *Timaeus*[22] and elsewhere in the *Corpus Platonicum*,[23] describes the Forms as being always the same and unchanging (ὡσαύτως ἀεὶ ἔχειν κατὰ ταὐτὰ ἔχειν or εἶναι), the Demiurge is literally described as a separate cause that undergoes changes. He is described as "reasoning and concluding";[24] he is also described as actively constructing the sensible cosmos.[25] Plato even makes a point of contrasting the Demiurge's active engagement in the creation of the sensible world with the state of rest that he settles into following the completion of these tasks.[26] Now whether Plato genuinely conceives of the Demiurge as a cause distinct from the Forms (as opposed to being one of or an aspect of the Forms) is subject to as much controversy as the literal-metaphorical question,[27] but it is crucial to evaluating his account of eternity. For when Plato turns to discuss eternity in *Timaeus* 37c–38b, it is only the

---

[22] Ancient defenders of the literal interpretation include Atticus, Plutarch, Harpocration, Hippolytus, and Philo of Alexandria (for sources and discussion, see M. Baltes, *Die Weltentstehung des platonischen Timaios nach antiken Interpreten*, vol. 1 (Leiden: Brill, 1976), 38–63), as well as Philoponus, who will be discussed later. Baltes, "Γέγονεν," 304, provides a list of modern literal interpretations, to which D. Sedley, *Creationism and Its Critics in Antiquity* (Berkeley: University of California Press, 2007), 99–106, may now be added.

[23] *Timaeus* 28a5–6; 29a1; 37b3; 38a2 (and see 35a1–2 and 41d6). E.g., *Phaedo* 78c6; d2–3; 79d5 (see 79d1–2 and 80b1–2); *Cratylus* 49d3–4; *Sophist* 248a12; 252a7–8; *Statesman* 269d5; *Philebus* 59b3–4; 61e2–3; *Republic* 479a2; e6–7; *Laws* 797b1–2.

[24] *Timaeus* 30b1: λογισάμενος οὖν ηὕρισκεν.

[25] See *Timaeus* 30b4–5: ψυχὴν δ' ἐν σώματι συνιστὰς τὸ πᾶν συνετεκταίνετο; 32b7–8: συνέδησεν καὶ συνεστήσατο οὐρανὸν ὁρατὸν καὶ ἁπτόν; 32c3–4: τοῦ συνδήσαντος; 32c7: συνέστησεν αὐτὸν ὁ συνιστάς; 33b1: ἐτεκτήνατο; 33b1–2: σχῆμα δὲ ἔδωκεν αὐτῷ τὸ πρέπον καὶ τὸ συγγενές; etc.

[26] *Timaeus* 42e5–6: Καὶ ὁ μὲν δὴ ἅπαντα ταῦτα διατάξας ἔμενεν ἐν τῷ ἑαυτοῦ κατὰ τρόπον ἤθει.

[27] Some reasons for considering the Demiurge to be either one of or an aspect of the Forms include the following. (1) He is described as ὄντος ἀεὶ θεοῦ at 34a8, and at 27d (see 48e) Plato might mean to divide all substances exhaustively into two categories, those that are ὄντα ἀεὶ and those that are γιγνόμενα, in which case the Demiurge would fall in the former category, which is described as νοήσει μετὰ λόγου περιληπτόν, ἀεὶ κατὰ ταὐτὰ ὄν, which looks like a fair characterization of the nature of the Forms. (2) The Demiurge appears to be called the "best" of τῶν νοητῶν ἀεί τε ὄντων (37a1), which again would seem to place him among the intelligible Forms (for alternative ancient

Paradigm that encompasses all of the Forms that is included in the domain of the eternal. Thus, if the Demiurge is not counted among the Forms, he is excluded from the domain of the eternal things, and the label "eternal" will apply only to the objects of thought, the Forms, but not to any thinking subject. Yet it is precisely by excluding the Demiurge from this domain that Plato, on the literal reading, is able to solve the reconciliation problem. For, with this distinction in hand, he may fully preserve the Forms as immutable and eternal causes of the sensible cosmos, since the Demiurge merely looks at them and models the cosmos after them, while he himself accounts for the changes necessary to explain how the cosmos came to be.

Since my examination of Plotinus will provide a context in which to explore a conception of eternity that can emerge when one thoroughly adheres to the metaphorical reading and denies that the Demiurge is distinct from the Forms, I will here investigate eternity as it presents itself on the literal reading. I will begin by looking at the passage in which Plato sets out his account of eternity:

> Now when the father who had begotten the universe thought of it in motion and alive, a shrine brought into being for the everlasting [ἀιδίων][28] gods, he rejoiced, and being well pleased he took thought to make it still more like its paradigm. So, just as the paradigm happens to be an everlasting [ἀίδιον] living thing, he

---

interpretations of this line, see F. M. Cornford, *Plato's Cosmology* (London: Routledge, 1937), 94 n. 2). (3) Not only is the Demiurge excluded from the list of causes at 50c7–d3, but here being (the Forms) is likened to a *father*, a term that was previously used to characterize the Demiurge (28c3). For a recent defense of the view that the Demiurge is *not* to be counted among the Forms, see Johansen, *Plato's Natural Philosophy*, 81–82.

28 Here I follow the standard practice of translating αἰών, αἰώνιος and ἀίδιος as "eternity," "eternal," and "everlasting," respectively, though this can be misleading, as these terms are commonly understood to be mutually exclusive—"everlasting" being predicated of what exists for all time (or duration) and "eternal" being predicated of what is beyond time (or duration). Plato, however, is willing to say both that time is "eternal" (37d7) and that the Forms are "everlasting" (29a3 and 37e5). Yet these terms cannot be synonymous here, either, since otherwise 37d1–3 would be too redundant. The suggestion explored here is that part of the meaning of αἰών and αἰώνιος is the completeness of a thing's life. See 39d7–e2, discussed below.

attempted to make this sensible universe as similar as possible in this respect. Now the nature of the living thing happened to be eternal [αἰώνιος], and it was impossible for him to wholly confer this unto what has been begotten. But he took thought to make a kind of moving image of eternity [αἰών], and simultaneous with his ordering of the heavens he created of eternity that abides in unity [μένοντος αἰῶνος ἐν ἑνὶ] an eternal [αἰώνιος] image moving according to number, and this number is what we have labeled "time." For days, nights, months and years did not exist before the heavens came to be, and he contrived their generation simultaneously with the constitution of the heavens. For these are all parts of time, and "was" and "will be" are generated forms of time that we unthinkingly and incorrectly apply to the everlasting [ἀίδιον] substance. For we say that it was, is and will be, but according to the true account only "is" is appropriate to the everlasting substance, while "was" and "will be" are appropriately said of the generation that proceeds in time. For these are both changes, but what is always the same and unchanging [τὸ ἀεὶ κατὰ ταὐτὰ ἔχον ἀκινήτως] cannot become older or younger through time, nor can it have been, nor can it now have come to be, nor will it be in the future. And in general none of the things that generation confers upon the things that are moving in the sensible realm are conferred upon it, but these have been generated as forms of time, which is imitating eternity [αἰών] and revolving according to number. And further, in addition to these things [we also say] things like this: what has come to be *is* what has come to be, and what is coming to be *is* coming to be, and further that what will come to be *is* what will come to be, and that what is non-existent *is* non-existent, and none of these locutions of ours is accurate. But perhaps right now is not the appropriate moment to go through all of these issues in great detail.[29]

---

29 *Timaeus* 37c6–38b5.

Many details in this passage are difficult, but one thing that is beyond dispute is that Plato is introducing αἰών as an alternative to time, and so we are justified in characterizing this αἰών as "timeless," at least in some sense of timelessness. Yet it is also made clear here (as well as in the sequel) that Plato is working with a very particular conception of time, and this raises questions about the exact sense of αἰών's timelessness. For time in the *Timaeus* is not synonymous with either duration or change (motion); rather it is something that supervenes on both of these.

Briefly recounting two stages of the *Timaeus* creation story that precede the introduction of time shows that change (motion) and duration exist independently of time. Before the Demiurge even begins to create the cosmos there is an initial state of affairs that we might call the *precosmic stage* and that is characterized by a "discordant and disorderly motion."[30] The Demiurge then begins to impose intelligible order by creating the body and the soul of the universe. This World-Soul is described as being not only spatially extended—it resembles a complex of eight concentric circles—but even as engaged in spatial motion, as each of these circles is revolving. It is these revolutions of soul that are to account for the movements of the planets and stars, once the latter are embedded into the former. At this point there is a cosmos (though it is not yet complete), but time still does not yet exist, so let us refer to this as the *pretemporal cosmic stage*. Both of these stages involve motion and thus duration, but neither of them is in time. Plato himself emphasizes as much at the start of the passage, when he draws attention to the fact that the universe is "in motion and alive."

But what is time for Plato if it is posterior to motion and duration? Time is said to come to be with the creation of the "wandering" stars, that is, the sun, the moon, and the five known planets (38c).[31] Their creation is meant to provide points of reference for the motions of

---

30 *Timaeus* 30a3–4: κινούμενον πλημμελῶς καὶ ἀτάκτως.
31 The fixed stars are introduced only after time has been generated, at 40a–b.

their orbits (which, again, already preexist as circles within the World-Soul), and these points of reference serve to make motion and change measurable. The system of measurement that results consists of the natural units determined by these moving bodies in relation to the earth and to each other. The most familiar units—day, night, month, year—are provided by the movements of the sun and moon, but there are also other, more obscure units determined by the less regular movements of the five planets (39b–d). Plato labels all of these units "parts" of time (37e), which raises a question about time as a whole, and his view appears to be that the whole of time is the so-called Great Year, which is consummated when the seven wandering stars return to their originally aligned positions (39d–e). After the consummation of the Great Year, the cycle of time presumably begins anew.[32] This suggestion that time for Plato is delimited and cyclical is controversial,[33] but in my view it offers the most promising angle for unpacking Plato's conception of αἰών, as I will show.

If, then, time is not tantamount to duration, eternity, though timeless, is not necessarily extradurational. Since, as noted, Plato routinely characterizes the Forms as being "always the same and unchanging," and the most straightforward way of understanding this expression would be that the Forms are everlasting and immutable (in a manner similar to Aristotle's Prime Mover), and since the domain of the eternal is limited to the Forms, the question now is whether Plato in this stretch of the *Timaeus* is looking to correct this straightforward understanding. It has often been thought that Plato here is indeed striving to articulate the notion of an extradurational eternity. The two main reasons given in support of this view are that Plato takes pains to emphasize that tensed language cannot be applied to the

---

[32] At *Timaeus* 38b Plato limits himself to saying that the universe will exist for all time, and leaves open the possibility that both the universe and time will be dissolved after the Great Year, but in 41a–b the latter possibility is ruled out, since the celestial bodies are said never to suffer destruction.
[33] The justification is that time is an image of αἰών by virtue of its completeness (39d8–e2) and it is the Great Year that is complete and determines "the complete number of time" (39d3–4).

Paradigm of Forms (*Timaeus* 37e3–38a8) and that the Paradigm does not grow "older or younger" than itself (*Timaeus* 38a3–4). Yet neither of these reasons is compelling. As was the case with Parmenides, Plato's point regarding tensed language might simply be that the use of "was" and "will be" would incorrectly suggest that the Forms are changing, when in fact the Form of F always was and always will be just what it *is*.³⁴ As for the latter reason, some have made the connection between this remark and a passage in Plato's *Parmenides*,³⁵ where Parmenides seeks to diagnose a contradiction in the theory of the Forms by arguing that existence is necessarily existence in time and that what exists in time is problematically becoming both older and younger than itself, and from this it has been concluded that Plato's intention here in the *Timaeus* is to reply to this argument by allowing the Forms to exist outside time. This might be right, but again to place the Forms outside time is not necessarily to place them outside duration. Moreover, this passage might rather be reiterating a point made in the *Symposium*, namely that sensible living things can achieve immortality only by reproduction, which Plato there describes as replacing an older with a younger self, and that the Paradigm's everlastingness is not dependent on such reproduction of another "self."³⁶ Thus, both of the main grounds for the extradurational interpretation can be called into question, and the case against the durationlessness of the Paradigm may be pressed further by pointing to several expressions Plato uses to describe it. He continues to describe the Paradigm

---

34 This point is made at greater length by Whittaker, "The 'Eternity' of the Platonic Forms," 140–141. The durational eternity of the Forms has also been maintained by Cornford, *Plato's Cosmology*, 98 n. 1 and 102; O'Brien, "Temps et éternité," 62–65. That the eternity of the Forms should be understood to transcend duration has been argued by H. Cherniss, *Aristotle's Criticism of Plato and the Academy*, vol. 1 (Baltimore: Johns Hopkins Press, 1944), 211, and *The Riddle of the Early Academy* (Berkeley: University of California Press, 1945), 5–6; Schofield, "Did Parmenides Discover Eternity?"; Tarán, "Perpetual Duration." An overview of the arguments both for and against durational eternity may be found in Sorabji, *Time*, 108–112, who concludes that "Plato did not decide between making eternity timeless and giving it everlasting duration."

35 *Parmenides* 141a–d.

36 *Symposium* 207a–e.

as "remaining" (μένειν) and as existing "always" (ἀεί) without offering us any alternative to the straightforward durational senses of "remaining" and "always."[37] Furthermore, Plato does not use the adjective "eternal" (αἰώνιος—a term he apparently coins himself) as an alternative to "everlasting" (ἀΐδιος), as he predicates both of them of the Paradigm.[38] Finally, the parallelism between Plato's use of the expressions "for all time" (τὸν ἅπαντα χρόνον) and "for all eternity" (πάντα αἰῶνα) at *Timaeus* 38c1–2 again suggests that eternity here is to be understood as a durational extension.[39]

It is reasonable, therefore, to conclude that Plato's conception of eternity is similar to Aristotle's, in that his eternal objects are timelessly everlasting and immutable. It remains for us to investigate what Plato has in mind with his enigmatic characterization of time as an *image* of eternity.[40] In fact, this characterization seems enigmatic only if one conceptualizes eternity as a mere negation of time, for example, as timelessness or durationlessness pure and simple. For how could anything be characterized as an image of its negation? Yet I have now established that there is some common ground between eternity and time after all, insofar as both involve duration, and Plato appears to point to this common ground in his explanation of the paradigm-image relation: "While the Paradigm is a being for all eternity, the [cosmos] has come into being and is and will be for all time right up to the end (διὰ τέλους)."[41] Yet duration appears to be only part of the

---

[37] Regarding μένειν, see *Timaeus* 37c6 and μόνιμος καὶ βέβαιος at 29b6. See also the use of μένειν of the Demiurge to refer to his state of rest after having executed his role in creation (*Timaeus* 42e5–6), which cannot mean that the Demiurge goes over into an extradurational state. Regarding ἀεί, see note 23 here and *Timaeus* 27d6, 35a1–2, 48e6. Here again it should be noted that the Demiurge, who is described in terms of a durational sequence of actions and who appears to fall outside the domain of eternity, is also characterized as existing "always."

[38] The terms ἀΐδιος and αἰώνιος are not synonymous in the *Timaeus*, since otherwise 37d1–4 would be indefensibly redundant, and elsewhere in the *Timaeus* (40b5) the term clearly has the sense "everlasting." See Sorabji, *Time, Creation and the Continuum*, 110, and O'Brien, "Temps et éternité," 63.

[39] This point has been made by O'Brien, "Temps et éternité," 63–64.

[40] Plato *Timaeus* 37d7; 38a7; 38b6–c3; 39e1–2.

[41] Plato *Timaeus* 38c1–3.

common ground shared by time and eternity, as emerges from the conclusion of Plato's discussion of the parts and the whole of time: "It is with respect to and for the sake of these [i.e., the parts of time and the Great Year] that all of the stars that possess turnings in their procession through the heavens [i.e., the sun, the moon, and the planets] were generated, in order that this [sensible living thing, i.e., the universe] might be as similar as possible to the complete and intelligible Living Thing by imitating its eternal [διαιωνίας] nature."[42] What deserves emphasis here is that the imitation of eternity is achieved not by time's going on indefinitely, although it does apparently go on forever via the repetition of the Great Year, but by possessing definite parameters. This suggests that the whole of time, that is, the Great Year, imitates eternity not simply by virtue of its duration but by virtue of its duration being *complete*. Unfortunately, Plato does not provide a fuller explanation of how the cosmos over the course of the Great Year captures the completeness of the Paradigm, though in the next section I will show Plotinus working out an admirable explanation for him.

Eternity in Late Antiquity

Whereas Plato's commitment to an extradurational eternity has been subject to debate, there is wide agreement that the Platonists of late antiquity did indeed hit upon a notion of durationlessness in their accounts of eternity. Here I shall pass over the remarks on eternity by several earlier Platonists,[43] in order to provide the much fuller account delivered by Plotinus, the founder in the third century AD of a new kind of Platonism (traditionally called "Neoplatonism"), the

---

42 Plato *Timaeus* 39d7–e2.
43 Notable here are above all the remarks by Plutarch and Philo. For a discussion of the relevant passages in Philo, see J. Whittaker, *God Time Being: Studies in the Transcendental Tradition in Greek Philosophy* (Oslo: Universitetsforlaget, 1971). For Plutarch, see J. Whittaker, "Ammonius on the Delphic E," *Classical Quarterly* 19 (1969): 185–192. Of interest is also Numenius's (*Fragments*, ed. É. des Places [Paris: Les Belles Lettres, 1973], Fr. 5) articulation of a conception of eternity as fixed and stable time.

development it deserves.[44] Plotinus's theory of eternity, which is found primarily in his treatise "On Eternity and Time" (*Ennead* 3.7 [treatise 45 in the chronological order]), is certainly heavily indebted to Plato (and Parmenides), with Plato's enigmatic description of time being an imitation of eternity forming the center point of Plotinus's investigation, but as I will show, Plotinus displays a striking degree of philosophical ingenuity in working out the metaphysics of the intelligible world, which in turn allows him to give a more detailed account of the concept of eternity.

In the first half of *Ennead* 3.7 Plotinus sets out to define eternity (the second half of the treatise is devoted to time), and although his investigation is generally successful, he does not deliver a single definition. Instead, he gives us a number of alternative formulations that collectively approach a definition, including these:

> This life, then, which belongs to that which exists and is in being, all together and full, completely without extension or interval, is that which we are looking for, eternity.[45]
>
> And eternity could be well described as a god proclaiming and manifesting himself as he is, that is, as being which is unshakeable and self-identical, and [always] as it is, and firmly grounded in life.[46]
>
> And if someone were in this way to speak of eternity as a life which is here and now endless because it is total and expends nothing of itself, since it has no past or future—for if it had, it would not now be a total life—he would be near defining it.[47]

---

44 Plotinus's theory of eternity and time has been the subject of many studies. See, for example, W. Beierwaltes, *Plotin, Über Ewigkeit und Zeit (Enneade III 7)* (Frankfurt: Klostermann, 1967); J. E. McGuire and S. K. Strange, "An Annotated Translation of Plotinus *Ennead* iii 7: On Eternity and Time," *Ancient Philosophy* 8 (1988) 251–271; S. K. Strange, "Plotinus on the Nature of Eternity and Time," in *Aristotle in Late Antiquity*, ed. L. Schrenk (Washington, DC: Catholic University of America Press, 1994), 22–53; A. Smith, "Eternity and Time," in *The Cambridge Companion to Plotinus*, ed. L. Gerson (Cambridge: Cambridge University Press, 1996), 196–216; and M. Wagner, *The Enigmatic Reality of Time* (Leiden and Boston: Brill, 2008).
45 *Enn.* 3.7.3.36–38, Armstrong translation.
46 *Enn.* 3.7.5.19–22, Armstrong translation.
47 *Enn.* 3.7.5.25–28, Armstrong translation.

The core idea behind these formulations is that eternity is to be identified with the life of the intelligible world, and a large part of the examination of eternity in *Ennead* 3.7 is aimed at explaining this notion of intelligible life. What emerges from this examination is a list of features, some of which are captured in the formulations above, that characterize intelligible life: it is all together,[48] in a unity;[49] it lacks extension;[50] it lacks parts;[51] it "remains,"[52] and it remains "the same";[53] it is not subject to past and future tense,[54] nor do "before" and "after" apply to it;[55] rather it is always in the present;[56] it is unlimited;[57] it involves both rest[58] and motion.[59]

Some background in Plotinus's metaphysics is necessary to make sense of these features, though only the briefest sketch is possible here.[60] Plotinus envisions a single principle from which everything that in any way exists, including matter, derives, and he calls this principle the One (and, less often, the Good). A second principle, Intellect, is derived from the One, and a third principle, Soul, is derived from Intellect. Together, these principles are commonly referred to as the

---

48 ὁμοῦ: *Enn.* 3.7.2.19–20; 3.7.3.11–12,19 and 37; 3.7.11.3; ἅμα: 3.7.3.18.
49 ἐν ἑνί: *Enn.* 3.7.2.32 and 35; 3.7.6.6; 3.7.11.4; 3.7.12.12; see 3.7.3.12.
50 ἀδιάστατον: *Enn.* 3.7.2.32–34; 3.7.3.15 and 37; 3.7.6.15–17 and 35; 3.7.11.53; 3.7.13.63. See αἰὼν οὐ συμπαραθέων οὐδὲ συμπαρατείνων: *Enn.* 3.7.1: 44–45; and see 3.7.6.24–26, where the text is uncertain but the underlying idea is lack of extension. See also 6.5.11.16–18.
51 ἀμερές: *Enn.* 3.7.3.19; 3.7.6.48–50; see 3.7.2.13; 3.7.4.8–11 and 37–38.
52 μένειν ἐν ἑνί: *Enn.* 3.7.2.35; 3.7.6.6; 3.7.12.12. See Plato *Timaeus* 37d6; μένουσαι ἐν τῷ αὐτῷ: 3.7.3.16 and 20–21; μένουσα δὲ ἀεὶ περὶ ἐκεῖνο καὶ ἐν ἐκείνῳ: 3.7.6.3; οὕτω μένον καὶ αὐτὸ τὸ μένον ὅ ἐστιν ἐνέργεια ζωῆς μενούσης παρ' αὐτῆς προς ἐκεῖνο καὶ ἐν ἐκείνῳ: 3.7.6.9–11; τοῦ ὡσαύτως καὶ μένοντος: 3.7.11.51.
53 ταὐτόν: *Enn.* 3.7.3.11; 3.7.5.21; ταὐτότητος: 3.7.3.26; 3.7.11.51; ὡσαύτως: 3.7.3.10 and 15; 3.7.6.8 and 14; 3.7.11.46 and 51; see ἀτρεμές: 3.7.5.21; 3.7.11.3.
54 *Enn.* 3.7.3.22–36.
55 *Enn.* 3.7.6.17–21.
56 *Enn.* 3.7.3.21–22.
57 ἄπειρον: *Enn.* 3.7.5.23–30.
58 στάσις: *Enn.* 3.7.2.20–36; 3.7.3.10; 3.7.11.46; see ἑστώς: 3.7.3.35; ἑστῶσαν: 3.7.11.4.
59 κίνησις: *Enn.* 3.7.2.27–28; 3.7.3.9–10; 3.7.11.28 and 49–51.
60 For an excellent introduction to Plotinus, see D. J. O'Meara, *Plotinus: An Introduction to the "Enneads"* (Oxford: Oxford University Press, 1993).

three "hypostases," though Plotinus does not himself characterize them by this term.[61] Soul is then responsible for generating matter and the entire sensible world, but it is the second principle that is of concern here. Plotinus famously combines Plato's intelligible world of the Forms and Aristotle's divine intellect (which he identifies with Plato's Demiurge) into a single principle with both subjective and objective aspects. Like Aristotle's intellect, it is a single entity that thinks itself, but unlike Aristotle's intellect, in thinking itself it is thinking the Forms. Eternity is the life of this principle, which is to say that eternity is intellection, and it is because one of Plotinus's major contributions to the philosophical tradition is his extended metaphysical analysis of intellection that he is in a much better position to give a detailed account of eternity, and in particular to explain why time, which seems so different from eternity, is nevertheless an imitation of it.

According to this analysis, intellection is what results from the two primary metaphysical processes, procession and reversion, that start from the first principle of all things, the One. Procession is a direct result of Plotinus's doctrine of double-activity. According to this doctrine, each thing can be effectively identified with a so-called internal activity. Plotinus's favorite examples here are fire and light. Fire, after all, is (among other things) essentially hot, so for fire to be is for it to be generating heat. But hot things cannot exist without spreading heat to their immediate surroundings. So here we have two activities: the fire itself (internal activity) and the heat sent off from the fire (external activity). This doctrine epitomizes two important Platonist principles that are together constitutive of procession. First, generation always follows from perfect existence.[62] In the example here, this means that the heat sent off from the fire is a necessary by-product generated by the

---

[61] See H. Dörrie, "Hypostasis: Wort und Bedeutungsgeschichte," *Nachrichten der Akademie der Wissenschaften in Göttingen* 3 (1955): 35–92. For a brief overview of Plotinus's use of the term "hypostasis," see M. Atkinson, *Plotinus: Ennead V.1 On the Three Principal Hypostases* (Oxford: Oxford University Press, 1983), 55–58.

[62] *Enn.* 3.8.5.6–8; 4.8.6.7–9; 5.1.6.37–38 and 7.37–38; 5.2.1.7–9; 5.4.1.26–36.

fire's achieving its own existence. Second, what is generated is always similar to but also inferior to its generator:[63] both the fire and the heat sent out from it are essentially hot, but the fire is hotter. "Procession," therefore, refers to this production of an inferior image that necessarily follows from a thing's achieving its own nature. "Reversion" refers to the process whereby the external activity then turns back and attempts to return to its source. The starting point of procession explains the necessity of reversion. For procession ultimately begins from the Good (the One), and so since all things desire the Good, all products of procession will necessarily turn back and strive toward it.[64]

Intellection may then be analyzed into the two aspects of procession from and reversion toward the One.[65] Considered from the perspective of procession alone there is only a *pre*-Intellect: it is still indeterminate, but it has the power and desire to receive determination and thus establish itself as a being.[66] Here Plotinus is following the Peripatetic model that likens intellection to sight:[67] just as the eye, in order to become an *actively seeing* eye, must receive determination (i.e., be affected by colors) from outside, so too must the (pre-) Intellect, in order to become an *actively contemplating* Intellect. Yet this attempt to see the One is not entirely successful on account of the utter simplicity of the One, and as a result the Intellect achieves only a refracted conception of the One.[68] It is this refraction that gives rise to the intelligible content of the Intellect, the Forms, and thus establishes intellection and the Intellect as such.[69] In this way intellection can be

---

[63] *Enn.* 3.8.5.24–25; 4.7.8³.9–11; 5.1.6.39 and 7.47–48; 5.2.2.1–4; 5.3.16.5–7; 5.5.13.37–38; 5.8.1.19–21; 6.7.17.4–6. See also A. C. Lloyd, "The Principle That the Cause Is Greater Than Its Effect," *Phronesis* 21 (1976): 146–156.

[64] This is a process that does not as easily lend itself to empirical illustration; the heat from a fire, for example, surely does not turn around and travel back to the fire.

[65] *Enn.* 3.8.8–11; 5.1.5–7; 5.2; 5.3.10–11; 5.6.5; 6.7.15–18.

[66] *Enn.* 2.4.2–5; 3.4.1.6–16; 3.5.9.53–5; 3.9 5; 4.4.2.6–8; 5.1.7.40–41; 5.9.4.10–12.

[67] *Enn.* 3.8.11.1–8; 3.9.5; 5.1.5.18 and 6.42–48; 5.2.1.19–20; 5.3.10–11; 5.4.2; 5.5.7.

[68] *Enn.* 3.8.8.32–8; 6.7.15.20–23 and 16.16–17.

[69] *Enn.* 6.7.17.32–34; see also 5.1.4.27–28; 5.1.7.27–30; 5.2.1.9–13; 5.9.5.12–13; 5.5.8.2–4.

understood both as an epistemic activity and as the productive activity that accounts for the generation of the Intellect and the Forms.

We may now paraphrase Plotinus's account of eternity as follows: eternity is the manner of presence of the plurality of Forms that results from this initial activity of procession from and reversion to the One, which is to say that it is the manner of presence of the Forms as they are contemplated by the Intellect. In what follows I shall aim to elucidate the two central characteristics of this account of eternity: (1) Plotinus's identification of the life or being of a thing with the manner of presence of that thing's formal principles, and (2) in what sense Plotinus takes this manner of presence to be durationless in the case of the Intellect.

We might best elucidate the first characteristic by focusing on a single Form such as the Form of Human Being as it is present at different ontological levels, since this shall provide a particularly clear illustration of the development that the theory of Forms undergoes in Plotinus's hands. Outside of the *Timaeus*, Plato's own commitment to Forms of natural kinds such as human being is questionable. In the *Parmenides* the young Socrates is even made to express his uncertainty about the existence of such Forms,[70] and even in the *Timaeus*, where such Forms are accepted, not only does the intended scope remain unclear,[71] but Plato does not make any attempt to connect the facts of human anatomy to the Form of human being. For when the generated gods undertake the task of creating the human body,[72] there is no indication that they are doing so by looking to the corresponding Form. By contrast, in Plotinus the theory of Forms takes a decidedly biological turn. Plotinus underscores that there are Forms corresponding to

---

70 *Parmenides* 130c1–4.

71 In *Timaeus* 39e–40a Plato suggests that there are only four kinds of Forms of living things corresponding to the four elements: heavenly bodies (fire), winged creatures (air), aquatic creatures (water), and terrestrial creatures (earth).

72 *Timaeus* 44d–47e and 69c–79a. The creation of the bodies of other living things is not discussed (see 90e–92c).

creatures of all kinds; his examples include human beings, horses, oxen, and dogs.[73] He even accepts a (or perhaps many) Form(s) corresponding to plants, which is a clear expansion of the scope of Forms over the *Timaeus*.[74] This is a first indication of Plotinus's metaphysical optimism. He is much more concerned than was Plato to establish the sensible world *in all of its detail* as an expression of the intelligible world. This same concern can be witnessed in how he relates the anatomical details of human and other animal bodies back to their respective Forms. Although Plotinus stops short of saying that there are Forms corresponding to parts of bodies, he allows that intelligible principles corresponding to these parts emerge as a Form is instantiated at lower ontological levels. For as I have shown, just as Intellect generates Soul and Soul in turn generates Nature, the lower power of Soul that is responsible for the creation of bodies in matter, so too are the contents of Intellect, namely the Forms, passed down to Soul and Nature, and as these contents descend, they become more complex. Thus, at the level of Nature one will find form-principles (Plotinus calls them *logoi*) for nearly all of the parts that make up the human body. But these contents also become diversified and individualized. It is not just a form-principle of nose that emerges; rather in one case a form-principle for a snub nose emerges and in another that for an aquiline nose.[75] The upshot of this development in the theory of Forms is that Plotinus can now say that to every individual human being there corresponds a form-principle that accounts for his or her particular features. To Socrates, for example, there corresponds a form-principle that accounts for his snub nose and bulging eyes. This admission of individual forms might seem counterintuitive if one does not bear in

---

73 See, e.g., *Enn.* 6.7.6–12. I have discussed the Forms of living things in much greater detail in Wilberding, "Intelligible Kinds and Natural Kinds in Plotinus," *Études Platoniciennes* 8 (2011): 53–73.

74 In the *Timaeus* plants are created simply as a means of sustenance for other living things, and not as part of the imitation of the intelligible world. See *Timaeus* 76e–77c.

75 See *Enn.* 5.9.12.4–12 and 5.7.

mind that these individual forms exist only at the level of Nature and so are not Forms in the strict sense.

We may now begin to understand what Plotinus means when he identifies the being and life of a thing with the manner of presence of its formal principles. The being or life of an individual body, such as that of Socrates, involves the actualization of all of the form-principles that determine the appearance and structure of that body, and this takes more or less its entire lifetime. The form-principle for growing a beard, for example, is already present in the seed but only begins to be expressed with the onset of puberty.[76] Thus, the manner of presence of the formal principles of Socrates's body, that is, the being and life of his body, is necessarily durational. Socrates's body, understood in this diachronic manner, is one durational instantiation of the Form of Human Being at this ontological level, but it is, taken all by itself, by no means a complete instantiation of the Form. For his body captures only one possibility within the Form of Human Being. The staggering diversity among human phenotypes can be fully captured only by a correspondingly large number of human individuals (again understood as diachronic individuals). Plotinus does not venture any guess as to the specific number of individuals needed to fully account for this diversity, but he does cautiously suggest that this number is limited and is significantly greater than the number of individuals alive at any one time: "But if it is precisely determined how many [individuals] there are, the quantity will be defined by the unrolling and unfolding of all the forming principles; so that, when all things come to an end [παύσεται], there will be another beginning; for the length that the universe ought to have, and all that it will pass through in its life [βίος], is established from the beginning in that which contains the forming principles."[77] The reference to the "end" of the universe's life must not be taken to mean that the universe at some point in time will cease

---

76 See *Enn.* 4.4.11.16–21. See Porphyry *To Gaurus on How Embryos Are Ensouled* 14.1 and 17.6.
77 Plotinus *Enn.* 5.7.3.14–18, Armstrong translation, revised.

to exist, since as Plotinus is quick to emphasize there will be "another beginning." What he rather appears to have in mind here is the "Great Year," when all of the wandering stars are realigned with the fixed stars.[78] I have shown that Plato's *Timaeus* already provided some reason for thinking that the Great Year determined the parameters of the universe's life, and that this had something to do with the capturing of the completeness of the Paradigm, though it was left unclear just how it was capturing that completeness. Plotinus appears to be taking up this suggestion and supplementing it with an account of completeness based on his more elaborate theory of Forms: the life of the universe defined by the Great Year is complete in the sense that it exhausts the information contained in all of the form-principles that derive from the Forms.[79] In the case of human beings, the universe completely instantiates the Form of Human Being over the course of the Great Year, since it takes this entire period to instantiate physically all of the possible combinations of the principles that derive from that Form.

This illustration is meant to show that there is something intuitive after all about Plotinus's identification of life and being with the manner of presence or actualization of formal principles, but it also shows that Plotinus is prepared to distinguish between the being and life of a thing on the one hand and its continuous existence on the other. In Socrates's case, the life of his body (that is, the actualization of all of its formal principles) and the duration of its continuous existence happen to coincide, but not in the case of the entire cosmos. For since the cosmos has always existed and always will exist,[80] its continuous existence encompasses an infinite number of Great Years and thus an infinite number of lives. I would like to capture this distinction by disambiguating between the duration required for a thing to achieve its

---

[78] See Armstrong's note at 5.7.2.23, and A. Petit, "L'éternel retour, un paradox plotinien," in *Études sur Plotin*, ed. M. Fattal (Paris: Editions L'Harmattan, 2000), 75–86. See also 4.3.12.12–19 and 4.4.9.
[79] Proclus approvingly gives a fuller articulation of this view in his *In Remp.* 2.11.24–16.3.
[80] See below.

life, which is to say the period in which it achieves its full completion by actualizing all of its form-principles (*constitutive* duration) and the duration of a thing's continuous existence (*existential* duration). Now we may say that the universe's constitutive duration is one Great Year, but its existential duration is bidirectionally infinite.[81] This distinction nicely sets the stage for our look at the second central characteristic of Plotinus's account of eternity: its durationlessness.

Plotinus goes to great length to emphasize and explain that in the Intellect the Forms are all present at once in a durationless manner, and his analysis of intellection puts him in a much better position to do this than his predecessors. I have already shown that one of the obstacles to crediting Parmenides and Plato with a conception of durationless eternity was that they both described their eternal beings as "remaining" (μένειν) without explicitly providing any alternative to the straightforward durational sense of "remaining." Plotinus's metaphysics of procession and reversion, however, does just this. For it provides, as it were, a distinct metaphysical axis perpendicular to that of duration. If the straightforward sense of "remaining" amounts to identity or changelessness over some (horizontal) duration, this metaphysical notion of "remaining" amounts rather to retaining a (vertical) ontological proximity to its cause. Thus, when Plotinus recasts Plato's description of eternity as "remaining in unity" (μένειν ἐν ἑνί; *Timaeus* 37d6) into his own metaphysical schema as remaining next to the One,[82] the "remaining" is now clearly non-durational, as the alternative is not changing over some duration but durationlessly proceeding further away from the One. Nevertheless, his remarks about the durationlessness of eternity are all too easily misunderstood. One possible misunderstanding would amount to simply identifying eternity and durationlessness. The error of this oversimplification can be

---

81 To be clear, this terminological distinction is not to be found explicitly in the *Enneads*, but it is implicit in Plotinus's claim that the length of the universe's life is one Great Year.
82 *Enn.* 3.7.6.4–8.

seen by considering the relation of eternity to the One. Plotinus makes clear that the One is no more durational than Intellect, yet it is *prior to* eternity,[83] the reason being that eternity is not simply durationlessness but a particular kind of unified presence of the plurality of Forms, and these Forms are not even present in the One. Nor may we simply identify eternity with the durationless activities of procession and reversion. For these activities are constitutive not only of Intellect but also of Soul, and eternity applies only to the former. This is because these activities give rise to a different manner of life and thought at the level of Soul, since it is more derivative and further removed from the One. Plotinus points to three basic differences between the kinds of thinking that occur at these two ontological levels. The first concerns the content of their respective thought: just as the original unity of the One gives way to the plurality of Forms in Intellect, so too does each of the Forms become further pluralized in Soul. The other two differences concern the thinking subject's relation to the contents of its thought. At the level of Intellect, the subject and the objects are identical, and the subject thinks all of the Forms all at once in a nondiscursive manner; at the level of Soul, the subject is not quite identical to its objects, and it must think them discursively, one after another.

It appears to be above all this nondiscursive manner of thinking that captures the core of the intelligible life. This life is, then, nondurational in the sense that Intellect comprehends all of the Forms all at once without requiring even the smallest duration to do so. But does this mean that the Intellect is entirely non-durational? This is to ask whether it makes sense to distinguish here again between the Intellect's

---

[83] *Enn.* 6.8.20.24–25. Post-Plotinian Platonists of late antiquity will speak of what is above the Intellect as being "pre-eternal" (προαιώνιος), e.g., Porphyry 223F, in A. Smith, *Porphyrius Fragmenta* (Stuttgart: Teubner, 1993), where I am inclined to accept the emendation suggested by Hadot and Segonds, as reported by Smith in the critical apparatus *ad loc.*; Proclus *Theol. Plat.* 1.51.4–11 (*Theologia Platonica* [*Théologie Platonicienne*], 5 vols., ed. H. D. Saffrey and L. G. Westerink [Paris: Les Belles Lettres, 2003]) (= Porphyry 232F in Smith, *Porphyrius Fragmenta*); Proclus *Theol. Plat.* 3.77.19–22; Proclus *Elem. Theol.* sec. 107. See Michael Psellus *Opusc. log.* 7.58–72 and 41.82–85 (*Philosophica Minora*, vol. 1, *Opuscula Logica, Physica, Allegorica, Alia*, ed. J. M. Duffy [Leipzig: Teubner, 1992]).

constitutional duration and its existential duration. If so, it would be possible to maintain that the Intellect is constitutionally durationless, insofar as it immediately comprehends all of the Forms, but is nevertheless existentially durational, insofar as it (timelessly) endures in the consummate completion of this activity. In this case the Intellect's relation to duration would be comparable to that of light. For light was widely held to be an activity that required no duration at all to illuminate even the largest of areas. The light of the sun and of the stars, for example, was held to illuminate the earth immediately.[84] And yet this light is continuously present in the universe. Light, then, could serve as a possible analogy to the Intellect, insofar as it is an activity that is constitutionally but not existentially durationless. If this is how Plotinus conceives of the Intellect, then his Intellect would be comparable to Aristotle's Prime Mover: an everlasting activity that is consummated at every moment. The question, then, is whether Plotinus's Intellect should be understood to be merely constitutionally durationless, or to be durationless both constitutionally and existentially.

One weighty reason for thinking that he did not intend his Intellect to be existentially durationless can be found in his argument for the bidirectional everlastingness of the cosmos, which he presents throughout the *Enneads*.[85] One particularly striking version of this argument is found in *Ennead* 5.8 (31).12.20–26: "For this reason those are not right who destroy the image-universe while the intelligible abides, and bring it into being as if its maker ever planned to make it. For they do not want to understand how this kind of making works, that *as long as* the higher reality gives its light, the rest of things can never fail: they are there *as long as it is there; but it always was and will be*. We are forced to use these terms in order to make our meaning clear."[86]

---

84 E.g., Porphyry 131F (in Smith, *Porphyrius Fragmenta*); Philoponus *In DA* 297.8–9; 327.1–6; etc.; *Contra Proclum* 18.3–13; 65,13–26. For Simplicius's criticism of this view, see *In Cat.* 308.28–34 and *In Phys.* 106.30–107.11.

85 See *Enn.* 2.3.18.19–22; 2.9.3.7–14; 2.9.7.1–2; 2.9.13.16–17; 3.2.2.17–18; 5.1.6.30–38; 6.6.18.46–47; 6.9.9.10–11. See 2.1.4.29–30.

86 *Enn.* 5.8.12.20–26, Armstrong translation, slightly revised with emphasis added.

The sensible universe always was and always will be because it is the automatic result of the activities of procession and reversion that derive from and establish Intellect. It is therefore crucial to this argument that the Intellect (and even the One) have some claim to everlasting endurance, that is, that it, too, in some sense, always was and always will be, and here Plotinus gives clear expression to this premise.[87] To be sure, Plotinus then signals that he is not entirely satisfied with saying that the Intellect "was" and "will be," but as he says, we are compelled to attribute a past and future to the Intellect if we want to give a clear articulation of the argument. This argument, then, provides strong grounds for concluding that the Intellect is in some sense enduring everlastingly and thus not completely durationless.

If, however, Plotinus does conceive of the eternal Intellect as enduring in this way,[88] he makes little effort to indicate this in *Ennead* 3 7. On the contrary, all of his efforts there appear to be focused on ensuring that we do not conceive of the Intellect in terms of duration at all, as is evidenced by the characterization he gives of eternity that was indicated above.[89] Some features of this characterization have certainly been taken over from Plato's account of eternity, which, as I have shown, is at least compatible with a durational understanding of eternity. Nevertheless, Plotinus seems to be interested in taking things a step further. I have already shown that he successfully deals with one of the obstacles to a durationless understanding of Plato's eternity by articulating an alternative to the straightforward durational sense of "remaining." In *Ennead* 3.7 he goes on to disambiguate an eternal sense of "always" (ἀεί) from its ordinary sense: "But when we use the word "always" and say that it [eternity] does not exist at one time but not at another, we must be thought to be putting it this way

---

87 See Plotinus's use of the past tense of the One at *Enn.* 6.8(39).16.30–31 and of the Intellect at 6.7.1.54–55, as well as his use of the future tense at 4.7(2).9.22–23 and 13.19–20.
88 Such an "enduring present" interpretation has been suggested by E. Stump and N. Kretzmann, "Eternity," *Journal of Philosophy* 58 (1981): 432.
89 See above notes 48–59.

for our own sake; for the "always" was perhaps not being used in its strict sense, but, taken as explaining the incorruptible, might mislead the soul into imagining an expansion of something becoming more and still more in order that it might never fail ... so the "always" must be taken as saying 'truly existing.'"[90] Moreover, Plotinus also goes beyond Plato by repeatedly emphasizing that eternity lacks extension.[91] Now it is perhaps possible to maintain that all such statements are aiming to establish only the constitutional durationlessness of the Intellect.[92] This would, in any case, help explain why Plotinus continues to make use of his cosmological argument even in treatises written after *Ennead* 3.7,[93] and why he frequently describes the Intellect not simply as αἰώνιος but also as ἀΐδιος and incorruptible.[94] Nevertheless, it would have been easy enough for Plotinus to say that the Intellect's activity is extended, in the sense that it is always being performed (existentially durational), yet unextended insofar as it is always complete (constitutionally durationless), and he never quite puts the matter in this way. Consequently, it appears best to conclude that Plotinus is seeking to remove eternity from duration entirely,[95] yet as a result we must also accept that Plotinus has to some extent undermined his own argument for the bidirectional everlastingness of the cosmos. For the force of that argument derived from conceiving of the intelligible and sensible worlds as existing side by side over an infinite duration,

---

[90] *Enn.* 3.7.6.21–34, Armstrong translation, slightly revised.

[91] See note 50 here. To this the following claims should be added: eternity is comparable to a point (*Enn.* 3.7.3.18–19 and 6.5.11.19, with the important clarification at 1.5.7.28); eternity is not of any size (3.7.6.47; 6.5.11.15).

[92] In this case Plotinus's point would be that the life of the Intellect is extensionless in the sense that no durational extension is required for it to achieve its full constitution; it is without future and past in the sense that, unlike Soul, it is not achieving its life by running through its contents discursively; and similarly for his other remarks.

[93] See *Enn.* 2.3(51).18.19–22 and 3.2(47).2.17–18.

[94] For ἀΐδιος, see *Enn.* 2.4.3.10; 2.4.5.24; 3.1.1.8; 3.2.14.9; 3.5.7.50; 3.7.6.1; 5.1.4.7; 5.1.6.38; 5.1.9.4; 6.5.11.17; 6.9.9.5. For incorruptibility, see 3.7.6.24 and 6.4.10.23.

[95] Even if on occasion he does slip into durational descriptions, see note 87 here and A. H. Armstrong, "Eternity, Life and Movement in Plotinus' Account of *Nous*," in *Le Néoplatonisme* (Paris: CNRS, 1971), 67–74, and Sorabji, *Time, Creation and the Continuum*, 114.

where the latter's existence is at every point necessitated by the former's. By rejecting this conception, Plotinus appears to be opening the door to alternative views on the everlastingness of the cosmos, which, as I will now show, some later Christian philosophers were very keen to explore.

The subsequent Platonic tradition develops and refines this Plotinian account in a number of ways that cannot be explored here,[96] though it will be instructive to see how two later Christian members of this tradition seek to reconcile the metaphysics of God's eternity with the world's creation. Augustine (354–430) grapples with the notion of eternity primarily in book 11 of his *Confessions*,[97] a treatise in which he also famously describes the formative influence that the Platonists had on his philosophical views and religious beliefs,[98] and this is particularly true of his conception of eternity, which he says consists in the whole being always in the present without any change.[99] Yet for Augustine this conception of God's eternity is entirely compatible with the world's having a datable starting point, and it is by exploiting Plotinus's own rejection of eternity's duration that Augustine aims to refute Plotinus's cosmological argument. For Plotinus, one problem with positing a datable starting point of creation was that it would

---

[96] For some discussion of later developments in Platonism, see P. Plass, "Timeless Time in Neoplatonism," *Modern Schoolman* 55 (1977): 1–19; M.-C. Galperine, "Le temps intégral selon Damascius," *Les études philosophiques* 3 (1980): 325–341; P. Hoffman, "Jambique exegete du pythagoricien Archytas: Trois originalités d'une doctrine du temps," *Les études philosophiques* 3 (1980): 307–323, and "Paratasis: De la description aspectuelle des verbes grecs à une définition du temps dans le néoplatonisme tardif," *Revue des études grecques* 96 (1983): 1–26; Sorabji, *Time, Creation and the Continuum*, 114–127; D. P. Taormina, *Iamblique, Critique de Plotin et de Porphyre, Quatre Études* (Paris: J. Vrin, 1993), 57–100; and C. Steel, "The Neoplatonic Doctrine of Time and Eternity and Its Influence on Medieval Philosophy," in *The Medieval Concept of Time: Studies on the Scholastic Debate and Its Reception in Early Modern Philosophy*, ed. P. Pasquale (Leiden: Brill, 2001), 3–31. S. Sambursky and S. Pines, *The Concept of Time in Late Neoplatonism* (Jerusalem: Israel Academy of Sciences and Humanities, 1971), offers a collection of some of the more important passages with English translations and notes.

[97] See also *City of God* 11.4–6.

[98] Augustine *Confessions* 7.9(13).

[99] Augustine *Confessions* 11.11(13): *totum esse praesens*; 11.13(15): *semper praesentis aeternitatis; City of God* 11.6: *in aeternitate autem nulla mutatio est*.

seem to entail the inactivity of the divine hypostases. Since the cosmos necessarily follows from their activity, if the cosmos did not always exist, then the hypostases were not always active, which is impossible. Augustine addresses this problem of God's inactivity—"What was God doing before he made heaven and earth?"—in chapter 10 of book 11, and in the chapters that follow he supplies his solution. He cautions that if anyone is surprised that God "abstained for unnumbered ages from this work before you [God] actually made it, he should wake up and take note that his surprise rests on a mistake. How would innumerable ages pass, which you yourself had not made? You are the originator and creator of all ages. . . .. You have made time itself. Time could not elapse before you made time. But if time did not exist before heaven and earth, why do people ask what you were doing then? There was no 'then' when there was no 'time.' "[100] Since for Augustine (as for Plato and Plotinus) time came to be together with the creation of the cosmos,[101] Augustine insists that Plotinus's problem does not arise: we are not forced to say that God was ever inactive because prior to the creation there was no "ever" to speak of. Of course, Augustine is appealing here to time rather than duration, but it seems fairly clear that he takes them both to be coextensive. For if he thought there could be duration without time, his argument would collapse.

John Philoponus, a sixth-century Christian philosopher whose commitment to Platonism involves some degree of ambivalence, explores a different path of reconciliation. In his treatise *Against Proclus on the Eternity of the World* Philoponus seeks to defend the Christian and, in his view, true Platonic view that the cosmos has not always existed, as well as the Christian (and by Philoponus's own admission un-Platonic) view that the cosmos will at some point cease to exist. Philoponus shows himself to be very aware of a number of problems

---

[100] Augustine *Confessions* 11.13(15). Translation from *Confessions*, trans. H. Chadwick (Oxford: Oxford University Press, 2009).
[101] Augustine *City of God* 11.6: *procul dubio non est mundus factus in tempore, sed cum tempore.*

that accompany these views, two of which we might single out for discussion here. The first concerns the beginning of time. Like Augustine, Philoponus subscribes both to the traditional philosophical view that time came to be together with the cosmos and to the Christian view that the cosmos has a datable starting point. Unlike Augustine, however, Philoponus is also concerned to some extent by the idea of time simply beginning. One source of his concern is a traditional argument, dating back to Aristotle,[102] against any beginning of time based on our normal linguistic habits, which, as one modern author puts it, "lead us, seemingly inexorably, to talk inconsistently of time before that beginning."[103] In response to this problem, Philoponus separates a kind of duration from time and accordingly develops a rather different conception of eternity:

> If eternity is the measure of the life of the Living Thing Itself, that is to say, of everlasting things [τῶν ἀιδίων], there is, one supposes, every necessity that eternity should not be a single point, but a kind of plane or extension, if I may put it so, which is co-extensive with the being of eternal things [τῶν αἰωνίων]. . . . I suppose one could say in regard to everlasting things [τῶν ἀιδίων] that, even though for eternity, which measures their being, there is no movement of a temporal interval, nevertheless it is certainly the case that a kind of self-uniform extension [παράτασίς τις ὁμοειδής] is thought of in connection with their being. For, as I have said, eternity is not a kind of point without parts and eternal things [τὰ αἰώνια] do not exist just as some point.[104]

---

102 *Physics* 251b10–11 and *Metaphysics* 1071b7–9. See Sextus Empiricus *Outlines of Scepticism* 3.141; Philo *De Aeternitate Mundi* 53.
103 W. H. Newton-Smith, "The Beginning of Time," in *The Philosophy of Time*, ed. R. Le Poidevin and M. MacBeath (Oxford: Oxford University Press, 1993), 168.
104 Philoponus *Contra Proclum* 114.24–115.13 (*De aeternitate mundi contra Proclum*, ed. H. Rabe [Leipzig: Teubner, 1899]), translation from *Philoponus. Against Proclus's "On the Eternity of the World 1–5,"* trans. M. Share (Ithaca, NY: Cornell University Press, 2005) with minor revisions. The reference to a point is directed at Plotinus; see note 91 here.

He ends his account of eternity by citing Plato's *Timaeus* 38b–c and concluding that his interpretation is true to Plato, for whom eternity is "a kind of single and uniform extension of the life of everlasting things, not [internally] divided by any differentiation, but always staying the same and remaining self-identical."[105]

Thus, rather than invoking the durationlessness of eternity as Augustine did, Philoponus returns to a concept of eternity that involves extension, that is, duration, in order to eliminate the inconsistency to which our linguistic practices drive us:[106] "This being so, when we say there *was* a point when there *was* no time and there *will be* a point when there *will be* no time,' we are not conceiving of 'was' and 'will be' in terms of time … rather, we claim that these 'temporal' terms, given the nature of the subject, are indicative of eternity."[107]

This attempt to accommodate our linguistic practices, however, forces Philoponus to confront the very problem that Augustine neatly sidestepped. For as Philoponus understands eternity, there is an infinite stretch of eternity prior to the generation of the cosmos (and time) over which God exists, and two undesirable conclusions would seem to follow from this understanding, namely that God was inactive

---

[105] Philoponus *Contra Proclum* 115.24–116.1, Share translation, slightly revised: μίαν τινὰ καὶ ὁμαλὴν τῆς ζωῆς τῶν ἀιδίων παράτασιν μηδεμιᾷ τεμνομένην διαφορότητι ἀλλ' ἑστηκυῖαν ἀεὶ τὴν αὐτὴν καὶ μένουσαν αὐτὴν ἑαυτῇ ἀπαράλλακτον.

[106] Although Plotinus preferred to think of eternity as intelligible life itself, for subsequent Platonists, including Philoponus's opponent here, Proclus, eternity becomes the *measure* of life, a view that might be legitimately inferred from Plato's *Timaeus*, since there measure appears to be a central feature of time and time is supposed to be an imitation of eternity. (For Proclus, see *In Remp.* 2.11.21–24: μέτρον δὲ ὁ πᾶς αἰών, μέτρον ἄρα καὶ ὁ πᾶς χρόνος. ἀλλ' ὁ μὲν τῆς τοῦ νοητοῦ ζῴου ζωῆς μέτρον, ὁ δὲ τῆς τοῦδε τοῦ κόσμου ζωῆς, ὁ χρόνος; *Plat. Theol.* 3.33.4–7; 3.72.16–18; *Elem. Theol.* 54 [see 52–53]; *In Parm.* 745.1–2; 1194.6–7; *In Tim.* 1.256.17–19; *In Tim.* 3.17.22–30.) This must be another motivating factor behind the reintroduction of a durational conception of eternity by Philoponus, who surely puzzled over how something utterly without extension could be a measure. (This, however, appears to have been Proclus's position, though he himself struggles to explain the manner in which eternity measures intelligible life. As Saffrey and Westerink remark (at Proclus *Theol. Plat.* 3.57.4): "Proclus nous a dit dans l'*In Tim.* III, p. 14.13–15, que la manière dont l'éternité est cause et mesure des essences, des puissances et des activités de tout, 'ce sont choses que je développe seulement dans les plus secrètes retraites de ma pensée.' Autrement dit, c'est l'object d'une méditation mystique."

[107] Philoponus *Contra Proclum* 116.1–9.

prior to creation and that at the point of creation God underwent a change. Philoponus proposes the following strategy for steering clear of these difficulties. First, he maintains that God remains the same and unchanging insofar as he is always willing the good.[108] This act of willing involves always willing the creation of all things, but with one qualification: he is always willing that all things come to be *when* it is best and most natural for them to come to be. As an example Philoponus points to the generation of Socrates. God has always been and always will be willing that Socrates come to be, but there is a natural time of Socrates's coming to be, namely after Socrates's father, Sophroniscus, has come to be.[109] Philoponus furthermore maintains that this same line of reasoning can be applied to the cosmos as a whole: "In the same way, even before the ordered state came to be, God was willing the ordered state to exist, but since it is a generated thing He was willing it to exist when it was able to exist. For none of the generated things are able to exist without beginning. . . . Therefore, it is not the case that if God was always willing the ordered state to exist, that it is thereby also necessary that the ordered state be everlasting."[110] Yet one might justly complain that these two cases are not sufficiently analogous. For in the case of Socrates, we have a ready explanation as to why Socrates was not generated earlier, namely that the conditions for his generation in the sensible world had to be met first, but in the case of the cosmos there can be no such appeal to preconditions in the sensible world. Thus, one is left with the impression that Philoponus has no satisfactory response to Parmenides's incisive question: "And what need would have driven it later rather than earlier, beginning from nothing, to grow?"[111]

---

108 Philoponus *Contra Proclum* 81.9–14.
109 Philoponus *Contra Proclum* 567.21–27. Plotinus certainly thinks that Socrates's body, as one possibility within the Form of Human Being, is generated only at a certain time, but he maintains that each Form is *always* in the process of instantiation. Philoponus denies this.
110 Philoponus *Contra Proclum* 567.28–568.5.
111 Parmenides 28B8.9–10. Translation from G. S. Kirk, J. E. Raven, and M. Schofield, *The Presocratic Philosophers* (Cambridge: Cambridge University Press, 1983).

Conclusion

This survey has shown how a handful of thinkers all defined eternity in terms of all things being present all at once, but that their accounts of eternity nevertheless differed from one another in significant ways. Due to constraints of space I have focused on examining how these accounts spelled out this feature of being "present all at once" in terms of the manner in which eternity and eternal beings are related to duration, but it would be fitting to conclude the survey by briefly calling attention to the diversity of opinions regarding the scope of "all things." For it is possible to witness this scope increasing over the centuries. What was originally in Parmenides limited to one thing, namely being, expands to include the plurality of intelligible Forms in Plato. Plotinus further broadens this scope in two ways, first by bringing together the demiurgic Intellect and the Forms, and second by expanding each Form to include a whole world of sensible possibilities. As I have shown, the Form of Human Being in some sense eternally contains a form-principle that accounts for Socrates's body, even though the sensible Socrates only exists for a limited time, and since Plotinus sees the sensible world as sequentially instantiating all of the possibilities contained in the intelligible world, he is even prepared to speak of the intelligible world containing the future.[112] While for Plotinus the future is present in the eternal Intellect only in this limited sense, in the early Christian tradition, where eternity is attributed above all to God, "all things" is understood to include both much more and much less than the Forms. It may be said to include much less than the Forms insofar as Christian thinkers carefully distinguish between (the eternity of) God and (the eternity of) the intelligible world. Augustine, for example, follows the tradition in having the intelligible Forms derive from a higher principle, but whereas Plotinus considered this higher principle, the One, to be beyond eternity precisely because there is no plurality of Forms there, for Augustine this

---

[112] Plotinus *Enn.* 6.7.1.48–57.

first principle, God, is the primary eternal subject.[113] Thus, it is not the Forms per se that are all together in God but some still more eminent versions of the Forms. Since, however, the possibility of a personal relationship to God is fundamental to Christian belief, early Christians also expanded the scope of "all things" to include all individual events and personal actions that take place in the sensible world.[114] As a result, when Christian philosophers such as Augustine and Boethius echo Plotinus's statement about the future being precontained in God's eternal present, they are referring to the future in a much fuller and much more straightforward sense.

Boethius, in fact, is a fitting figure with which to end this survey of ancient philosophers' views on eternity, as his famous account of eternity as God's mode of being was itself to serve a mediating role in the history of philosophy: it is heavily inspired by its Greek sources and becomes itself a major source of influence for many medieval discussions of eternity. Boethius's discussion of eternity is found above all in *The Consolation of Philosophy* 5.6 and *On the Sacred Trinity* 4, and the following classic passage from the former may serve to give a fair indication of the scope of influence:

> Eternity therefore is the whole and perfect possession all together of a life which cannot end, which becomes clearer from a comparison with temporal things. For whatever lives in time progresses as something present from what is past to what is future, and there is

---

113 Augustine calls the intelligible universe the "heaven of heavens" (*caelum caeli*), but also a *creatura aliqua intellectualis* (*Confessions* 12.9(9)), *spiritalem uel intellectualem illam creaturam* (12.17(24)), an *intellectualis natura*, and a *mens rationalis et intellectualis* (12.15(20)). Like Plotinus's Intellect, its activity of contemplation is nondiscursive (12.13(16)), and it achieves its being by turning back to the first principle (*On Genesis Against the Manicheans* 1.2(3); see *Confessions* 12.11(12¹). Because it is (timelessly) created by God, it is not "coeternal" (*coaeterna*) with God, but it is eternal (*Confessions* 12.9(9); 12.11(12–13); 12.12(15) 12.15(19); etc.). Moreover, unlike Plotinus, Augustine places emphasis on the fact that this *mens* is, as a created being, not by its very nature immutable, though it does in fact not undergo any change thanks to its relation to God (12.12(15) and 12.15(21)).

114 In fact, this expansion is already found in Post-Plotinian Platonists of late antiquity such as Iamblichus and Proclus. For a brief discussion and references, see R. Sharples, *Cicero "On Fate" and Boethius "The Consolation of Philosophy" IV.5–7, V* (Warminster: Aris and Philipps, 1991), 25–29.

nothing placed in time which could embrace the whole extent of its life equally. It does not yet grasp tomorrow, and it has already lost yesterday. Even in today's life you do not live more than in the moving and transitory moment. So what is subject to the condition of time is not yet such as rightly to be judged eternal [*aeternus*], even if, as Aristotle believed of the world, it never began to exist, and does not cease, but has its life stretched out with the infinity of time. For even if its life is infinite, it does not include and embrace the whole extent of that life all together, since it does not yet possess the future and it already lacks the past. So that which embraces and possesses equally the whole completeness of a life which cannot end, and for which there is not some of the future missing nor some of the past elapsed—that is rightly held to be eternal [*aeternum*]. And it must be in possession of itself and always present to itself, and must have present to itself the infinity of moving time.[115]

Some features already familiar from the Platonic accounts explored above are on full display here. Notably, eternity is contrasted to time and characterized by completeness and life. Elsewhere, we also find other familiar characterizations of God's mode of eternal being: he is said to be "always" and "remaining," and time is said to be an imitation of eternity.[116] As I have now shown in my examinations of earlier Platonists, these features are prima facie compatible with a number of possible conceptualizations of eternity, and so it should perhaps come as no surprise that scholars have reached no consensus on Boethius's own understanding of eternity. Some attribute to him an atemporal durational conception of eternity, comparable to that

---

115 Boethius *Consol.* 5.6.4–8 (Moreschini ed., ll. 9–29), translation from Sorabji (*Time, Creation and the Continuum*, 113–114).
116 For "always" (*semper*), see *Consol.* 5.6.8 (Moreschini ed., l. 28) and *Trin.* 4. (Moreschini ed., ll. 231–48). For "remaining," see *Trin.* 4. (Moreschini ed., l. 244: *permanens*); *Trin.* 4 (Moreschini ed., l. 245: *consistens*); *Consol.* 5.6.12 (Moreschini ed., l. 49: *manentis*). For imitation, see *Consol.* 5.6.12 (Moreschini ed., ll. 39–47).

of Philoponus above;[117] others see a more Plotinian extradurational eternity in these discussions.[118]

Since both of these conceptions have already been adequately explored in the previous sections, I shall return to the point introduced above about the extension of the scope of "all things" to include individual events and actions. Boethius was well aware that this expansion naturally raises new problems of its own, foremost among which is a dilemma that he articulates in book 5 of his *The Consolation of Philosophy* regarding the causal relation between God's knowledge and actions in the sensible world.

Since God eternally knows what our future actions will be, it would seem to be the case that our future actions are predetermined by God and that free will in human beings is a mere illusion. Boethius employs the concept of eternity to work out an ingenious solution to this problem. Since "God's condition is always eternal and present," God "in his simple act of knowing considers all things as if they were happening now."[119] Thus, God is no more determining our future actions than the casual observer is determining the actions she happens to witness in the present.[120] Yet Boethius's analogy to the casual observer suggests that Boethius's rejection of predetermination rests on his embracing the other horn of the dilemma, namely that our actions are somehow the causes of God's knowledge. Even if we put aside for the moment the question of how actions that have yet to be performed can be causes of present knowledge, there remains the more fundamental objection that God should be the sole cause of his own knowledge. As

---

117 E.g., Stump and Kretzmann, "Eternity," and B. Leftow, *Time and Eternity* (Ithaca, NY: Cornell University Press, 1991), 112–146. In J. Marenbon and D. E. Luscombe, "Two Medieval Ideas: Eternity and Hierarchy," in *The Cambridge Companion to Medieval Philosophy*, ed. A. S. McGrade (Cambridge: Cambridge University Press, 2003), Marenbon also appears to endorse a durational interpretation (52–53).

118 E.g., Sorabji, *Time, Creation and the Continuum*, 119–120.

119 Boethius *Consol.* 5.6.15 (Moreschini ed., ll. 63–64). Translation from Sharples, *Cicero "On Fate" and Boethius "The Consolation of Philosophy" IV.5–7, V*.

120 Boethius *Consol.* 5.6.18–19 (Moreschini ed., ll. 70–74).

Boethius puts it: "How backward this is, that the outcome of temporal things should be said to be the cause of eternal foreknowledge!"[121] It would seem, then, that while Boethius admirably articulates some of the problems surrounding God's knowledge of the particulars in the sensible world and explores some promising solutions, he ultimately falls short of presenting an adequate resolution to the dilemma and bequeaths these problems to his medieval philosophical successors, whose own engagement with these issues will be examined in the next chapter.

Abbreviations

AMMONIUS *In de Int.*   Ammonius, *In Aristotelis De interpretatione commentarius* (Commentary on Aristotle's *On Interpretation*)

BOETHIUS *Consol.*   Boethius, *De consolatione philososophiae* (*Consolation of Philosophy*)

BOETHIUS *Trin.*   Boethius, *De sancta trinitate* (*On the Sacred Trinity*)

JOHN PHILOPONUS *In DA*   John Philoponus, *In Aristotelis De anima libros commentaria* (Commentary on Aristotle's *On the Soul*)

JOHN PHILOPONUS *Contra Proclum*   John Philoponus, *De aeternitate mundi contra Proclum* (*Against Proclus On the Eternity of the World*)

MICHAEL PSELLUS *Opusc. log.*   Michael Psellus, *Opuscula Logica, Physica, Allegorica, Alia* (Logical, Physical, Allegorical and Other Works)

OLYMPIODORUS *In Meteor.*   Olympiodorus, *In Aristotelis Meteora Commentaria* (Commentary on Aristotle's *Meteorology*)

PLOTINUS *Enn.*   Plotinus, *Enneads*

---

121 Boethius *Consol.* 5.3.15 (Moreschini ed., ll. 46–48).

PROCLUS *Theol. Plat.* Proclus, *Theologia Platonica* (*Platonic Theology*)

PROCLUS *Elem. Theol.* Proclus, *Elementatio theologica* (*Elements of Theology*)

PROCLUS *In Parm.* Proclus, *In Platonis Parmenidem commentaria* (Commentary on Plato's *Parmenides*)

PROCLUS *In Remp.* Proclus, *In Platonis Rem Publicam commentarii* (Commentary on Plato's *Republic*)

PROCLUS *In Tim.* Proclus, *In Platonis Timaeum commentaria* (Commentary on Plato's *Timaeus*)

SIMPLICIUS *In Cat.* Simplicius, *In Aristotelis Categorias commentarium* (Commentary on Aristotle's *Categories*)

SIMPLICIUS *In Phys.* Simplicius, *In Aristotelis Physica* (Commentary on Aristotle's *Physics*)

Reflection

ETERNITY AND ASTROLOGY IN THE WORK OF VETTIUS VALENS

Dorian Gieseler Greenbaum

Vettius Valens (c. 120–185 CE) wrote an important treatise on the practice of astrology (*Anthology*) that was transmitted to the Arabs and became a useful text for astrologers in the medieval period. Though hardly a philosopher, Valens had a reasonable knowledge of the philosophical trends of his day and was versed in classical Greek authors, such as Plato, Euripides, and Aristotle. Thus there is some justification for examining Valens as representative of an astrologer's understanding of eternity. In this case study, I examine Valens's use of the Greek word αἰών. When Valens employs αἰών and its derivatives, they do not always mean "eternity" in the sense of something everlasting or always existing, but sometimes connote a long period of time, that is, an age. However, Valens does insist that astrology itself possesses qualities that can be considered eternal in Aristotle's sense of a heavenly eternity.

An important passage in which Valens uses the word αἰών (*Anthology* IX, 2.2) is as follows:

> For just as the all-seeing Sun, with untiring motion, whirls round with the cosmic revolution and rides through a time of great age, he by turns reinstates the dances of the stars with their varying

courses and [then] detaches them. He causes seasonal turnings, proper times and phases, beginning from where he left off and leaving off from where he begins; and enchanting and arousing the souls of men, he is responsible for reputation, action, and every kind of advance. Likewise also the Moon, fortune of the cosmos, waxing and waning by the Sun's power, makes [her own] phases and changes of weather, and in ripening fruits for men she becomes co-responsible for life. In the same way, too, it will be necessary to investigate, for every birth, in what parts of the cosmos the Lots of Fortune and Daimon fell out . . .

ὄνπερ γὰρ τρόπον ἐπὶ τοῦ κοσμικοῦ περιπολίσματος ὁ παντεπόπτης "Ἥλιος ἀκαμάτοις φοραῖς δινούμενος καὶ μακροῦ αἰῶνος χρόνον διιππεύων τὰς τῶν ἀστέρων χορείας ἀλλεπαλλήλοις δρόμοις ἀποκαθίστησι καὶ ἀποχωρίζει, τροπάς τε καὶ καιροὺς καὶ φάσεις ποιούμενος, ἀρχόμενος ὅθεν ἔληξε καὶ λήγων ὅθεν ἄρχεται, τὰς δὲ ψυχὰς τῶν ἀνθρώπων θέλγων καὶ διεγείρων, αἴτιος δόξης καὶ πράξεως καὶ πάσης προκοπῆς τυγχάνει, ὁμοίως δὲ καὶ ἡ Σελήνη, τύχη τοῦ κόσμου ὑπάρχουσα καὶ ὑπὸ τῆς ἡλιακῆς δυνάμεως αὐξομειουμένη, τὰς φάσεις ποιεῖται καὶ τὰς τῶν ἀέρων μεταβολάς, καὶ τοὺς καρποὺς πεπαίνουσα τοῖς ἀνθρώποις ζωῆς παραιτία γίνεται, τῷ αὐτῷ τρόπῳ καὶ ἐπὶ πάσης γενέσεως σκοπεῖν δεήσει τὸν κλῆρον τῆς τύχης καὶ τοῦ δαίμονος ἐν ποίοις μέρεσι τοῦ κόσμου ἀπερρύησαν . .[1]

This highly poetic passage, possibly derived from an earlier verse original,[2] introduces the topic of the Lots of Fortune and Daimon, two positions in the natal chart that are responsible for, among other things, determining the length of a life along with its actions and reputation. Without becoming mired in the intricacies of astrological technique, calculating these lots involves the Sun and Moon, two

---

[1] In Vettius Valens, *Anthologiarum libri nctem*, ed. David Pingree (Leipzig: B. G. Teubner, 1986), 318.17–28, my translation.
[2] See Stephan Heilen, "Some Metrical Fragments from Nechepsos and Petosiris," in *La poésie astrologique dans l'Antiquité*, ed. I. Boehm and W. Hübner (Paris: De Boccard, 2011), 23–93.

essential components of an astrological chart. In astrology, the Moon is associated with fortune (τύχη) and the body; the Sun with mind, soul, and spirit (δαίμων). Fortune and Daimon have a long and intertwined association in Hellenistic culture.[3] (See table 1a.1.)

Circles and Starry Dances

The Sun circling through the "cosmic revolution" initiates planetary cycles that arise in relationship to it. Its daily cycle gives us day and night; its movement through the yearly cycle gives the seasons marked by solstices and equinoxes.

Valens invokes the metaphor of the dances of the stars in combination with the word αἰών. This is not his invention: Plato uses the metaphor of the starry dance in *Timaeus* (40c–d), but it originates in the fifth century BCE (e.g., Philolaos, Euripides).[4] The stars in a choral dance, representing a regular and repeating motion, subsequently become a common trope.[5] Plato uses it to describe the planets, which meet and separate, reverse course, and move forward (*Timaeus* 40c).

Valens too alludes to this dancing together and separation of the planets. The *Epinomis* (982d–e) and Philo Judaeus (*De opificio mundi* 78) also use the imagery of the stars' choral dance.[6] The Middle Platonist Apuleius (*De Platone et eius dogmate* 1.10.9–14)[7] connects the choric dance of the planets to time. Komorowska has demonstrated Valens's exposure to Middle Platonism.[8] So Valens employs a well-known image, whether it was obtained directly from Plato or not.

---

[3] See Dorian Gieseler Greenbaum, "The Daimon in Hellenistic Astrology: Origins and Influence" (Ph.D. diss., Warburg Institute, University of London, 2009), and eadem, *The Daimon in Hellenistic Astrology: Origins and Influence* (Leiden/Boston: Brill, 2015).
[4] See James Miller, *Measures of Wisdom: The Cosmic Dance in Classical and Christian Antiquity* (Toronto: University of Toronto Press, 1986), 23, 28–37, 46–49.
[5] Miller, *Measures of Wisdom*.
[6] Miller, *Measures of Wisdom*, 41, 59.
[7] Miller, *Measures of Wisdom*, 258.
[8] Joanna Komorowska, *Vettius Valens of Antioch: An Intellectual Monography* (Kraków: Ksiegarnia Akademicka, 2004).

TABLE IA.I *Astrological Attributes of the Sun, Lot of Daimon, Moon, and Lot of Fortune*

Sun	Lot of Daimon	Moon	Lot of Fortune
Breath	Soul	Body	Human bodies
Soul	Intentional mind	Mother	Actions in life
Movement	Character	Conception	Sufferings of soul
Mind	Power	Fortune	Companionship
Light of the mind	Worth	Breath	Reputation
Intentional mind	Reputation	An eye	Fortune
Spirit (*daimōn*)	Religious rites	*Pronoia* (providence, foreknowing)	
Action	Intended plans		
Vision, an eye	Advice		
Oracular response of gods	Mental activity		
Soul's sense-perception			

Sources: Vettius Valens, *Anthologiarum libri novem*, ed. David Pingree (Leipzig: B. G Teubner, 1986), I, 1; II, 20; IV, 4 and 25; IX, 2; Antiochus of Athens (ap. Rhetorius), "Thesauroi," in *Catalogus Codicum Astrologorum Graecorum I*, ed. F. Boll and F. Cumont (Brussels: Henri Lamertin, 1893), 160; Antiochus of Athens (ap. Rhetorius), "On the Seven Planets in an Epitome from Antiochus," in *Catalogus Codicum Astrologorum Graecorum VII*, ed. F. Boll (Brussels: Henri Lamertin, 1908), 127; Paulus Alexandrinus, *Elementa Apotelesmatica*, ed. Emilie Boer (Leipzig: B. G. Teubner, 1958), chap. 23; Paulus Alexandrinus and Olympiodorus, *Late Classical Astrology: Paulus Alexandrinus and Olympiodoru. with the Scholia from Later Commentators*, trans. Dorian Gieseler Greenbaum (Reston, VA: ARHAT, 2001), 41–42, 103, 108; Olympiodorus, *Eis ton Paulon <Heliodorou>. Heliodori, ut dicitur, in Paulum Alexandrinum Commentarium*, ed. Emilie Boer (Leipzig: B. G. Teubner, 1962), chap. 22; Rhetorius, "Selected Chapters," in *Catalogus Codicum Astrologorum Graecorum VIII, pars IV*, ed. F. Cumont (Brussels: Henri Lamertin, 1921), 122.20–23.

The Sun leads and regulates the choral dance because it embodies and sets both primary and secondary motion. These correspond to the two Platonic motions of Same and Other, respectively. As leader of the planetary dance, the Sun physically represents the cycle of the seasons and repeating cycles that eventually result in "great ages," one example of which is Plato's Great Year.⁹ In astrology, different cycles of the Sun, Moon, and planets create "maximum," "middle," and "minor" years assigned to each planet; minor years are based on recurrence cycles with the Sun. Valens gives the Sun preeminence here because of its leadership role, and because its position is essential for the Lots of Fortune and Daimon, which can determine both lifespan and reputation.

In *Timaeus*, the existence of the sun, moon, and planets defines and protects the numbers of time (*Timaeus* 38c); their orbits, revolutions, and cycles allow a way for time to be measured. Similarly in Valens, the Sun and Moon cause the Lots of Fortune and Daimon to exist, determine, and provide life and reputation.

A Great Age (μακρὸς αἰών)

The use of αἰών and χρόνος in the same phrase is unattested elsewhere in this form.¹⁰ Perhaps Valens means just a "long span of time"—or the juxtaposition deliberately recalls the difference between time and eternity in the *Timaeus*. Perhaps Valens means the idea of a great age, even something like Plato's Great Year (*Timaeus* 39d–e), where at its basic astronomical level the eight celestial revolutions of the starry sphere and the planets, sun, and moon meet at the same place, but also join harmonically and philosophically.¹¹ With the circle of the Sun moving through the cosmos, coupled with the Sun's leadership of the circles of the planets, Valens describes a hierarchy of two systems moving cyclically together, with the planets' cycles dependent on the Sun's authority.

---

9 See Godefroid de Callataÿ, *Annus Platonicus: A Study of World Cycles in Greek, Latin and Arabic Sources* (Paris: Peeters, 1996).
10 Heilen, "Some Metrical Fragments from Nechepsos and Petosiris," 60.
11 See de Callataÿ, *Annus Platonicus*, 7–9.

Interestingly, Valens uses the phrase μακρὸς αἰών elsewhere in the *Anthology*, at IX, 8.35: "[Nature] ... serves the cosmos, rousing it from sleep and cycling it round in a Great Age" (ἡ φύσις ... ὑπηρετοῦσα δὲ τοῖς κοσμικοῖς καὶ ἐξ ὕπνου ἐγειρομένη καὶ εἰς μακρὸν αἰῶνα κυκλοστρεφουμένη).[12]

Interpreting μακρὸς αἰών here as a kind of great age is plausible, with its idea of a cycle forming such an age. This later quotation further supports a "great age" interpretation in the earlier passage. Here, too, the cosmos is "roused from sleep" just as the Sun in the earlier passage "arouses the souls of men." Again a cosmic image is juxtaposed with a solar one.

Turns of Fortune

The passage quoted above (IX, 2.2) contains several wordplays on the Greek word τροπή, "turn, turning," which cannot always be fluently rendered into English. Turning is an important function in astronomical cycles (e.g. the solstices), as it is in dancing, and it does not seem coincidental that τροπή (or idiomatic forms deriving from the same verb τρέπω) appear three times. In fact, the first words of the passage (ὄνπερ γὰρ τρόπον), which mean "in the same manner," translate literally as "by the very same turn." Next, the Sun makes "turnings" (τροπάς), its apparent turns north and south at the solstices, causing the seasons. Finally, Valens uses τῷ αὐτῷ τρόπῳ, "in the same manner," literally "by the same turn," to insert the calculation of the lots into the turning process.

The Moon, too, is part of the process of cyclical change. While the Sun regulates the planets in their choral dance and creates both the day and the year, the Moon, by contrast, gives us the month (its own personal dance with the Sun) and, as Valens emphasises, affects weather and the production of food to sustain life. As the

---

12 Valens, *Anthology* IX, 8.35 (*Anthologiarum libri novem*, 330.13).

astrological ruler of the physical body and the closest luminary to the earth, the Moon symbolically demonstrates, through its waxing and waning, the vagaries and changeability of life. It is associated with both good and bad fortune. Unlike the Sun, though, the Moon is only "co-responsible," since it is also dependent on the Sun. However both astrological lots, the Lot of Fortune, "the Moon's lot," and the Lot of Daimon, "the Sun's lot" (Paulus Alexandrinus, *Elementa Apotelesmatica*, chap. 23),[13] are important for Valens, especially for determining length and quality of life.[14]

Chronos versus Aiôn in the Anthology

Astrology deals with *chronos* on a fundamental level: the mundane passage of time and allotments of it are given at birth to every human. (The planet Saturn, by way of a Greek pun, rules time—Kronos/Chronos.) The prediction of eclipse cycles was well established by Valens's time. Astrological techniques assigned different planets to rule different times of life.[15]

Yet for Valens, astrology is more than *chronos*; it is "a holy and immortal theory."[16] Those who study "dances of the gods and mysteries" (i.e., astrologers) enter into "an immortal place" and attain a "god-like glory" and "eternal honor and fame."[17] The knowledge of heavenly bodies is "divine and revered" and inspires

---

13 Paulus Alexandrinus, *Elementa Apotelesmatica*, ed. Emilie Boer (Leipzig: B. G. Teubner, 1958), 49; Paulus Alexandrinus and Olympiodorus, *Late Classical Astrology: Paulus Alexandrinus and Olympiodorus with the Scholia from Later Commentators*, trans. Dorian Gieseler Greenbaum (Reston, VA: 2001), 42.

14 Valens will demonstrate this in the remainder of Chapter 2 of Book IX. See the discussion of these lots so used in Greenbaum, *The Daimon in Hellenistic Astrology*, 331–335; see also the rich commentary of Stephan Heilen, "Hadriani Genitura"—*Die astrologischen Fragmente des Antigonos von Nikaia* (Berlin/Boston: De Gruyter), 2 vols., here vol. 2, 1158–1182, whom I thank for emphasizing their importance in this chapter and spurring me to further investigation.

15 See Valens, *Anthology* IV,1–10, 17–25 (*Anthologiarum libri novem*,).

16 Valens, *Anthology* IV, 11.9 (*Anthologiarum libri novem*, 163.20–21).

17 Valens, *Anthology* IV, 11.14 (*Anthologiarum libri novem*, 164.13–15); *Anthology* IX, 6.7 (*Anthologiarum libri novem*, 345.5).

Valens to keep his life pure and his soul "immortal."[18] These descriptions mirror Aristotle's description of the heavenly αἰών as "immortal and divine" (*De caelo* 1.279a28).

Valens even swears by the cycles of the planets, Sun, and Moon: "I adjure you, my most honored brother, and those being initiated into the mysteries with this composition, by the starry vault of heaven and circle of the twelve zodiac signs, by both the Sun and Moon, and the five wandering stars through which all life is driven, by both Providence itself and holy Necessity, to keep these things in secret."[19]

Thus, Valens finds an approximation of eternity both in studying astrology and the doctrine itself. Though the body may be mortal, Valens ardently believes that the theory and practice of astrology is a path for the soul to the immortal, divine, and eternal.

---

18 Valens, *Anthology* VI, 1.15 (*Anthologiarum libri novem*, 232.6–8).
19 Valens, *Anthology* IV, 11.11–12 (*Anthologiarum libri novem*, 163.25–29); similarly at *Anthology* VII, 1.3 (*Anthologiarum libri novem*, 251.18–22) and VII, 6.231–234 (*Anthologiarum libri novem*, 281.4–18) (*Anthologiarum libri novem*, 231–234).

Reflection

ETERNITY AND WORLD-CYCLES

Godefroid de Callataÿ

According to Plato's definition in the *Timaeus*, time should be conceived of as "a moving likeness of eternity," and the planets, or more exactly the planetary spheres, as the "instruments of time." The *Timaeus* also provides us with the mathematical problem of the "Great Year," which was to become for many centuries the locus classicus in terms of world-cycles.[1] Much like the hands of an analog clock, the seven planets of the traditional geocentric representation (namely, counting in the ascending order from the Earth: the Moon, the Sun, Mercury, Venus, Mars, Jupiter, and Saturn) move along the sphere of the fixed stars at different speeds. The "Great Year"—which Plato rather calls the "Perfect Year"—is defined as the period of time necessary for all seven planetary spheres to come back into conjunction with one another and with the eighth sphere, namely the sphere of the fixed stars. The *Timaeus* presents the problem as not absolutely unsolvable. Thus, after having dealt with the lunar (one month) and solar (one year) revolutions, Plato states: "None the less it is possible to grasp that the Perfect

---

[1] Godefroid de Callataÿ, *Annus Platonicus: A Study of World Cycles in Greek, Latin and Arabic Sources*, Publications de l'Institut Orientaliste de Louvain, 47 (Paris: Peeters, 1996).

Number of time fulfills the Perfect Year at the moment when the relative speeds of all the eight revolutions have accomplished their courses together and reached their consummation, as measured by the circle of the Same and uniformly moving" (*Timaeus* 39d).² Plato nowhere gives a solution to the problem, although it is plausible that he left in his writings—especially in some of his most famous myths (the Atlantis myth in *Timaeus/Critias*; the myth of the two alternating revolutions of the world in the *Politicus*; the myth of Er at the end of the *Republic*)—indications enabling one to reconstruct the "perfect cycle" as well as its principal divisions in the form of ages, floods, and conflagrations.

The Great Year as a grand conjunction of the planets is found in a wide variety of ancient sources, all of them connecting it with the idea of a complete recommencement of the heavenly motions and most of them suggesting as well a certain recurrence in the order of things, cycle after cycle, in the world here below. Apart from a few skeptical and Epicurean thinkers who doubted the existence of such a phenomenon, resolute opposition to the theory was largely found among Christians. Origen (third century CE) and Augustine (fourth–fifth centuries CE), for instance, expressed the view that such a periodicity was wholly incompatible with the uniqueness of Jesus Christ's incarnation.³ On the opposite side, the Stoics and, in general, all those who believed in the power of astrology endorsed the theory with enthusiasm and did not hesitate to push the argument to its ultimate consequences, that is, to take just one typical example, by assuming that the same Socrates would be born again and teach his students an infinite number

---

2 F. M. Cornford, *Plato's Cosmology* (London: 1937), 117.

3 Godefroid de Callataÿ, "Platón astrólogo: La teoría del Gran Año y sus primeras deformaciones," in *Homo Mathematicus, Actas del Congreso Internacional sobre Astrólogos Griegos y Romanos celebrado en Benalmádena (Málaga, 8–10 de Octubre 2001)*, ed. A. Pérez Jiménez and R. Caballero (Charta Antiqua, 2002), 317–324.

of times in the future. As a result, Plato's original conception came to be increasingly distorted in the sense of an extreme astral determinism, by which it was postulated that absolutely every event, from the tiniest to the most significant, would eternally return the same again, cycle after cycle, in this world of coming-to-be and passing-away. Although ridiculed by the Christians for its rigidity, the conception of a cyclical restoration of the entire universe, which the Stoics used to call "apocatastasis," became extremely popular in late antiquity. The Great Year was to become one among various symbols of this "eternal return" of things. It was even confused by some with the cycle of the phoenix's life span.

A further series of distortions and confusions were made in Abbasid times, when the theory made its entry into the Arab-Muslim world on the occasion of the unprecedented movement of translation of scientific works that took place in Baghdad and other seats of learning of the Middle East between the eighth and the tenth centuries CE. Plato's cycle was at this point amalgamated with the precessional period, a cycle whose origin lies in the fact that the axis of the earth is not absolutely fixed in space but rotates, like a spinning top, around the poles of the ecliptic. Technically speaking, the precession of equinoxes is the name given to the slow motion (one degree in about seventy-two years or approximately twenty-six thousand years for a complete revolution, according to the modern value) of the vernal point (that is, the point of intersection of the celestial equator and the ecliptic) with respect to the sphere of the fixed stars. The importance of this movement is considerable: by affecting the coordinates of all points of the heavenly globe, it forced astronomers to introduce a ninth sphere into their systems. The Greek astronomer Hipparchus (second century BCE) is generally credited with its discovery. Nevertheless, it was Ptolemy's approximate value, as mentioned in the *Almagest* (namely, one degree in one hundred years, or thirty-six thousand years for the whole period), that was transmitted as some sort of standard to posterity. Although the two theories should not have anything to do with one another in fact, it is understandable that Plato's Great Year and the movement of equinoctial precession were confused with one another, since both could claim to be "the greatest cycle" in the Universe. Macrobius in late antiquity already offers us an example of the confusion, albeit with a different numerical value from the one mentioned in Arabic sources. It is not possible to decide with precision who in Islam was responsible for the amalgamation between the two periods, but it appears that Farghānī (the Alfraganus of the Latin tradition) in the ninth century CE was instrumental in popularizing this misconception. To add to the general confusion, Abū Maʿshar (the Albumasar of the Latins) and other authors of the same period helped introduce a series of new millennial cycles originally elaborated by Indian astronomers in the fifth and sixth centuries CE and eventually transmitted to the Islamic world through Sassanid Persia. Integrating the Platonic notion of an original conjunction of all the planets, which in their case is invariably fixed on Aries zero degrees (that is, the starting point of the zodiac), these Indo-Iranian cycles were assumed to consist of truly gigantic numbers of years, ranging from 360,000 for the Yuga to 4,320,000,000 in the case of the kalpa.[4]

Under the ever-growing influence of astrology, other important elements inherited from the Greeks were related to the notion of a conjunctional world-cycle lasting thirty-six thousand years. In particular, the historian and geographer Masʿūdī (tenth century CE) made much use of Aristotle's assumption that seas and mainlands periodically interchange on the surface of the earth, to the effect that every nine thousand years a shift should occur between the quarters of the earth's surface. The Ikhwān al-Ṣafāʾ (or Brethren of Purity), a mysterious group of people who wrote an encyclopedia of philosophy and science at about the same

---

4 D. Pingree, *The Thousands of Abū Ma'shar* (London: Warburg Institute, 1968).

time as Mas'ūdī, took a unique position on the history of world-cycles. Strongly committed to astrology, they sought as well to put geological observations—such as, for instance, the discovery of marine fossils or sediments at a great distance from the sea—in line with long periods in the supralunar world. Like Mas'ūdī, they believed that the transformations of the earth's surface were caused by the thirty-six-thousand-year cycle, although in their case it is clear that this precessional period was thought to be different from the Indian 360,000-year conjunctional cycle. In their *Epistle on Minerals* they put forward a theory of geological changes that is striking for its intrinsic coherence and anticipation of modern ideas, to the point that a recent expert has praised it as having no equivalent until the grand vision of James Hutton in the eighteenth century.[5] Secular erosion, stratified layers of marine sediments, uplifting of these layers into new mountains while the sea covers again the former mainlands: never had the chain of the geological principles at work on the surface of our planet been so coherently conceived, nor would it be again until the rise of modern geology.

Turning to the medieval Latin west, we observe that the theory followed on the same general pattern, the influence of Arab authors being especially noticeable on authors like Pietro d'Abano (thirteenth to fourteenth centuries). Once again, the main resistance came from Christians, who did their utmost to reject the Great Year doctrine as sheer heresy. In 1277 Stephen Tempier, the bishop of Paris, solemnly included among the list of 219 tenets officially condemned by the Roman Church the doctrine

---

5 F. Ellenberger, *Histoire de la géologie, I: Des anciens à la première moitié du XVIIe siècle* (Paris: Éditions Lavoisier, 1988). Godefroid de Callataÿ, "World Cycles and Geological Changes According to the Ikhwān al-Ṣafā'," in *In the Age of al-Fārābī: Arabic Philosophy in the Fourth/Tenth Century: Proceedings of the Conference Held at the Institute of Classical Studies and the Warburg Institute (London, 19–21 June 2006)*, ed. P. Adamson, Warburg Institute Colloquia, 12 (London: Warburg Institute, 2008), 179–193.

"that when all heavenly bodies have come back to the same point, which takes place every thirty-six thousand years, the same effects now in operation will come back." A major blow was inflicted on the theory of Great Years when the French canon Nicole Oresme (fourteenth century CE) decided to fight against its partisans by using their own weapons, that is, the language of numbers. Thus, he dedicated several of his works to refute the theory of grand conjunctions by demonstrating, on purely mathematical grounds, that the periods of the planets were not commensurable with one another. Another rejection of the theory was finally provided by Francesco Piccolomini in the sixteenth century. It must be said that the Great Year doctrine had by then already gone largely out of fashion, since it could not find any place in the new, heliocentric representations of the Renaissance.

Reflection

THE ETERNALITY OF LANGUAGE IN INDIA

Andrew Ollett

In 2004, the European Space Agency launched the space probe *Rosetta*. Mounted on this probe was a nickel disk three inches in diameter that had texts in thousands of languages etched on its surface. The goal of the "Rosetta disk," like the "Golden Records" that were launched with the Voyager spacecrafts in 1977, was to leave a permanent monument of humankind's existence. The assumption underlying efforts such as these is that human language is *eternal*, at least in the sense that the signs etched onto the "Rosetta disk" or the sound waves imprinted on the "Golden Records" will continue to signify something thousands, millions, or billions of years from now. If other intelligent beings ever encounter these objects and make sense of them, it will be because there is a sense to be made. The inspiration for this project comes, of course, from the Rosetta Stone: Champollion's 1822 decipherment was a recovery of a relationship between signifier and signified that had existed for more than two thousand years.

The idea that language could be eternal, however, runs up against several other aspects of the way we generally understand it. Language changes, and thus lacks the permanence that we might expect eternal things to have. Language is also a product of human culture, and there is a sense in which created things simply cannot be eternal: even if they have a permanent existence in the future, which is generally doubtful for the creations of human beings, they cannot have had a permanent existence in the past.

In ancient India, the question of whether and how language could be eternal was urgent, because entire worldviews stood and fell on the answer. A number of thinkers, but especially Buddhists, maintained that language is conventional: the relationship between signifier and signified (*śabda* and *artha* in Sanskrit) is a contrivance that serves the purpose of human communication but possesses no inherent stability. This view accords with Buddhism's basic skepticism regarding the ultimate existence of phenomena and its doctrine that all things are in constant flux. Another group included Mimamsakas—specialists in the interpretation of the Vedas, the sacred texts of Hinduism—who took the counterintuitive position that language is eternal. They were in some sense required to take this position, since they claimed that the Veda's authority—on which their entire worldview depended—derives from the fact that the Veda is itself uncreated and eternal, which is only possible on the assumption that the Sanskrit language in which it consists is equally eternal.

The Mimamsakas' position seems like an apologetic maneuver to preserve the authority of the Vedas at any cost. But between the sixth and seventh centuries CE, Mimamsakas strengthened their position with an arsenal of philosophical arguments and thereby produced a compelling vision of language as a stable and self-contained system that was inherently meaningful and therefore intelligible. This vision would dominate Indian thought for centuries, and the Mimamsakas' philosophical intervention is closely related to the intellectual and cultural dominance of the Sanskrit language itself. The key figure in this period is Kumarila Bhatta (seventh century), who wrote a series of commentaries on the basic texts of Mimamsa, the system of Vedic interpretation.

One of his goals was to defend this system against the recent attacks of a philosophically sophisticated Buddhism, represented by the works of Dharmakirti (also seventh century).

Kumarila draws a strong distinction between the *formal* constituents of language and the individual utterances in which they are instantiated. Most of Kumarila's arguments in this direction would find acceptance in a modern linguistics classroom: words are composed of phonemes, which are finite and self-similar across their infinite articulations; and words themselves, as fixed sequences of phonemes, have a formal existence that endures in between the moments when they are "manifested" by speech. Kumarila compares the word reflected in several utterances to the sun reflected in several pots of water. Similar arguments establish the fixed character of *meaning* as well: for Kumarila, words do not signify individual things but classes. Classes, which are also called "forms" and are quite similar to Platonic forms, are thought to be eternal. The word "cow" will still have a *sense* even if, in a dystopic cow-less future, it can no longer be assigned a *reference*. This account fits well with Mimamsa's broader goal of interpreting the Vedas. First, the Vedas were orally transmitted from teacher to student among families of Brahmans spread all over India: the words of the Vedas were never identified with one authoritative instance of recitation and even less of transcription but were thought to be (and for the most part actually were) identical across all such instances. Second, Mimamsakas did not believe that the Vedas referred to states of affairs in an increasingly distant past; for them, the *meaning* of the Vedas consisted in their injunctions to perform rituals (such as "one who desires heaven should sacrifice") and hence was oriented toward a future in which those rituals were to be performed again and again.

The idea that words, meanings, and the relation between them are fixed would seem to contradict the observable facts of linguistic diversity and linguistic change. People use the same words in different meanings and express the same meanings in different words. Rather than admit that the relationship of signifier and signified was a convention that human speakers constantly renewed among themselves, as the Buddhists thought, Kumarila asserted that there is an *originary* relationship between signifiers and signifieds underlying the communicative practices of human beings. Those practices may be haphazard and degenerate, but it remained possible to "purify" them—that is in fact the meaning of "Sanskrit"—and restore the originary relationship. This is a provocative idea. It implies first of all that what was eternal was not language in general but the Sanskrit language in particular. Second, it implied not only that Sanskrit would exist eternally in the future but that it had existed eternally in the past as well.

It was disarming in its simplicity. Kumarila's predecessor, Shabara, had discussed the possibilities that the Vedas were created by human beings, as the Buddhists believed, or by God, as some other Brahmans maintained. His dismissal began modestly: "There is another way." According to Shabara, we observe that the Vedas are transmitted from teacher to student over generations, and we in fact have no reason to believe that it was ever otherwise. This is the sociocultural equivalent of what is called the uniformitarian assumption in the natural sciences: the present is the key to understanding the past. Kumarila tried to tip the balance by arguing that there were no viable alternatives: without the supposition of an originary connection between word and meaning, there would have been no way for language to ever get off the ground. Suppose a person in the distant past made some sounds, like Quine's "gavagai": how would any other person know what exactly was meant by them or, indeed, that anything was meant at all? For Kumarila, it was too much to ask us to believe that any language, least of all the complex and specific register of philosophical Sanskrit that he employed, got started in guessing-games like this.

Once we are driven to suppose that language has always existed, the supposition that the Vedas are eternal—which is what Kumarila wanted to prove in the first place—is not far off. Mimamsakas viewed the Vedas as language in its pure form: faithfully transmitted from time out of mind, without any of the defects that characterize the discourse of careless and mendacious humans. But Kumarila drew a still more polemical implication. To the extent that his Buddhist and Jain opponents recognized "eternal truths," such as nonviolence, and derived such truths from their own sacred texts, they contradicted themselves. If there is anything eternal in discourse (*vyavahāra*), it can only be because discourse is based on the eternal relation between a signifier and a signified. If you admit the eternality of this relation, argued Kumarila, you no longer have sufficient grounds for denying the authority of the Vedas; if you persist in denying this relation's eternality, your moral certitude disappears from under your feet, along with any eternally abiding discursive truths.

The eternality of language, as propounded by Mimamsakas, offers scope for comparative reflection. It might at first recall the Christian idea of the eternality of the Word (*logos*) or the idea of the inescapable linguisticality of all understanding present in several streams of twentieth-century thought. But it differs radically in its specificity. Christian theologians never claimed that the words of their scriptures had existed since the beginning of time, but this is precisely what the Mimamsakas asserted of the Vedas. And what enabled Gadamer, for example, to claim that all understanding is linguistically mediated was the idea, now associated with the work of Noam Chomsky, of language as part of the inner architecture of the human mind; Mimamsakas conceived of language as radically external to human minds and even to human existence. And as much as Kumarila's arguments seem incompatible with our own worldview, the challenges they posed—challenges that Buddhists and Jains felt compelled to answer for centuries afterward—have not quite disappeared.

CHAPTER TWO

Eternity in Medieval Philosophy

Peter Adamson

Eternity is a central theme in medieval philosophy. In particular, the question of the eternity of the world shows up as a major topic in some of the most important philosophical works written in the medieval period.[1] One thinks immediately of al-Kindī's *On First Philosophy*, al-Ghazālī's *Incoherence of the Philosophers*, Maimonides's *Guide for the Perplexed*, and, if you still aren't impressed, Thomas Aquinas's *Summa Theologiae*. On the face of it, it's hard to see why this should be so.

---

[1] Throughout this chapter I will use the term "world" to refer to the physical universe. In a medieval context, "world" therefore comes close to meaning everything other than God. There is, however, the complication of nonphysical created entities such as souls and angels or celestial intelligences. With some exceptions (for instance al-Rāzī), medieval philosophers normally take souls to be created along with the bodies they inhabit. (Avicenna argues explicitly for this, as I have discussed in "Correcting Plotinus: Soul's Relationship to Body in Avicenna's Commentary on the *Theology of Aristotle*," in *Philosophy, Science and Exegesis in Greek, Arabic and Latin Commentaries*, ed. Peter Adamson, Han Baltussen, and M. W. F. Stone [London: Institute of Classical Studies, 2004], vol. 2, 59–75.) So souls do not play much role in the eternity debate, although they do arise in the context of arguments about actual infinity: if souls live on forever after their bodies die, and the world is eternal, then there must now be an actually infinite number of souls. (For this argument see below on al-Ghazālī.) Celestial intelligences or angels, by contrast, might well be taken as eternal even if the physical universe (the "world") is not.

After all, philosophers of the Jewish, Christian, and Muslim faiths are generally committed to the absolute omnipotence of God. Let us suppose, rather inadequately, that this means God can do anything that can be done. Then, if we assume that the world *can* be eternal—and why not?—God could have created an eternal world. Likewise, if the world *can* begin existing after not existing—and again, why not?—then God could have created a noneternal world. This means it is (or was) up to God whether or not the world is eternal. It's not clear why anything further would turn on the question.

Of course, this deflationary conclusion has followed only because we twice said "and why not?" Perhaps there is some reason why the world cannot be eternal. In that case, God's hand is forced: if he is to create a world at all, he must create a noneternal world. If on the other hand the world cannot begin after not existing, God must either create an eternal world or no world at all. Medieval arguments over the world's eternity turned precisely on this question, as philosophers tried to find principled reasons for constraining God's creative activity in one direction or another. Some, including Maimonides and Aquinas, claimed instead that there are no such reasons. For them, philosophy cannot say whether the world is eternal. To know for sure we must learn the answer from revelation.

Yet we are still in the dark about why these philosophers found the issue so important. One might seek another kind of deflationary answer by referring to the Greek sources on which they drew. As has been shown in chapter 1 here, the world's eternity was already hotly debated in the late ancient period. But some of the heat of that debate was stoked by an apparent conflict between Plato and Aristotle. The *Timaeus* seems to teach that God (the Demiurge) created the world with a beginning in time—though late ancient Platonists spared no ingenuity in avoiding that reading of the dialogue—whereas Aristotle could hardly have been more emphatic that the world is eternal. This tension was still felt in the medieval period, when there was still discussion of Plato's harmony with Aristotle. Still, the potential clash of authorities does not seem to have been a prime mover for the eternity debate.

Perhaps we can get closer to the motives underlying the medieval discussion by considering the ancient dispute involving Proclus, Philoponus, and Simplicius. This was a confrontation between two pagan philosophers who defended the world's eternity and a Christian philosopher who denied it. The religious tensions underlying the debate are evident from Simplicius's attitude toward Philoponus and from the fact that theologians prior to Philoponus had already been assaulting the Aristotelian position from a Christian perspective. Yet it would be too simple to conclude that adherents of revealed faiths were bound to reject the world's eternity, perhaps because of scriptural authority. After all, several prominent Muslim philosophers accepted it, and this was precisely what goaded al-Ghazālī into writing his *Incoherence*. Rather, it seems that there were strong philosophical motives driving the medieval heirs of Philoponus, like al-Kindī, Saʿādia Gaon, and al-Ghazālī, to say that God did not create an eternal world.

As I will try to show in this chapter, there were at least two main motives. The first was the desire to uphold the belief that God creates freely. To say that the world is not eternal is to deny that God's relation to the world is a necessary one. Ironically, the opponents of eternity were led by this motive to restrict God's action in the other direction. As already noted, a robust denial of the world's eternity seems to require saying that God *cannot* create an eternal world—since otherwise he did, for all we know. (I here assume that the question cannot be settled empirically, though that issue does arise occasionally, as I shall show.) This puts a burden of argument on the eternity deniers. But the pro-eternity camp has its own problem. They think the world *must* exist eternally, typically because the existence of God is in itself enough to guarantee an eternally existing world. This opens them to the charge that they make God an automatic cause, giving rise to creation the way fire gives rise to heat—whereas adherents of these three faiths would normally want to speak of God's creation as generous and freely given.

The second motive for denying the possibility of an eternal world relates to the claim that God himself is eternal. For the pro-eternity camp, that is if anything a reason to embrace the eternal world hypothesis: an eternal cause should have an eternal effect. But this runs afoul of the broad consensus that God is unlike, perhaps even wholly unlike, the things he creates. (This is grounded in scriptural texts like Qur'ān 42:11: "[God] is the Creator of heaven and of earth . . . there is nothing like unto Him [laysa ka-mithlihi shay'un].") Just as God can do everything and knows everything, whereas his creatures are limited in action and wisdom, so God's eternity may be contrasted with a temporal limitation on the side of his creation. To put it more succinctly, eternity should be God's prerogative alone, just like his other perfections. The alternative is, apparently, to say that God and the world are on a par, as far as eternity goes. Those philosophers who endorsed the world's eternity (or the possibility of its eternity) were thus obliged either to admit that God is after all comparable to created things in this respect or to explain how God is eternal in a way superior to the eternity of what he makes. Thus the debate over the eternity of the world involves reflection on the eternity of God and, more broadly, on the meaning of eternity itself.

1 Al-Kindī and His Circle

In Islam, the most famous representative of the anti-eternity side of the debate is certainly al-Ghazālī. As I will show, he seems to have been driven primarily by the first motive of securing God's freedom. Yet the first major Muslim thinker to deny the world's eternity was motivated more by the second issue of God's uniqueness. This was al-Kindī (d. after 870), who denied that the world is eternal despite being an unstinting adherent of Hellenic philosophy. It would not have been obvious for al-Kindī to adopt this stance. His primary philosophical influence was Aristotle, and writings by Plotinus and Proclus had been translated in his circle. These figures all explicitly endorsed the necessary and eternal existence of the world. Al-Kindī certainly knew Aristotle's stance on the question, having had access to versions of the *Physics* and *Metaphysics*, among other texts, but he passes over this particular feature of the Aristotelian corpus in silence.

As for Plotinus and Proclus, the Arabic versions of their writings produced in the so-called Kindī circle do modify the original texts on the topic of eternity, as on many other topics. But they do not do so in order to eliminate the commitment to the world's eternity. To the contrary: an original section inserted at the end of the first section (*mīmar*) of the *Theology of Aristotle*, the largest text representing the Kindī-circle version of Plotinus's *Enneads*, raises the question of how God relates to the world temporally. The author of this insertion writes:

> The Ancients were forced to speak of "time" regarding the beginning of creation simply because they wished to describe how things were brought to be. Thus they were forced to include time in their description of [this] bringing to be, and of creation, which was not in time at all . . . so as to distinguish between primary, superior causes and secondary, inferior causes. For when one wishes to ascertain and explain the cause, one is forced to mention time, because a cause is of course "prior" to its effect. So one gets the impression that the priority is time, and that every agent performs its act in time. But this is not the case: that is, not every agent performs its act in time, and not every cause is prior to its effect in time.[2]

This looks very much like the ancient, so-called non-literal readings of the *Timaeus* that had Plato speaking of a beginning (*archê*) of time only in the sense of causal priority rather than temporal priority. So we might

---

2 'Abdurraḥmān Badawī, ed., *Plotinus apud Arabes* (Cairo: Dirāsa Islamiyya, 1947), 27. All translations are mine unless otherwise indicated.

suspect a link to a Greek exegetical tradition.[3] But whatever its provenance, the passage insists that God is *not* temporally prior to his creation.

Nonetheless, the Arabic versions of Plotinus, and of Proclus, do take steps to avoid the consequence that God and His creation are on a par by both being eternal. In a maneuver typical of the Arabic Plotinus, the epithet "eternal" is both ascribed to the First Principle (frequently called "the Creator") and denied of him. As I have pointed out elsewhere, the Arabic version of the *Enneads* refers to God as "eternal" without any prompt from the original text.[4] Elsewhere in the text, eternity is associated with the level of intellect (Greek *nous*, Arabic *ʿaql*), whereas God is above both time and eternity.[5] We find this also in the *Book of the Pure Good*, a redaction of the Arabic version of Proclus's *Elements of Theology* (later known in Latin as the *Liber de Causis*, that is, *Book of Causes*). For instance:

> Every true being [anniyya] is either higher than eternity [al-dahr] and prior to it, or together with eternity, or posterior to eternity, yet above time. The being that is prior to eternity is the First Cause. For it is the cause for [eternity]. The being that is together with eternity is the intellect, for it is the stable[6] being. The being that is posterior to eternity, yet above time, is the soul. For it is on the lower horizon of eternity and above time. The proof that the First Cause is prior

---

[3] On the passage from the *Theology* see also Cristina D'Ancona, "La Teologia neoplatonica di 'Aristotele' e gli inizi della filosofia arabo-musulmana," in *Entre Orient et Occident: La philosophie et la science gréco-romaines*, ed. Ulrich Rudolph and Richard Goulet (Vandoeuvres: Fondation Hardt, 2011), 135–195.

[4] For instance at Badawī, *Plotinus apud Arabes*, 119. See also Peter Adamson, *The Arabic Plotinus: A Philosophical Study of the "Theology of Aristotle"* (London: Duckworth, 2002), 145.

[5] Badawī, *Plotinus apud Arabes*, 143.

[6] Here there are conflicting readings of the text. The Latin *Liber de Causis* reads *esse secundum*, "second being," which reflects an underlying Arabic text *al-anniyya al-thāniyya*. That is also the reading printed in Bardenhewer's edition: Otto Bardenhewer, ed., *Die pseudo-aristotelische Schrift Ueber das reine Gute* (Freiburg im Breisgau: Herder, 1882). However, Richard C. Taylor proposes emending the text to *al-anniyya al-thābita* (which is palaeographically almost identical in Arabic), "the stable being." See his note 4, p. 12, in St. Thomas Aquinas, *Commentary on the Book of Causes* (Washington DC: Catholic University Press, 1996). Both readings would make sense philosophically: the intellect is indeed stable and unchanging—hence eternal—yet it is also second, after the First Cause or God.

to eternity in itself is this: that the being in [eternity] is acquired. We say that every eternity is a being, but not every being is an eternity. Thus being is of greater extent than eternity. The First Cause is above eternity because eternity is caused by it. Whereas the intellect conforms to eternity because it spreads out along with it, without altering or changing. But the soul is connected to eternity from below, since it is a lower effect of the intellect, but [connected] to time from above, because it is the cause of time.[7]

This is one of numerous passages in the Arabic Proclus materials that betrays influence from the Arabic Plotinus. The more complex hierarchy of Proclus is assimilated to the Plotinian scheme of One, Intellect, and Soul, and even the image of soul as being on a "horizon" is Plotinian. Distinctively Proclan, however, is the notion that God must transcend eternity if he has ultimate effects that lack eternity. God is the source of all being and shares his own features with whatever he creates. From this it follows that, if God were eternal, everything would be eternal; but this is not the case. Thus he is above eternity.

The works of al-Kindī are deeply influenced by these texts, so we might expect him to take up a similar view. And we do indeed find echoes of the Arabic Plotinus and Proclus in al-Kindī. He certainly agrees with these texts that God is beyond time. In his *On First Philosophy*, he remarks that God "is first in time, because he is the cause of time" (sec. 1.5)—notice that this echoes the passage just cited from the *Book on the Pure Good* but makes God, and not soul, the cause of time.[8] Al-Kindī's way of putting the point might make it sound as though God temporally precedes other temporal things or even, bizarrely, that God temporally precedes time itself.

---

7 *Book of the Pure Good*, proposition 2, in Bardenhewer, *Die pseudo-aristotelische Schrift*, 61–63.
8 Section numbers and translations for all Kindian works are taken from Peter Adamson and Peter E. Pormann, trans., *The Philosophical Works of al-Kindī* (Karachi: Oxford University Press, 2012).

But arguments later on in *On First Philosophy* make it clear that time applies only to things that can move (sec. 6.8), namely bodies. Since God is incorporeal, God is not subject to time, a conclusion repeated later on (sec. 20.3). Likewise, al-Kindī repeats the idea found in the Arabic Plotinus that God creates "without time" (*bi-ghayr zamān*).[9] This occurs in a digression on divine creation in *On the Quantity of Aristotle's Books*, a survey of the Aristotelian corpus. Here al-Kindī is explaining the difference between divine and non-divine agency and says that God's action is unique in needing neither material substrate nor time (sec. 6.8). Rather, God simply wills that something exist, and it exists. For al-Kindī this is the meaning of the Qur'ān's repeated statement that "when God wishes something, He says to it 'be!' and it is."[10]

In saying all this, al-Kindī is thus taking up the late antique idea that God is eternal in the sense of timeless and is repeating the more specific idea of the Arabic Plotinus that God creates without time. But he does not agree with these texts when they say that God transcends eternity as well as time. Indeed, as I have argued elsewhere, his arguments against the world's eternity are intended to show that eternity is reserved for God alone.[11] Turning back to *On First Philosophy*, we find al-Kindī repeatedly equating God with "the eternal." He insists that only the First Cause can be eternal: the eternal is that which cannot fail to be, because it has no cause (sec. 5.1). Here al-Kindī is applying a notorious Aristotelian line of argument to God. Aristotle tried to show (in *de Caelo* 1.12, for instance) that not only does necessity imply eternity, but the reverse also holds. For Aristotle, then, what did not begin and will never be destroyed *could not* have begun and *cannot* be destroyed. Similarly, al-Kindī explains that the First Cause, being eternal, cannot be destroyed or change (secs. 5.2–3).

---

9 Badawī, *Plotinus apud Arabes*, 114.
10 Qur'ān 36:78–82.
11 Peter Adamson, *Al-Kindī* (New York: Oxford University Press, 2007), chap. 4.

Thus al-Kindī adopts a simpler metaphysics than that of the translations executed in his own circle. The *Theology of Aristotle* and *Book of the Pure Good* envision a temporally everlasting physical world caused by soul. Eternity proper is ascribed to the cause of soul, namely the intellect. This intellect is in turn caused by the First Cause, a Creator who is beyond even eternity. Al-Kindī, by contrast, presents us with a physical cosmos that is subject to time and whose cause is a timelessly eternal God.[12] Nor does this exhaust his departures from the Greco-Arabic sources. In the face of both Aristotelian and Platonist authority insisting that the world is eternal in the sense of "everlasting," al-Kindī lavishes great attention on proving that the world is not eternal in even this sense. Here, as numerous scholars have discussed, al-Kindī is again drawing on a Greek source, but a Christian one: John Philoponus.[13] For instance he uses Philoponus's argument that if the world had no beginning, an infinite amount of time must already have been traversed to reach the present moment; but this is impossible (*On First Philosophy* 8). Ironically, another text displaying al-Kindī's close association of divinity with eternity is a brief refutation of the Christian doctrine of the Trinity.[14]

It seems that al-Kindī is missing a trick here. He has made sufficiently clear that God is eternal not just by existing forever but by being timeless. This is eternity in the sense already established by Plotinus, Proclus, and other Platonists. If he simply wishes to safeguard God's uniquely transcendent status, wouldn't it be enough to show that whereas the world is eternal in the sense of everlasting, God is eternal in the sense of being timeless? Al-Kindī never betrays any awareness of a distinction between these two senses of eternity, which in part explains his failure to adopt (or at least say why he does not adopt) this solution. However, there may be a deeper explanation for

---

12 Actually, al-Kindī does sometimes seem to endorse that hierarchy. For instance *On First Philosophy* 19.6 speculates that intellect could be considered the first multiple effect of God—an obvious, if less than emphatic, allusion to the Plotinian system.

13 See especially Herbert A. Davidson, "John Philoponus as a Source of Medieval Islamic and Jewish Proofs of Creation," *Journal of the American Oriental Society* 89 (1969): 357–391.

14 See *On the Trinity*, in Adamson and Pormann, *Philosophical Works of al-Kindī*.

the apparent oversight. The question of whether anything other than God is "eternal" (*azalī*) was a much-disputed one in al-Kindī's lifetime. The ʿAbbāsid caliphs, with whose courts he associated, imposed an infamous "test" (*miḥna*) in the early to mid-ninth century, requiring adherence to the dogma that the Qurʾān is not created. In the disputes that raged over this question, it was fundamental to assume a dichotomy between "created" and "eternal." It was perhaps with this dispute in mind that al-Kindī not only assumed the same dichotomy in *On First Philosophy* but also sought to push God's eternity as far away as possible from the status of creatures. For, on his view, the created is temporally limited while God is timelessly eternal. There is no room in this stark opposition for a created but everlasting world.

2 Saʿādia Gaon

For a strikingly similar handling of the eternity question, we may turn to the Jewish thinker Saʿādia Gaon (d. 942).[15] As Herbert Davidson pointed out decades ago, Saʿādia, like al-Kindī, takes inspiration from Philoponus.[16] For instance, both infer the temporal finiteness of the world from its spatial finiteness,[17] and from the fact that it is composed.[18] There has as yet been little attempt to go beyond this observation to reflect on how and why al-Kindī and Saʿādia both use this

---

[15] I here draw on his philosophical magnum opus, *Kitāb al-Amānāt wa 'l-Iʿtiqādāt*, ed. Samuel Landauer (Leiden: Brill, 1880), English version in Samuel Rosenblatt, trans., *The Book of Beliefs and Opinion* (New Haven, CT: Yale University Press, 1948); cited by page numbers from the Landauer edition, which are indicated in angled brackets in the Rosenblatt translation. For general overviews of his thought see Israel Efros, *Studies in Medieval Jewish Philosophy* (New York: Columbia University Press, 1974), and Sarah Pessin, "Saadya," in the online *Stanford Encyclopedia of Philosophy*, http://plato.stanford.edu/entries/saadya.

[16] Davidson, "John Philoponus as a Source." See also Davidson, *Proofs for Eternity, Creation and the Existence of God in Medieval Islamic and Jewish Philosophy* (New York: Oxford University Press, 1987).

[17] *On First Philosophy* sec. 6; al-Kindī's other short treatises against the world's eternity also depend principally on the idea that the world is finite in size. For the argument in Saʿādia, see *Book of Beliefs*, 32–33.

[18] *On First Philosophy* sec. 7, *Book of Beliefs*, 34.

source. The two men were not contemporaries—Saʿādia was born after al-Kindī's death—and if one wanted to name a Jewish author strongly influenced by al-Kindī one would choose Isaac Israeli, who drew extensively on the Kindian corpus.[19] Yet the parallels between al-Kindī and Saʿādia go beyond their adoption of Philoponus's arguments, even as they differ to some extent in their reactions to Philoponus.

To begin with the differences, Saʿādia is much more detailed in his consideration of the whole issue. He not only presents a series of arguments against the world's eternity, but announces that he has engaged in independent reflection on whether the arguments are sound. Thus, for instance, he discusses possible alternatives to the view that the world is limited in size, considering that there could be an infinite number of worlds scattered throughout infinite space (as had been suggested by the Epicureans, though Saʿādia does not mention any proponents of this view by name). He also shows knowledge of ancient sources not used by al-Kindī in the context of the eternity debate, for instance the *Timaeus*. Saʿādia takes Plato (also not mentioned by name) as an upholder of the view that God creates the world out of something, namely "indivisible parts" comparable to "dust." This turns out to refer to the triangles out of which the faces of the five elemental polyhedra are made.[20]

But for the purposes of this discussion, the most notable difference between the two is that Saʿādia endorses Philoponus's claim that the heavens are not made of an indestructible fifth element.[21] Instead, they are fiery. Saʿādia even repeats Philoponus's point that Aristotle (yet again, none of the protagonists of the Greek debate are named) was wrong to

---

19 As shown in Alexander Altmann and Samuel Miklos Stern, *Isaac Israeli: A Neoplatonic Philosopher of the Early Tenth Century* (Oxford: Oxford University Press, 1958).
20 *Book of Beliefs*, 41. It should be noted that al-Kindī, too, knows the polyhedron theory of the *Timaeus*. He in fact wrote a treatise on it, titled *Why the Ancients Related the Five Geometric Shapes to the Elements*. In sharp contrast to Saʿādia, al-Kindī greets the theory with an uncritical analysis based on number theory and fails to connect the theory to the eternity question. Incidentally, it is also worth noting that al-Kindī wrote a lost treatise against the theory of "parts that cannot be divided," so he would agree with Saʿādia in rejecting any theory on which God made the world out of atoms. See the list of works in Adamson and Pormann, *Philosophical Works of al-Kindī*, sec. 158.
21 *Book of Beliefs*, 58–61.

suppose that the heavens cannot be made of fire, given that fire naturally moves straight upward and not in a circle. Rather, fire moves up only because it is within air; once it reaches its natural place above air, it moves in a circle. By contrast, al-Kindī not only accepted the fifth element theory but wrote a treatise designed to prove it, precisely on the grounds that the four sublunary elements (including fire) move rectilinearly by nature.[22] One might wonder how al-Kindī could endorse the orthodox Aristotelian view despite his firm opposition to the world's eternity. The answer is that he considered God's will to be required even for the continued existence of the otherwise indestructible heavenly body. It is "stable in its state for the days of its duration which its Creator, the great and exalted, allotted to it—until he destroys it [so that it is again] just as when it began."[23] Thus does al-Kindī elide the whole debate between Philoponus and Aristotle, with a nonchalance Saʿādia evidently did not feel he could afford. Note, however, that Saʿādia too remarks that created things subsist only for as long as God has decided.[24]

Despite this fundamental disagreement between the two thinkers regarding the fifth element, there are equally fundamental points of agreement. Both al-Kindī and Saʿādia are emphatic about the method to be used in discussing the world's eternity: the topic should be approached not using sensory observation but through intellectual reflection. As Saʿādia puts it, this is a matter too subtle for the senses, and he calls for "discovery through what is intelligible" (*min ṭarīq al-istidlāl bi-l-maʿqūl*), using the faculty of thought (*fikr*).[25] He repeatedly castigates upholders of the world's eternity for assuming

---

22 See *On the Nature of the Celestial Sphere*, translated in Adamson and Pormann, *Philosophical Works of al-Kindī*.
23 *On the Nature of the Celestial Sphere* sec. 13 (Adamson and Pormann translation).
24 *Book of Beliefs*, 70.
25 *Book of Beliefs*, 30; see 41, where he says that his conclusions have been achieved "by means of theoretical inquiry" (*bi-ṭarīq al-naẓar*). These remarks should be understood in light of the methodological section that begins the work, where Saʿādia speaks of inquiry that proceeds from necessary principles to establish necessary conclusions, in contrast to inquiry based on sensation (see 9). For the same sentiments in al-Kindī see my discussion in Adamson, *Al-Kindī*, 90. It should however be noted that Saʿādia does not deploy the kind of axiomatic, Euclidean style of argument found in al-Kindī.

that the only possible kind of agency is the one we see exercised by created beings. He also turns the empiricist tendencies of these unnamed (and in some cases probably hypothetical) opponents against them: we can never use the senses to experience something's being eternal, so by this standard we could as well infer that nothing is eternal.

The discussions in al-Kindī's *On First Philosophy* and Saʿādia's *Book of Beliefs* have remarkably parallel structures. After a methodological introduction to the problem of eternity, arguments drawn from Philoponus are used to prove that the world was created ex nihilo. It is then observed that the world cannot have created itself since nothing can do this.[26] Thus the world must have an external creator, namely God, who alone is eternal. Finally, both thinkers move on to discuss at length the oneness of God and the extent to which divine oneness is compatible with the possession of attributes. These parallels are, perhaps, to be explained not only by the use of a similar range of sources but also by the fact that both philosophers are upholding the contemporary theological idea that eternity is the privilege of God alone.

I have already shown this idea at work in al-Kindī. In Saʿādia it emerges in passages like the following. Here he is in the midst of arguing that the world is not made out of anything but created ex nihilo:

> It is clear too that whatever thing we suppose existing things to be created out of, we will be compelled to infer that this thing is eternal. But if it is eternal then [this thing] and the Creator are equal in eternity, and necessarily, he will not be in a position to create things from it,[27] nor will it accept his command, so as to be affected as he wishes, or be shaped as he wills, unless we join to them a third cause in our suppositions, which would distinguish between the two so that through it, this one becomes the maker [ṣāniʿ] and that one the made.[28]

---

26 *On First Philosophy* sec. 9, *Book of Beliefs*, 37–38.
27 This translates Landauer's Arabic text. Rosenblatt seems to think of emending from *min-hu ashyāʾ* to *min-hu shayʾ*, since he translates "to create any part of it." But this misunderstands the argument, as I will go on to explain.
28 *Book of Beliefs*, 39.

Here Saʿādia seems to be assuming that if two things are both eternal, neither can be causally prior to the other such that one is a cause and the other an effect. To make something other than God eternal is to postulate that God has an equal or peer, over which he could not exercise control. It would not be until Avicenna that a philosopher showed how to get around this problem and to assert that the world is both eternal and caused by God. But a contemporary of Saʿādia was nonetheless happy to assert that there are eternal principles apart from the Creator. This contemporary was a man known for adopting unusual philosophical views: al-Rāzī.

3 Al-Rāzī

Abū Bakr al-Rāzī (d. 925) is not usually mentioned as a significant figure in the debate over the eternity of the world, but he ought to be. He not only had a strikingly original view about eternity and its relationship to God and the world but also brought an unexpected Greek source into the fray. Al-Rāzī did engage with Aristotle's physics and even with Aristotle's discussions of the world's eternity. Al-Rāzī was primarily a doctor though, not a philosopher, and the Hellenic author who looms largest in his corpus is not Aristotle but Galen. Al-Rāzī had access to Galen's *On Demonstration*, which is now lost, apart from the fragments and testimonia preserved by a variety of authors. Al-Rāzī himself provides one of the most interesting reports, which quotes Galen on the world's eternity.

The report is found at the beginning of al-Rāzī's *Doubts about Galen*,[29] in which al-Rāzī applies to Galen the same rigorous standards Galen prided himself on applying to his own predecessors.[30] Although

---

29 Abū Bakr Muḥammad ibn Zakariyyāʾ al-Rāzī, *Kitāb al-Shukūk ʿalā Jālīnūs*, ed. Mehdi Mohaghegh (Tehran: Society for the Appreciation of Cultural Works and Dignitaries, 1993).
30 On this point see Lutz Richter-Bernburg, "Abū Bakr al-Rāzī and al-Fārābī on Medicine and Authority," in *In the Age of al-Fārābī: Arabic Philosophy in the Fourth/Tenth Century*, ed. Peter Adamson (London: Warburg Institute, 2008), 119–130.

there are many examples of Galen's methodological strictures in his extant corpus, the best example may have been *On Demonstration*.[31] Al-Rāzī praises this work of Galen as "the best and most useful book apart from the books revealed by God."[32] Despite this strikingly positive introduction, al-Rāzī goes on to complain about the work's contents at some length. He quotes Galen as follows: "Galen had earlier denied but then concluded, in the fourth chapter of *On Demonstration*, that the world does not corrupt. He said, 'If the world were corruptible, then the [celestial] bodies, the distances between them, their magnitudes, and their motions would not persist in one and the same state, and, moreover, the sea-water from before our time would vanish. But not a single one of these ever departs from its state or alters, as the astronomers have observed for thousands of years. Therefore, it necessarily follows that, since the world does not age, it is not susceptible to corruption.'"[33]

Al-Rāzī is not impressed. For one thing, he complains, this contradicts what Galen has said elsewhere. In both *On My Own Opinions* and *On Medicine* Galen denies that we can know whether or not the world is eternal. For another thing, Galen fails to rule out the possibility that the world could be suddenly destroyed, rather than (so to speak) dying of old age. For instance a building whose foundation is removed will be destroyed all at once, as will a tree when it is uprooted. Intriguingly al-Rāzī points out that some purveyors of religion (*mutadayyinūn*) think the world will end in exactly this way.[34] Besides, if the universe is deteriorating very slowly, this might be imperceptible even across many generations.

---

31 On which see Benjamin Morison, "Logic," in *The Cambridge Companion to Galen*, ed. R. J. Hankinson (Cambridge: Cambridge University Press, 2008), 66–115.

32 Al-Rāzī, *Kitāb al-Shukūk*, 3, l. 13.

33 Al-Rāzī, *Kitāb al-Shukūk*, 3, ll. 17–21; translation, with modifications, from Jon McGinnis and David C. Reisman, *Classical Arabic Philosophy: An Anthology of Sources* (Indianapolis: Hackett, 2007), 51. On this passage see also Pauline Koetschet, "Galien, al-Rāzī, et l'éternité du monde. Les fragment du traité *Sur la Démonstration* IV, dans les *Doutes sur Galien*," *Arabic Sciences and Philosophy* 25 (2015): 167–198.

34 Al-Rāzī, *Kitāb al-Shukūk*, 5, l. 3.

As this discussion in al-Rāzī shows, issues of philosophical methodology are never far from the surface in the debate over the world's eternity. In fact he is probably misrepresenting Galen's purposes here in *On Demonstration*. Galen's approach seems to have been highly dialectical, seeking to show flaws in the arguments of other thinkers that fell short of demonstrative status.[35] So he was probably not, as al-Rāzī tries to make out, arguing positively in favor of the world's eternity. Indeed al-Rāzī all but admits this, imagining an objector who says that Galen was "unable to say for sure that the world is incorruptible, but rather just looked to see where one would need take premises from."[36] I take this to mean that Galen's own discussion was merely suggesting how someone should argue if they do want to establish the world's eternity, without necessarily claiming that the argument would be decisive.

An unusual feature of Galen's proposed argument is that, instead of arguing that the world *cannot* start to exist or be destroyed, it proceeds on the basis of empirical observation. If the world were corruptible, it would surely deteriorate over time, but then we would see signs of that deterioration. Al-Rāzī responds by showing the limitations of empirical observation for this topic, given the huge time scale involved (similarly, he mentions that the sun is so large that changes in size would be undetectable to us). This is an example of something I have already mentioned: al-Rāzī turning Galen's own methodology against him. Galen likewise emphasized that sense-experience cannot provide certainty about some philosophical issues, such as the question of whether there is void surrounding the cosmos, as was claimed by the Stoics.[37]

---

35 See Riccardo Chiaradonna, "Le traité de Galien *Sur la démonstration* et sa postérité tardo-antique," in *Physics and Philosophy of Nature in Greek Neoplatonism*, ed. Riccardo Chiaradonna and Franco Trabattoni (Leiden: Brill, 2009), 43–77. This is not an isolated example: as I have argued elsewhere, al-Rāzī engages in a similar distortion regarding the topic of void in *On Demonstration*. See Peter Adamson, "Galen on Void," in *Philosophical Themes in Galen*, ed. Peter Adamson, Rotraud Hansberger, and James Wilberding (London: Institute of Classical Studies, 2014), 197–211.
36 Al-Rāzī, *Kitāb al-Shukūk*, 6, ll. 5–6.
37 Galen, *De Placitis Hippocratis et Platonis* [On the opinions of Plato and Hippocrates], ed. Philip De Lacy (Berlin: Akademie, 1978–1984), 9.6.21.

Al-Rāzī's critique of Galen suggests that sense-experience simply cannot settle the issue of the world's eternity. We know from other texts that al-Rāzī thought the issue could indeed be settled, but by other means. His approach revolves around a critique of Aristotle's physical theory. In a work called *On Metaphysics*,[38] he shows a fairly detailed knowledge of the Aristotelian arguments in favor of eternity. He first mentions the argument in *Physics* 8 that a first motion is impossible because a previous motion would be needed to bring it about.[39] In a way al-Rāzī agrees with Aristotle that such a regress can be avoided by positing an unmoved mover: motion and body are "jointly originated" (*ḥādithān*) by a divine act that involves no change in God.[40] Another attempted proof, al-Rāzī goes on, is to be found in *Metaphysics* book 12. This argument is that there cannot be a first moment of time; for to say that time begins to exist is to say that it starts existing *after* not existing, so that there must be time *before* it exists.[41] Against this, al-Rāzī responds that it must be possible for a period of time to have a "starting point" that is not itself part of time, since otherwise even a given day, like Sunday, could not begin. But if this is possible, then there can be a non-temporal starting point that provides a "before" for the temporal span of the world.[42]

Although this critique of Aristotle reveals something of al-Rāzī's view on the world and its relation to time, it does not tell us the whole

---

38 I here assume that this is to be treated as a genuine work of al-Rāzī. On the work see Giulio A. Lucchetta, *La natura e la sfera: La scienza antica e le sue metafore nella critica di Rāzī* (Bari: Milella, 1987).

39 Abū Bakr Muḥammad ibn Zakariyyā' al-Rāzī, *Rasā'il falsafiyya (Opera philosophica)*, ed. Paul Kraus (Cairo: Paul Barbey, 1939), 128, ll. 3–6; see Aristotle, *Physics* 8.1.251a17–28.

40 Al-Rāzī, *Rasā'il*, 128, ll. 7–9.

41 Al-Rāzī, *Rasā'il*, 128 ll. 11–16; see Aristotle, *Metaphysics* 8.6.1071b6–11. Intriguingly al-Rāzī adds that this is similar to a proof he knows from Proclus.

42 What al-Rāzī is envisioning here is more or less just an Aristotelian instant (*nun*), which is also non-extended and not a part of time. So in fact their disagreement boils down to the question of whether such an instant must be a division between past and future or whether it can simply mark the beginning of time with no preceding past.

story.⁴³ The whole story is well worth telling, because it involves al-Rāzī's most celebrated philosophical idea: his distinction between "absolute" (*muṭlaq*) and "relative" place and time. Here another passage from *On Metaphysics* is relevant. Responding to Aristotle's argument against the possibility of a first moment, al-Rāzī says: "One may say to him, 'you claim that if time is originated, there must be [another time] prior to [the moment of its origination]. But then how do you differ from someone who claims that if place exists in the world, it must be in [another] place—thus either abolishing place or saying that every place is in another place, to infinity?'"⁴⁴ Here we see, not for the first time and not for the last, a parallel being drawn between temporal extension and spatial extension. In this case, al-Rāzī has put his finger on a sensitive spot in Aristotle's theory. Already Aristotle's student Theophrastus worried about whether the cosmos as a whole is in a place. It would seem not, given that for Aristotle there is nothing outside the cosmos, not even empty void—yet place is defined as the inner bound of a containing body.⁴⁵

It now looks as though there will be a problem both for Aristotle, who wants to say that the world is finite in spatial magnitude, and for an opponent of Aristotle who denies the world's eternity and thus wants to say that it is finite in temporal magnitude. Both face the difficulty that there should be something "outside" the magnitude in question. Al-Rāzī himself solves the problem through the expedient of positing not only the time and place relative to the world and its

---

43 For reasons of space I leave out a third argument in *On Metaphysics* in which al-Rāzī argues against the Aristotelian inference from circular celestial motion to eternity. This passage is complicated by the fact that the main target of al-Rāzī's critique is not Aristotle but a version of the Aristotelian view found in Thābit b. Qurrā, a ninth-century philosopher and scientist. On him see Roshdi Rashed and Marwan Rashed, eds., *Sciences and Philosophy in 9th Century Baghdad: Thābit Ibn Qurrā (826–901)* (Berlin: de Gruyter, 2009).

44 Al-Rāzī, *Rasā'il*, 129, ll. 6–9.

45 There is further discussion of this issue in al-Rāzī's *On Metaphysics*, at *Rasā'il*, 132–134, which similarly begins by drawing a parallel between a spatial boundary and a first motion. For the issue in the Aristotelian tradition see Richard Sorabji, *The Philosophy of the Commentators, 200–600 AD: A Sourcebook in Three Volumes* (London: Duckworth, 2004), vol. 2, sec. 13b.

parts but also absolute time and place. He compares absolute place to a "container" or "vessel" for the world, refers to it as "void," and says that it is unlimited in extent.[46] Thus he has the resources to solve the difficulty concerning place: the world is in a place, namely void or absolute place, whereas this outer place is not in a place because it is unlimited.

Al-Rāzī's explanation of how time can be finite is similar: the time of the world is finite, but before this world-relative time, there was absolute time. Indeed there has always been absolute time. For this reason al-Rāzī refers to it not only as "duration" (*mudda*) but also as "eternity" (*dahr*).[47] Just as relative place is the place of a given body and depends on that body for its reality (this is why it is "relative" or "supervenient": *muḍāfī*), relative time measures a motion and does not exist without motion.[48] Al-Rāzī has done something rather subtle with Aristotle's core physical notions here. He does not exactly reject the Aristotelian conceptions of time and place, but he makes them merely relative, because of their dependence on body and motion. Furthermore, he insists that without absolute time and place, relative time and place cannot exist. But the reverse is not true: there have always been absolute time and absolute place. Thus al-Rāzī dignifies them with the status of eternal principles, along with God, soul, and matter.

The eternity of absolute time guarantees that there is never, strictly speaking, a first moment of time. There is however a first moment of *relative* time, something al-Rāzī has explained by appealing to the notion of a beginning point that itself has no temporal duration. This allows him to escape from Aristotle's argument against time beginning. He has also insisted that there is no valid inference from God's immutability to an eternal world. Al-Rāzī confronts a third argument for eternity, though,

---

46 See for instance al-Rāzī, *Rasā'il*, 245, ll. 16–17, a report of his view found in Ibn Ḥazm: "according to them void is place with nothing placed in it, and the celestial sphere, according to them, exists in the void, since void has no limit, in the sense of a surface [min ṭarīq al-misāḥa]."
47 For instance at al-Rāzī, *Rasā'il*, 195.
48 Al-Rāzī, *Rasā'il*, 198, ll. 15–17.

which will turn up again in Avicenna and al-Ghazālī. This argument proceeds not from God's immutability but from his rationality. The problem is that, as al-Rāzī puts it, "it cannot be that God, being wise, begins to act, after having refrained from doing so for an infinite duration" (*mudda*).[49] To begin an action suddenly, without the appearance of some new reason for action, is characteristic of a "frivolous" agent, not a wise one like God. Al-Rāzī accepts the premises of this argument and infers from it that God cannot be the sole cause for the beginning of the universe. He argues that there must be an "ignorant" or "foolish" agent involved as well. This, he claims, is the eternal soul, which is the sort of agent that can spontaneously begin to act for no good reason.[50]

Clearly al-Rāzī has put a great deal of thought into safeguarding the temporal beginning of the universe. He not only raises and refutes at least three arguments in favor of the world's eternity but also postulates both absolute time and an eternal, foolish soul in order to explain how the world can begin to exist. Admittedly, he does not do so only to avoid admitting that the world is eternal; for instance, the idea of an ignorant soul allows him to address the problem of evil. Nonetheless, his cosmology is clearly designed in part to show how the world could be non-eternal. This raises the question of why al-Rāzī was so desperate to secure this result. He never gives an argument in favor of the world's beginning but contents himself with refuting arguments in favor of the world's eternity.

A full answer to this question is beyond the scope of this chapter. But there are at least two relevant factors, in my view. First, his arguing partners were mostly theologians, not philosophers inspired by Aristotle.[51] His dialectical situation was one in which his opponents accepted that the world begins to exist after not existing, so he simply

---

49 Reported in Fakhr al-Dīn al-Rāzī, *al-Maṭālib al-'āliya*, ed. A. Ḥ. al-Saqqā, 9 vols. (Beirut: Dār al-Kitāb al-'Arabī, 1987), vol. 4, 407.

50 On this see not only the report in Fakhr al-Dīn cited in the previous note but also, e.g., al-Rāzī, *Rasā'il*, 311–312.

51 Marwan Rashed, "Abū Bakr al-Rāzī et la *kalām*," *Melanges de l'Institut Dominicain d'Etudes Orientales du Caire* 24 (2000): 39–54.

assumes this and shows what must follow. Second, al-Rāzī was deeply influenced by Plato's *Timaeus*. Unlike late ancient Platonists from Plotinus onward, whose eternalist reading of this dialogue is echoed in the first chapter of the *Theology of Aristotle* (see section 1 of this chapter), al-Rāzī adhered to the sort of reading we find in Platonists like Plutarch. On this reading, the *Timaeus* says that the Demiurge and the Receptacle are eternal, but there is a temporal "beginning" (*archê*, see *Timaeus* 28b) for the cosmos that the Demiurge fashions within the Receptacle. Galen is again relevant here, as al-Rāzī would have known the *Timaeus* through Galen's paraphrase and commentary on it. Galen's *On Demonstration* criticized the Aristotelian theory of time—perhaps in part because of Galen's own engagement with the *Timaeus*—and this criticism seems to have been a partial inspiration for al-Rāzī's theory of absolute time.[52] Thus al-Rāzī occupies a unique place in the medieval debates over eternity. Drawing principally on Galen and through Galen on Plato, and arguing principally with theologians rather than philosophers, al-Rāzī embraces a temporally finite cosmos but ascribes eternity to time, place, matter, and soul, as well as God.

4 Miskawayh and Avicenna

Thus far, I have looked at three philosophers, all of whom believed that the physical cosmos is generated in time (even if al-Rāzī did accept the eternity of principles apart from God). Yet in the medieval Islamic tradition, philosophy came to be strongly associated with a belief in the world's eternity. This is due in part to Aristotle, of course, but it is above all a sign of Avicenna's enormous impact on philosophy in the Islamic world. Even before Avicenna, some philosophers had endorsed the world's eternity. Notable here is al-Fārābī, whose system

---

52 See Peter Adamson, "Galen and Abū Bakr al-Rāzī on Time," in *Medieval Arabic Thought: Essays in Honour of Fritz Zimmermann*, ed. Rotraud Hansberger and Charles Burnett (London: Warburg Institute, 2012), 1–14.

anticipates Avicenna's in many respects. But Avicenna provides a new and powerful rationale for the eternity of the world, which is inextricably linked to a new and powerful way of thinking about God.

Still today, the most famous Muslim philosophers—the ones who are, for instance, usually touched on in courses on medieval philosophy—are those who defended the world's eternity: al-Fārābī, Avicenna, Averroes. But we should not be misled by this into supposing that al-Kindī and al-Rāzī were exceptional in rejecting that thesis. Consider, for example, a contemporary of Avicenna, the historian and philosopher Miskawayh (d. 1040). It would be hard to make a case that he was a great thinker. But precisely for that reason, he can be used (with caution) as a barometer of which ideas were accepted by mainstream, intellectually conservative philosophers in the early eleventh century. It is thus significant that in his epistle *On Soul and Intellect*, he confidently rejects the eternity of the world.[53] In this epistle, Miskawayh quotes and refutes an unnamed opponent. The opponent is reminiscent of the empiricist eternalists attacked by Saʿādia but is clearly a real, not hypothetical, antagonist.[54] This opponent asserts that we should content ourselves with what we can learn from sensation and not go beyond this to suppose the existence of immaterial causes. Natural processes in the world (in particular life) can be explained with reference to fire, an idea that inevitably calls the Stoics to mind, while cognitive processes are explained in terms of light. The world as a whole or, more specifically, what the opponent calls "the essence of the macrocosm," is itself a first cause and requires no further explanatory principle.

Miskawayh is unsurprisingly appalled by this atheistic sentiment, and for the sake of refuting it points out that the opponent's view

---

53 For translation and discussion see Peter Adamson and Peter E. Pormann, "More than Heat and Light: Miskawayh's Epistle on Soul and Intellect," *Muslim World* 102 (2012): 478–524; I quote from this version in what follows. For the Arabic text see ʿAbdurraḥmān Badawī, *Dirāsāt wa-nuṣūṣ fī l-falsafa wa-l-ʿulūm ʿinda l-ʿArab* (Beirut: al-Muʾssasa al-ʿArabīya, 1981), 57–97.

54 Compare especially *Book of Beliefs*, 57–58, where Saʿādia imagines an objector who says that fire's innate tendency to burn makes it superfluous to postulate further causes.

implies the eternity of the world. This is because "whatever has no cause and no prior reason exists always, that is, it is unoriginated [lam yazal]. For the eternal [al-azalī] is that which has no before and nothing prior to it. What is like this does not subsist through anything else, nor is its existence due to any of the four causes."[55] Notice the mutual implication assumed here: an uncaused cause must be eternal, and what is eternal must be uncaused. This signals Miskawayh's agreement with al-Kindī, which is no coincidence. As he goes on to refute the eternity of the world, he makes use of materials from *On First Philosophy*, especially the argument from composition.[56] He thus repeats al-Kindī's fundamental conclusion that the eternal is not only "absolutely first" but also free of all multiplicity.[57]

I take this to show that Avicenna was not simply following philosophical orthodoxy when he affirmed the eternity of the world. If anything, he was arguing against a rather well-entrenched view, which has a good deal of intuitive appeal: if something is eternal, it has no cause. Of course Avicenna was not the first philosopher to deny this. Aristotle made his God the cause for an eternal cosmos—albeit a cause of motion rather than existence[58]—and Plotinus saw the world as eternal yet dependent on immaterial causes. Yet Avicenna does bring something new, by using modal notions to capture the unique status of God as an uncaused cause, while accepting that some of God's effects are eternal.[59] His fundamental insight is that we should not see eternity, causal primacy, and necessity as inextricably linked. It is tempting to think that there is such a link, because necessary existence

---

55 Badawī, *Dirāsāt*, 77–78.
56 As I have shown in "Miskawayh's Psychology," in *Classical Arabic Philosophy: Sources and Reception*, ed. Peter Adamson (London: Warburg Institute, 2007), 39–54.
57 Badawī, *Dirāsāt*, 78.
58 Ammonius had argued that this is a distinction without a difference: God causes the cosmos to exist by making it move, since this means bestowal of actuality.
59 I am sympathetic to Wisnovsky's argument that Avicenna's innovation was inspired, at least in part, by emerging notions in Islamic speculative theology (*kalām*), which made eternal divine attributes dependent on God's essence. See Robert Wisnovsky, *Avicenna's Metaphysics in Context* (London: Duckworth, 2003).

seems to imply both eternity (after all, how could something eternal ever fail to exist?) and uncausedness (after all, why would something necessary need a cause?). Against this, Avicenna contends that there are actually two ways of being necessary. What exists without being caused is necessary-through-itself, whereas what is caused to exist is necessary-through-another. Thus, we can also have two kinds of eternal things: the eternal-in-itself and the eternal-through-another.[60] To put the same point differently, an uncaused cause will be intrinsically necessary and eternal and can pass on that necessity and eternity, so that what it causes will have these features extrinsically.

To understand Avicenna's treatment of eternity in greater detail, we must turn to his philosophical theology. Famously, this begins from a proof for the existence of God—or rather, a proof for the existence of a necessarily existent being. As I have argued elsewhere, Avicenna does not simply assume that a necessary existent would have all the traditional features associated with God.[61] In fact, in the metaphysical sections of texts like the *Healing* (*al-Shifāʾ*) and the *Pointers* (*al-Ishārāt*) he devotes more space to inferring the divine attributes from the central trait of necessity than to his proof that there is a necessary existent. He argues that features such as uniqueness, simplicity, immateriality, goodness, generosity, and knowledge are implied by necessity of existence. More or less the same goes for eternity and being uncaused, except that these features may simply be part of the meaning of necessity: "the true nature [ḥaqīqa] of necessity of existence is nothing but the very guarantee of existence [nafs taʾakkud al-wujūd]."[62]

---

60 Averroes alludes to this way of putting the distinction in his *Incoherence of the Incoherence*, his response to al-Ghazālī: Averroes, *Tahāfut al-Tahāfut*, ed. Maurice Bouyges (Beirut: Dar el-Machreq, 1930), 6.

61 See Peter Adamson, "From the Necessary Existent to God," in *Interpreting Avicenna: Critical Essays*, ed. Peter Adamson (Cambridge: Cambridge University Press, 2013), 170–189. For the proof that there is a necessary existent, see, e.g., Toby Mayer, "Avicenna's *Burhān al-Siddiqīn*," *Journal of Islamic Studies* 12 (2001), 18–39.

62 Avicenna, *The Metaphysics of the Healing*, ed. and trans. Michael E. Marmura (Provo: Brigham Young University Press, 2005), sec. 1.7.6; Marmura's section numbers, with my translations.

The upshot of this is that Avicenna does accept the following inference: "If something is uncaused, it is eternal." But he rejects the converse inference: "If something is eternal, it is uncaused." This is in contrast to all the philosophers we have looked at so far, who assumed that the implication between "uncaused" and "eternal" was a mutual one. This still leaves Avicenna with some work to do, however. To deny the second inference is merely to say that something *could* be both eternal and caused. But this does not prove that anything eternal is *in fact* caused, or vice versa. Perhaps God, who is necessary, uncaused, and eternal, creates a world that is neither necessary nor uncaused nor eternal. That is precisely the position that al-Ghazālī and many subsequent philosophers and theologians will take. How does Avicenna rule it out and insist that eternity is passed on from God to his creation?

His argument comes late in *The Metaphysics of the Healing*, in section 9.1. The thrust of this section is that a temporally finite creation is incompatible with God's necessity. There is an irony here: whereas al-Kindī, Saʿādia, and Miskawayh had denied the eternity of the world to safeguard God's unique possession of eternity, Avicenna thinks that God's unique possession of necessary existence requires us to embrace the eternity of the world. His reasoning has its roots in ancient treatments of divine causation. For instance, in his arguments in favor of the world's eternity, Proclus had argued that God cannot change so as to begin exercising causal efficacy.[63] Avicenna agrees. Reminding his reader that "the existent necessary in itself is a necessary existent in all of its aspects, and cannot start to have a feature it did not [yet] have," and that "the intrinsic cause [al-ʿilla li-dhātihā] necessitates its effect, so that if it endures it necessitates the effect enduringly [dāʾiman]," he states that this would already be enough—meaning by this, apparently, enough to show that God's effect will be eternal (9.1.2).

---

63 See Proclus, *On the Eternity of the World*, trans. Helen S. Lang and A. D. Macro (Berkeley: University of California Press 2001).

Avicenna forges on nonetheless, and it quickly becomes clear that he is trying to explain not one but two things. First, of course, the eternity of God's created effect, but also the reason why some things in the created world are *not* eternal. It's obvious why this is a problem for Avicenna: if God is the first cause and eternally passes on necessary existence to his effect, then won't every further link in the causal chain initiated by God also be eternal? After all, each cause will be caused eternally, and the presence of a genuine cause will yield its effect "without delay" (9.1.5). For anything to happen non-eternally, we need some factor that can introduce change: "On the agent's side, [this] would be for instance a volition [irāda] that would necessitate the act, or a nature that would necessitate the act, or an instrument, or a time. On the recipient's side, [this] would be for instance a readiness that had been absent [istiʿdād lam yakun]. Or, on both sides it might be, for instance, one of them reaching the other. Clearly all of this would be through some motion [ḥaraka]" (9.1.9). What Avicenna means here is that an eternal cause can give rise to new effects if it does so by eternally *moving*. In particular, it can do so by achieving a previously absent proximity to that which it affects ("one of them reaching another"). He has in mind the way the celestial bodies move over things in the sublunary realm and thus produce modifications in that lower realm, as when the sun heats the earth.

This is something God cannot do. He is immaterial and cannot get "closer to" or "further away from" anything. More generally, he never produces an effect he has not already been producing: "When a single essence remains as it has been in all respects, with nothing having been made to exist from it previously, and it is still like this now, then now too nothing will be made to exist from it. If it comes to make something exist from it, then an intention or volition [irāda] has arisen or ceased in the essence, or a capacity and power, or something else of this sort, that had not yet been present [lam yakun]" (9.1.13). Avicenna anticipates the objection that God could eternally will a temporally finite effect. He rejects this possibility, on the basis that a volition to produce something will immediately be efficacious unless there is some obstacle or reason to wait. But in this case, there can be no obstacle or reason to wait, nor can there be any change in God's power to act. Thus, God's volition will give rise to its immediate effect whenever God exists, that is, eternally. It is only through intermediaries, and in particular moving intermediaries, that God can indirectly give rise to anything noneternal. This is the second time that I have mentioned what we might call the "why not earlier" argument.[64] We first saw it in al-Rāzī, who invoked a foolish eternal causal principle (the soul) to explain how the world can begin at some arbitrary time despite God's perfect wisdom. Now, in Avicenna, we have seen the more obvious response to the problem: since every time is equivalent with respect to God's causal agency, God must exercise that agency at every time.

5 Al-Ghazālī

This brings us finally to the most famous text on the world's eternity written in the Islamic world: al-Ghazālī's *Incoherence of the Philosophers*.[65] After some opening remarks, he begins this work as follows:

> Philosophers disagreed about the past eternity [qidam] of the world. The majority opinion, among both the ancient and recent [philosophers], has settled on saying that it is eternal, and that it has always existed [lam yazal mawjūdan] along with God the exalted, and is caused by him and concurrent with him, rather than coming after him in time, as effect is concurrent with cause and light with the sun. They claim that God's priority to [the world] is like the priority of cause to effect, that is, a priority of essence and order, but not a

---

64 The basic thought can be found as far back as Parmenides, in fragment DK 28B8 9–10, quoted by Wilberding in the previous chapter.
65 Al-Ghazālī, *The Incoherence of the Philosophers*, ed. and trans. Michael E. Marmura (Provo: Brigham Young University Press, 1997). Marmura's section numbers, with my translations.

temporal priority. It is reported that Plato said the world is brought to be and originated [muḥdath]; but then, some of [the philosophers] interpreted his statement by denying that he was committed to the origination of the world. Galen, toward the end of his life in the book called *On His Own Opinions*, adopted suspension of judgment [tawaqquf] regarding this question, holding that one cannot discover whether the world is eternal or created. Perhaps he showed that he could not know, not because of any shortcoming on his part, but rather because this question is in itself impenetrable to the mind. But this is something of an anomaly among their school [madhhab], given that all of them teach that it is eternal in the past, and that in general, one simply cannot conceive of anything coming to be originated from eternity, with no intermediary. (sec. 1.1–3)

In the previous section I expressed some skepticism about al-Ghazālī's historical claim. He singles out Plato and Galen—referring to the same text quoted by al-Rāzī—as exceptions to a philosophical consensus in favor of eternity. While Plato adhered to the view al-Ghazālī himself would endorse, that the world is created with a first moment in time, Galen was agnostic on the issue.[66] As I have shown, this is probably the correct interpretation of Galen's remarks about the world's eternity. But al-Ghazālī passes over the fact that many philosophers in the Arabic tradition had rejected the eternity of the world, in order to portray his refutation of Avicenna as a confrontation with philosophy itself.

Thus, he launches into an extensive examination of the "philosophers'" arguments for the world's eternity, considering in section 1 whether the world has eternity in the past (*qidam*), and in section 2 the possibility that the world will exist forever into the future. Al-Ghazālī

---

66 Later (sec. 2.7–11), al-Ghazālī takes issue with the same Galenic argument mentioned by al-Rāzī, that we would see signs of decay in the cosmos if it were not destined to exist forever. He simply condemns this as a bogus argument, without discussing Galen's possible intentions in providing it.

sees an asymmetry here. Whereas he insists that the world cannot always have existed, he concedes that it might possibly be held in existence forever once it has been made (2.4). Why accept this asymmetry? As I suggested at the outset of this chapter, one might suppose that an omnipotent God could create the world for as long as he wishes. If this means that he can maintain the world's existence eternally into the future, why could he not already have created it eternally in the past? Al-Ghazālī does provide some reason for thinking this is impossible. Like al-Kindī, and Philoponus before him, al-Ghazālī contends that a world that has already existed eternally will constitute an actual infinity, something the "philosophers" themselves reject. In defending this inference he gives the example of the souls of the infinity of humans who have already lived and died (1.30). This is an improvement on the Kindī version of this argument, which refers to the infinity of past times that have already elapsed. An Aristotelian might respond to al-Kindī that so long as the times are not simultaneously present, they do not form an actual infinity but only a potential one. (Indeed this is precisely Averroes's counterargument in his *Incoherence of the Incoherence*.) But that reply is not available in the case of the infinity of immortal, immaterial souls that would now already exist if the world were eternal.

For the most part, though, al-Ghazālī occupies himself not with arguing positively that the world was created but with undermining arguments to show that it is eternal. He singles out the "why not earlier?" argument as the strongest consideration put forward by the philosophers. Admittedly, he damns it with faint praise, remarking that "what they say on other metaphysical questions is more feeble than what they say in this case, because here they can exploit various tricks of the imagination [funūn min al-takhyīl], which they cannot do elsewhere" (1.11). What he means is that our imaginations almost compel us to suppose that if God creates the world with a first temporal moment, there must be earlier moments at which God does not create. In which case, why did he wait for an infinite time before creating, despite already having willed to create? As al-Ghazālī says, closely echoing Avicenna's arguments in the *Healing*, one would suppose that an "instrument, capacity, purpose, or nature" must have arisen that previously was absent (1.9; see 1.16) that would "preponderate" the world to be created rather than not being created. But no such preponderating factor can arise, since before the world exists there is only God, and God never changes.

Al-Ghazālī has two ways of responding to this challenge. His first response is to welcome the consequence that God would need to choose an arbitrary time to create the world. A capacity for arbitrary choice is precisely what we mean by God's attribute of "volition" (*irāda*). As al-Ghazālī puts it: "The world began to exist where it did, as it did, and when it did, through volition [irāda]. Volition is an attribute whose role is to distinguish something from what is like it. If this were not its role, then it would be enough to have power [qudra]. But, since the relation of power to the two contraries was the same, and there needed to be something to select one thing in preference to what was like it, it is said that in addition to power, there was an attribute whose role is to select something in preference to what is like it" (1.41). This observation is a clue to the polemical purposes of the *Incoherence* in general. Just as al-Ghazālī here seeks to uphold God's possession of *irāda* against the idea that God causes the world necessarily and eternally, so later on, in section 3, he tries to show that if God causes in this way he does not count as an "agent" (*fāʿil*). The sixth discussion revolves around the problem of divine attributes in general; discussions sections 11, 12, and 13 concern the question of how God has the attribute of "knowledge"; and so on. The *Incoherence* is, to large extent, a challenge to Avicenna's systematic attempt to derive the traditional divine attributes from the central feature of necessity and to argue that if anything, the traditional attributes presuppose a non-necessary creative act. The eternity of the world, or so I would argue, arises in this text only as a subsidiary issue. Al-Ghazālī thus has good reason to develop the "why not earlier?" argument at such length. It allows him to make what is for him a central point concerning divine freedom.

Al-Ghazālī's second response renders his first response somewhat problematic. This is to say that there were, in fact, no times or moments before the world was created, because "time and duration" are created together with the world. He alludes to this only briefly in responding to the "why not earlier?" argument (at 1.36). But he develops the point further as he answers a second philosophical argument, to the effect that if the world is not eternal, there must have been an infinity of "empty" time before the world began to exist—but time is the measure of motion (1.80). Thus, the philosophers argue, God's priority to the world is causal or "essential" rather than "temporal." But as al-Ghazālī sees, the philosophers' point can be turned against them: if time is indeed the measure of motion, and motion is created, then time too is created. This idea was expressed already by al-Kindī, who made God "prior in time" in the sense of being the "cause of time" (*On First Philosophy* 1.5). However, God's timelessness, such a key issue in late antique philosophical theology (and, as I will explain below, in Latin medieval philosophy), did not really come to the fore in Arabic treatments of eternity before al-Ghazālī.

In this context, the question of timelessness does not concern God's foreknowledge or mode of existence, as in Boethius, for example. Rather, al-Ghazālī worries about the meaning of the claim "God was, while the world was not" (*kāna Allāh wa-lā ʿālam*) (1.84). Like other thinkers grappling with the notion of timelessness, he warns against being misled by tense. We should take the claim not to involve anything about the past—which would imply a time before the creation of the world—but rather the sheer fact of one thing's being without another. Cleverly, he picks out this fact as the common content shared by both "God was while the world was not" and "God will be while the world will not" As al-Ghazālī puts it, "the fundamental notion conveyed by the two expressions is the existence of one essence and non-existence of the other" (1.86). We cannot help but associate this with a time, but that is a product of misleading supposition (*wahm*).[67]

---

[67] Marmura translates *wahm* as "estimative faculty," a faculty Avicenna discusses in his theory of the internal senses. But t should be born in mind that Avicenna also makes *wahm* responsible for misleading (if all but irresistible) false impressions. That seems to be the sense of the term here.

Again showing his keen instinct for dialectical argument, al-Ghazālī points out that the philosophers are in an exactly analogous situation with regard to spatial extension. Just as those who uphold the world's creation must deny that there is time before the moment of creation, so the Aristotelians must deny that there is any place outside the outermost heaven. For they hold that the world is a perfect sphere, with nothing beyond it, not even void. He puts the point succinctly: "if one may affirm an 'above' with no 'above' above it, then one may affirm a 'before' with no genuine 'before' before it, apart from that in imagination and supposition, just as with 'above'" (1.87). Furthermore, the philosophers must face spatial arguments similar to the "why not earlier?" proof. It seems clear that the world could be very slightly bigger or smaller, but this is a matter of indifference (1.100). Likewise, the direction of heavenly rotation seems to be indifferent (1.62)—if the heavens went from west to east, then all the same results would be achieved, albeit spatially inverted. Thus the philosophers too are committed to accepting arbitrariness in creation.

Of course, al-Ghazālī would in fact welcome some arbitrariness in divine causation—this will allow him to secure the attribute of volition. Ironically, he can more easily insist on this in the spatial cases he forces on the philosophers—the size of the world, or direction of rotation—than in the case of time. For, in rejecting the existence of time prior to creation, he seems to have undermined his own carefully argued point that God arbitrarily selects a moment for the world to begin. If there is no time prior to the world, then no such selection occurs. As far as I can see, the only way to save the argument would be to say that it is a matter of indifference how long the world has already existed prior to a given event—for instance, how long the cosmos existed before you began reading this book. Yet this seems to suppose that adding or removing events to the set of events that have already occurred is indifferent, which is implausible.

Before leaving al-Ghazālī, I should note one final aspect of his polemic against Avicenna. A careful reader of his opponent, al-Ghazālī realizes that Avicenna was trying to achieve two things in 9.1 of the *Metaphysics of the Healing*. He not only wanted to show that the world is eternal, but to explain how it is that anything temporal can emerge from an eternal cause. Avicenna sought to achieve this by positing eternal motion. The motion of the heavens yields the possibility of non-eternal things, thanks to the heavenly bodies' varying proximity to the matter they influence.[68] Al-Ghazālī, though he understands Avicenna's idea well (as is clear, for instance, at *Incoherence* 1.72–73), is dismissive of the whole strategy. He complains that on this theory, we can never resolve the problem of how an eternal thing generates a temporally limited thing, because the eternal is "stable" (*thābit*), that is, unchanging (*Incoherence* 1.76). I think this underestimates the effectiveness of Avicenna's solution. If the heavenly motion has never begun, then God can arguably cause it to happen without changing. But one might find the critique more persuasive if one understands eternity to mean timelessness rather than everlasting existence. The question would then be, not how God causes motion without changing, but how he can give rise to something that is subject to time without himself being subject to time.

6 Maimonides and Aquinas

In this final section, I will show that this idea of God's timelessness underlies the position of Thomas Aquinas in the heated debates over the world's eternity that took place in Paris, especially toward the end of his life. Its role is not at first obvious, though. Aquinas takes his chief inspiration for the idea of divine eternity from Boethius,[69] whereas his position on the world's eternity is normally, and rightly,

---

68 For a similar maneuver in Averroes see Ruth Glasner, *Averroes' Physics: A Turning Point in Natural Philosophy* (Oxford: Oxford University Press 2009), 79.

69 Albeit that, as Shanley has emphasized, Aquinas "makes the Boethian definition fit his position rather than vice-versa." See Brian J. Shanley, "Eternity and Duration in Aquinas," *Thomist* 61 (1997), 537.

seen as heavily influenced by the great Jewish thinker Maimonides. With Maimonides and Aquinas, we finally seem to arrive at a clear statement of the position I suggested at the beginning of this chapter: that God's untrammeled power means that he could create either an eternal or non-eternal world—in which case there is no way for us to know whether the world has always existed, and always will exist, unless God further sees fit to inform us of his will on the matter.

Maimonides's handling of the eternity question in his *Guide for the Perplexed* is one of the most contentious and often-discussed aspects of a contentious and often-discussed work.[70] It is clear enough that Maimonides voices the view just sketched: that there can be proof neither of the world's eternity nor of its non-eternity, given that God can choose between either option, so it is only from Revelation that we learn that the world was in fact created with a first moment of time. What is less clear is whether he asserts this as his genuinely held opinion. Some interpreters have suggested that in reality Maimonides does accept the world's eternity. A sufficiently attentive reader will see that the surface profession of the world's non-eternity is undermined by clues scattered throughout the text.[71] Without getting too far into this debate here, it is worth mentioning one strong argument in favor of this reading and a reason for questioning this argument.

The argument is as follows. When Maimonides proves the existence of a God who is immaterial and is one, he explicitly depends on the premise that the world is eternal.[72] He cautions the reader that this

---

70 Arabic text in Maimonides, *Dalālat al-Ḥā'irīn*, ed. Solomon Munk and Issachar Joel (Jerusalem, 1929); for English versions see Maimonides, *The Guide of the Perplexed*, trans. Shlomo Pines (Chicago: University of Chicago Press, 1963); Maimonides, *The Guide for the Perplexed*, trans. Michael Friedländer (New York: Dover, 1956).
71 See for instance Herbert A. Davidson, "Maimonides' Secret Position on Creation," in *Studies in Medieval Jewish History and Literature*, ed. Isadore Twersky (Cambridge, MA: Harvard University Press, 1979), 16–40; Shlomo Pines, "The Philosophic Purport of Halachic Works and the Purport of the Guide of the Perplexed," in *Maimonides and Philosophy*, ed. Shlomo Pines and Yirmiyahu Yovel (Dordrecht: Kluwer, 1986); Shlomo Pines, translator's introduction to Pines, *Guide of the Perplexed*.
72 See *Guide* 2, introduction.

is a premise he accepts "because one can on its basis demonstrate our sought for conclusion." While this might suggest that he does not in fact subscribe to the premise, he fails to provide demonstrations of these key points—the existence (*kawn*) of God, his incorporeality (*lā jisman*), his being one (*wāḥid*)—on any other basis. Thus one might well wonder whether he in fact subscribes to the eternity of the world, because he believes that otherwise these points are indemonstrable. However, I find plausible a rival proposal recently made by Daniel Davies.[73] He points out evidence showing that for Maimonides, the desired conclusions concerning God are simply *obvious* if God created the physical cosmos with a first moment of time. His intention would thus be to show that even if one makes the less favorable assumption of the world's eternity, these conclusions can still be shown to follow.

Be all that as it may, at the surface level Maimonides's treatment of the world's eternity is an attempt to show that considerations offered so far in the eternity debate have been inconclusive. He sets out various arguments for and against the proposition that the world is eternal and shows that none of the arguments are successful. His most telling point against Aristotelian pro-eternity arguments is simply that divine causation cannot be assumed to work the way physical causation does. Thus, for instance, physical causes always operate on preexisting material, but this need not be the case for God.[74] Partially on this basis, Maimonides rules out one of the three possible positions on the question, namely (what he takes to be) Plato's: the world as we know it is not eternal, but it was fashioned from eternal, preexisting matter. He reads a number of rabbinical commentators on the Torah as having held the same position.[75] This leaves two options that are apparently genuinely possible: Aristotle's position that the world

---

73 Daniel Davies, *Method and Metaphysics in Maimonides' Guide for the Perplexed* (New York: Oxford University Press, 2011), 49.

74 *Guide* 2.17. See Davies, *Method and Metaphysics*, 30.

75 On this see Kenneth Seeskin, *Maimonides on the Origin of the World* (Cambridge: Cambridge University Press, 2005), 16.

has always existed more or less as it does now, and what seems to be Maimonides's own view, that the world was created ex nihilo with a first moment of time.

One might assume that Maimonides is quietly chastising Aristotle for claiming to have proven the world's eternity when no such proof can in fact be given. But, famously, Maimonides shows his respect for Aristotle by insisting that Aristotle was well aware that his proofs are not decisive. He cites three passages, from the *Physics, On the Heavens*, and *Topics*, that supposedly show Aristotle expressing doubt concerning the world's eternity.[76] The most compelling evidence Maimonides offers is from the *Topics*, where Aristotle says that dialectical problems include those "in regard to which we have no argument because they are so vast, and we find it difficult to give our reasons, e.g. the question whether the universe is eternal or not; for into questions of that kind too it is possible to inquire."[77] Maimonides's irenic attitude toward Aristotle will be inherited by Aquinas in most of his discussions of the question, though his *Commentary on the Physics* is less generous toward Aristotle on this score.[78]

A powerful argument that no demonstration is possible in respect of the world's eternity is that God's will determines the duration of the world—and God wills contingently. But Maimonides faces a now familiar difficulty on this very point. If God goes from willing, or causing, the nonexistence of the cosmos to willing, or causing, its existence, won't God change? Maimonides answers by drawing an analogy between God and the Active Intellect (as propounded by Aristotle and al-Fārābī, says Maimonides). As immaterial causes, they are both efficacious without any actualization of a preceding potentiality (*Guide* 2.18). But Maimonides realizes that this analogy threatens to undermine his

---

76 *Guide* 2.15. See Seeskin, *Maimonides on the Origin*, 92.
77 *Topics* 104b13–17, translation by W.A. Pickard Cambridge, in J. Barnes (ed.), *The Complete Works of Aristotle* 2 vols (Princeton: Princeton University Press, 1984), vol. 1, 174.
78 John F. Wippel, "Did Thomas Aquinas Defend the Possibility of an Eternally Created World?" *Journal of the History of Philosophy* 19 (1981), 21 n. 2.

own position. The effects of the Active Intellect occur at some times rather than others because of changes on the side of the recipient, that is, the matter. For example, an animal is formed when matter is suitably prepared. Once the matter is ready, the form comes automatically from the Active Intellect, since it is always emanating all the forms, which are taken on by whatever matter is in the appropriate condition. This is very like Avicenna's picture of divine causation: divine emanation is eternal and unchanging, and variation in the effects is caused by heavenly motion.

Of course this sort of account is no help in explaining why the world comes to exist after not existing. One cannot appeal to changes in a recipient if the recipient does not yet exist. So even if Maimonides has shown that an immaterial cause can be efficacious without going from potentiality to actuality, he has not yet solved a different, and by now familiar, problem. If God's wisdom leads him to will a created universe, and this wisdom and will are eternal, why is the universe not likewise eternal? He poses this objection explicitly but thinks it is a fairly feeble one: "For just as we are ignorant of his wisdom which necessitated that there are nine celestial spheres, no more and no fewer; and that the number of stars is what it is, neither more nor fewer, neither greater nor lesser; in the same way, we are ignorant of his wisdom in making the universe exist, after not doing so only a little beforehand. The universe is consequent upon his eternal and unchanging wisdom, but we are wholly ignorant of the rules [*qānūn*] and decrees of that wisdom" (*Guide* 2.18). As Kenneth Seeskin has emphasized, Maimonides appeals to certain evident features of the universe that are (as far as we can tell) arbitrary.[79] What difference could it make if there were one more star, or one fewer spheres? The strategy is like that of al-Ghazālī, who drew a comparison between the temporal extent of the universe and cosmological features, such as its spatial dimensions and the direction of the heavenly rotation.

---

79 See Seeskin, *Maimonides on the Origin*, at, e.g., 120, 167.

Maimonides sharpens the dialectical move by turning the philosophers' equation between eternity and necessity against them. If they are right that the eternal is necessary and vice versa, then the apparently arbitrary cosmological features of the universe show that the universe is characterized by contingency and therefore *not* eternal.[80] In light of this, one might wonder whether Maimonides has compromised his stated position that we can prove neither the world's eternity nor its non-eternity. Doesn't the argument just considered prove that it is not eternal?[81] No: Maimonides argues only that as far as we know or could know, the world is non-eternal. Certain cosmological features of the world do *seem* to be non-necessary, and what is not necessary is not eternal. However, this may only be a symptom of our own cognitive limitations, what Maimonides calls our "ignorance" (*jahl*) in the passage cited above. Despite our impression that certain cosmological features of the world are contingent, it could turn out that they are in fact necessary and eternal. We cannot know God's will or the wisdom behind it. Perhaps, if we did have the divine perspective on things, we would see that the apparently contingent is necessary after all (for example because God sees that it is best, and God necessarily makes whatever is best).

When we turn to Aquinas, we find that the epistemic limitations of humankind are again crucial. It is well known that Aquinas followed Maimonides in thinking that the world is not eternal but might have been created as eternal; we know the truth only through revelation. But what is the force of this "might"? Does it mean that it was genuinely "up to" God whether or not to create an eternal world, so that both options were *in fact* possible? Or does it mean only that *as far as we know*, both options were possible? The latter, epistemic

---

80 Thus we have a *modus tollens*: if the world is eternal, its cosmological features are necessary; but its cosmological features are arbitrary, not necessary; therefore the world is not eternal. The reason he focuses on cosmological features is of course that these are "permanent," in the sense that they obtain whenever the world exists.

81 Compare Seeskin, *Maimonides on the Origin*, who at 172 and 181 (referring to *Guide* 2.19) avers that this argument from the "particularization" of contingent factors is "nearly" a proof.

possibility is compatible with the factive impossibility of an eternal world; the impossibility would be for reasons beyond our ken.[82] As argued in a useful study by John Wippel, Aquinas seems to have subtly modified his position on this issue during his career.[83] For most of his life,[84] he explicitly endorsed only the epistemic interpretation of this claim. That is, he argued that we cannot demonstrate the non-eternity of the universe without positively asserting that its eternity is in fact possible.

But debates over the question became more and more heated and politicized during his lifetime.[85] Others, like John Pecham and Bonaventure, asserted the demonstrable impossibility of an eternal world.[86] Ultimately, Aquinas was provoked into explicitly claiming that the world could in fact be eternal, although it of course is not.[87] This comes late in his life in a very short work called simply *On the*

---

82 Of course Aquinas can rule out that there is (unbeknownst to us) a factive necessity of the world's being eternal, since for him it is false that the world is eternal.
83 Wippel, "Did Thomas Aquinas Defend the Possibility of an Eternally Created World?"
84 Wippel considers the *Sentences* commentary, *De Potentia*, *Summa contra Gentiles*, and *Summa Theologiae*.
85 This story is well told in Richard C. Dales, *Medieval Discussions of the Eternity of the World* (Leiden: Brill, 1990).
86 On Bonaventure's position and its comparison to that of Aquinas, see Bernardino Bonansea, "The Impossibility of Creation from Eternity According to St. Bonaventure," *Proceedings of the American Catholic Philosophical Association* 48 (Washington, DC: 1974), 121–135, and "The Question of an Eternal World in the Teaching of St. Bonaventure," *Franciscan Studies* 34 (1974), 7–33; Francis J. Kovach, "The Question of the Eternity of the World in St. Bonaventure and St. Thomas: A Critical Analysis," *Southwestern Journal of Philosophy* 5 (1974): 141–172; Peter van Veldhuijsen, "The Question on the Possibility of an Eternally Created World: Bonaventura and Thomas Aquinas," in *The Eternity of the World in the Thought of Thomas Aquinas and His Contemporaries*, ed. Josef Wissink (Leiden: Brill, 1990), 20–38; Albert Zimmermann, "*Mundus est aeternus*? Zur Auslegung dieser These bei Bonaventura und Thomas von Aquin," in *Die Auseinandersetzungen an der Pariser Universität im XIII. Jahrhundert*, ed. Albert Zimmermann (Berlin: de Gruyter, 1976), 317–330.
87 The literature on the eternity of the world in Aquinas himself is of course large. See for instance Jan A. Aertsen, "The Eternity of the World: The Believing and the Philosophical Thomas. Some Comments," in Wissink, *The Eternity of the World*, 9–19; Norman Kretzmann, *The Metaphysics of Creation: Aquinas's Natural Theology in Summa contra gentiles II* (Oxford: Oxford University Press, 1999), esp. chap. 5; John F. Wippel, "Thomas Aquinas on God's Freedom to Create or Not," in Wippel, *Metaphysical Themes in Aquinas II* (Washington, DC: Catholic University Press, 2007), 218–239; and for comparison to Maimonides, Avital Wohlman, *Thomas d'Aquin et Maïmonide: Un dialogue exemplaire* (Paris: Cerf, 1988).

Eternity of the World.[88] Aquinas's rationale is, again, the one I proposed above: God is omnipotent, so unless there is something intrinsically impossible in an eternal world, he is free to create one. One might well wonder how Aquinas can go beyond (what I take to be) the Maimonidean claim that *as far as we know* such a world is possible to the stronger claim that *in fact* such a world is possible. The answer is that createdness ex nihilo and eternity are compatible notions. To put it in the terms I used above in discussing Avicenna, one cannot say that "eternal" implies "uncaused." Interestingly, Aquinas seems to take the point from Avicenna. Already in Aquinas's commentary on the *Sentences*, written in the 1250s, he remarks: "even if it be supposed that heaven has existed from eternity, yet it is true to maintain that heaven is out of nothing, as Avicenna proves."[89]

Why is it that Aquinas, while holding firmly to the non-eternity of the world, is so forthcoming about admitting the possibility of its eternity? One reason, surely, is the one already mentioned—that denying this possibility seems to amount to a restriction on divine power.[90] But there is, as I anticipated at the beginning of this section, another factor at work here: God's timelessness. If we cast our minds back to the issue that worried figures like al-Kindī, we will recall that a major impetus toward denying the world's eternity had always been the need to distinguish radically between God and world. A "created" versus "eternal" dichotomy is ideally suited for such a distinction, because the world can be safely located on the "created" side with the "eternal" side reserved for God alone. But Aquinas had access to a source unavailable to al-Kindī and the other figures I have examined from the Arabic tradition, namely Boethius. Certainly, as I have shown, al-Kindī, al-Ghazālī, and others did suggest that God is somehow timeless, if only because he is

---

[88] James Wiesheipl, "The Date and Context of Aquinas' *De aeternitate mundi*," in *Graceful Reason*, ed. Lloyd P. Gerson (Toronto: Pontifical Institute, 1983), 239–271. For a translation see Cyril Vollert, Lottie H. Kendzierski, and Paul M. Byrne, trans., *St. Thomas Aquinas, Siger of Brabant, St. Bonaventure: On the Eternity of the World* (Milwaukee: Marquette University Press, 1964), 18–24.

[89] *Commentary on the Sentences*, 5.2.2, cited by Aertsen, "The Eternity of the World," 17.

[90] This is the rationale emphasized by Wippel, "Did Thomas Aquinas Defend the Possibility of an Eternally Created World?," 32.

unmoving, and time is the measure of motion. But Boethius had considered God's timelessness at greater length and identified timelessness, rather than eternal temporal duration, as the true sense of "eternity."

It is this, I contend, that freed Aquinas to assert the possibility of a temporally everlasting world so boldly. Such a world would in any case not compromise God's unique status as "eternal" if, that is, "eternal" is understood as meaning "timeless."[91] This point becomes explicit in *On the Eternity of the World*, where Aquinas cites Boethius's *Consolation*. Boethius remarks that some people think Plato is committed to the co-eternity of world and God. They are wrong, because "it is one thing to be carried through an endless life, which is what Plato attributed to the world, and quite another to embrace the whole presence of endless life all at once" (*Consolation* 5.6). Having quoted this with approval, Aquinas goes on to say: "hence it is clear that the difficulty feared by some does not follow, that is, that the creature would be on a par with God in duration [aequaretur Deo in duratione]. Rather we must say that nothing can be co-eternal with God, because nothing can be immutable save God alone."[92]

For Aquinas, then, the Boethian notion of timeless eternity defused to some extent the issue of the world's eternity. This was not the case for all his contemporaries, as shown by the inclusion of propositions regarding the eternity of the world in the notorious condemnations laid down by the church authorities of Paris in 1270 and 1277. But Aquinas faces other problems instead. Aside from the question of what "timeless eternity" even means,[93] Aquinas must cope with the dilemma

---

[91] One might speculate that Maimonides's freedom to defend an agnostic position had a similar basis: given his thoroughgoing negative theology, it would not be permissible in his view to describe God as "eternal," unless this were understood as a mere concealed negation. Thus, even if the world is "eternal" in the sense of "everlasting" this will not put the world on a par with God.

[92] Translation from Aquinas, *On the Eternity of the World*, sec. 11, in Vollert et al., *St. Thomas Aquinas, Siger of Brabant, St. Bonaventure*, 23.

[93] Again, a subject much discussed in the secondary literature. In addition to works already cited, see e.g. David B. Burrell, "God's Eternity," *Faith and Philosophy* 1 (1984): 389–406; Brian Davies, *The Thought of Thomas Aquinas* (Oxford: Oxford University Press, 1992), chap. 6; Christopher Hughes, *A Complex Theory of a Simple God* (Ithaca, NY: Cornell University Press, 1989), chap. 4; Nikolaus Wandinger, "Der Begriff der 'Aeternitas' bei Thomas von Aquin," *Zeitschrift für Katholische Theologie* 116 (1994): 301–320.

I discussed in section 5 of this chapter. How can something timeless produce, or even relate, to the temporal? This is simply an instance of a broader phenomenon, already familiar from late antique Platonism. Created things have their natures by participating in higher principles (in this case God), yet those higher principles are meant to transcend the natures of their effects. On one interpretation, Aquinas solves this further problem by seeing God's eternity more as a negation (a consequence of immutability) than as a preeminent possession of all that is expressed temporally in his effects (as would be the case for Plotinus and, indeed, Boethius).[94] This though is only one reading of Aquinas among several, and later medieval thinkers continued to ponder the meaning of God's eternity.[95] The distinction between timeless eternity and temporal everlastingness may have solved the eternity of the world debate to Aquinas's satisfaction. But as will be shown in subsequent chapters, the relation of the timeless to the temporal has remained a perennial, not to say unending, source of philosophical perplexity.

---

94 See Shanley, "Eternity and Duration in Aquinas," 530, 541, 545.
95 See e.g. Richard A. Cross, "Duns Scotus on Eternity and Timelessness," *Faith and Philosophy* 14 (1997): 3–25.

Reflection
"ETERNALISTS" AND "MATERIALISTS" IN
ISLAM: A NOTE ON THE *DAHRIYYA*
Hinrich Biesterfeldt

Dahriyya is an Arabic collective noun that a tenth-century encyclopedia of Arabic-Islamic technical terms defines as "those who believe in the eternity of the course of time."[1] The basic noun from which *dahriyya* is derived is *dahr*, a Qur'anic word denoting (a long period of) time. Sura 45, verse 24 talks about the contention of the unbelievers, "deafened, their hearts sealed and blinded" by God, who maintain: "There is only our life here and now, we die and we live (here), and we are destroyed only by Time [al-dahr]."[2] It is probable that this verse is at the root of (1) the negative connotation of *dahr*, *dahrī* ("eternalist"), *dahriyya* or *ahl al-dahr* ("eternalists, eternalism") in Muslim theology, (2) its frequent contamination of *dahr* with the meaning "materialist, materialism," and (3) the fact that we know about the development and actual use of these terms mainly from Muslim heresiography. No original *dahriyya* testimonies

---

1 Al-Khwārazmī, *Mafātīḥ al-ʿulūm*, ed. Gerlof van Vloten (Leiden: Brill, 1895), 35: *al-dahriyya: alladhīna yaqūlūna bi-qidami l-dahri*.
2 On the pre-Islamic notion of *dahr*, "all-devouring time," see Hellmut Ritter, *The Ocean of the Soul: Man, the World and God in the Stories of Farīd al-Dīn ʿAṭṭār* (Leiden: Brill, 2003) (German original: *Das Meer der Seele* [Leiden: Brill, 1978]), 43–44, with references.

survive, next to no individual *dahrī* is known by name, and the doctrine of the *dahriyya*, as the very varied sources report on it, are far from clear-cut. Obviously the *dahriyya* has never constituted a "school" or a philosophical "faction," yet the great number of polemics from perhaps already the eighth down to the twelfth century against *dahriyya* views attests to its presence in Muslim theology and philosophy.

One of the most vivid accounts of the arguments of the *dahriyya* comes from ninth-century Baghdad, the place and time at which its representatives may have commanded the greatest attention.[3] Its contemporary author is al-Jāḥiẓ (d. 869), one of the most famous prose writers in classical Arabic. He is not a heresiographer but a litterateur, a Muʿtazilite in the sense of showing a keen interest in a rationalist approach to the theological and philosophical issues of his time, and a brilliant master of disputation and polemic. In his *The Book of Living*, he writes:

> For he [the dahrī] it is who denies divine lordship [rubūbīya]; who holds commanding right and forbidding wrong to be inconceivable; who rejects the possibility of apostleship; who makes clay [ṭīna] primordial matter; who repudiates reward and punishment; who knows not licitness or illicitness; who avers that the world contains no proof which indicates a craftsman and his handiwork, a creator and his creation. It is he who makes the celestial sphere (which does not know the difference between itself and someone else, which does not distinguish between new and old and between good and bad, which is not capable of increasing its motion or decreasing its revolution, of alternating rest with motion, of coming to a halt for

---

[3] See Patricia Crone, "Dahrīs." In *The Encyclopaedia of Islam*, 3. referenceworks.brillonline.com/entries/encyclopaedia-of-islam-3. Comprehensive information on the sources for *dahriyya* thought is collected in Ignaz Goldziher and Amélie-Marie Goichon, "Dahriyya," in *Encyclopaedia of Islam*, new ed., vol. 2 (Leiden: Brill, 1993), 95–97; Mansour Shaki and Daniel Gimaret, "Dahrī," in *Encyclopaedia Iranica* (Costa Mesa, CA: Mazda, 1993), vol. 6, 587–590; Crone, "The Dahrīs According to al-Jāḥiẓ."

the blink of an eye or of turning from its course) the origin and the destruction of all this good craftsmanship, of subtle and momentous matters, of these wondrous instances of wisdom and perfect arrangement [tadābīr], of novel composition [al-taʾlīf al-badīʿ] and wise disposition [tarkīb], all according to an identified pattern and a known order, at the extremes of subtly fine wisdom and perfect fabrication!⁴

In this passage, al-Jāḥiẓ combines the most frequent suspicions against the *dahriyya* that Muslim theologians of the ninth century entertained: its denial of God and his commands and retribution and of his sending prophets to mankind and its blindness to the wisdom of his creation (the Islamic "argument from design")—indeed denial of creation altogether, which for Muslims constitutes God's fundamental act that sets time in motion. Instead, the *dahriyya* postulates "clay" as primordial or eternal matter,⁵ and it assigns the management of the world, which together with the creation of the world was traditionally seen as a role belonging exclusively to God, to "the celestial sphere." Al-Jāḥiẓ's extensive account of that sphere's incapacities—lack of self-conscience, ignorance of temporal stages,⁶ and of ethical positions, inability to determine its own movement—shows how crucial this part of the *dahriyya* argument is for him. "How could the heavenly bodies, which did not have the ability to vary their own movements, be the regulators of everything?"⁷

---

4 I follow Montgomery's translation of the book title and of the passage in question: James E. Montgomery, *Al-Jāḥiẓ: In Praise of Books* (Edinburgh: Edinburgh University Press, 2013), 263–264 (= al-Jāḥiẓ, *K. al-Ḥayawān*, ed. ʿAbdassalām Muḥammad Hārūn [Cairo: Maktabat Muṣṭafā al-Bābī al-Ḥalabī, 1938–45], vol. 7, 12–13).
5 "Primordial" is *qadīma* in Arabic, which expresses the same idea as al-Khwārazmī's *qidam* (see note 1 here). In al-Jāḥiẓ's time, some Arabic translations use *ṭīna* to render Greek *hylē*; see Gerhard Endress, "Ṭīna," in *Encyclopaedia of Islam*, new ed. (Leiden: Brill, 2000), vol. 10, 530.
6 Crone, "The Dahrīs According to al-Jāḥiẓ," 65, translates "that which appears in time and the eternal" (*al-ḥadīth wa-l-qadīm*).
7 Crone, "The Dahrīs According to al-Jāḥiẓ," 66, with a reference to earlier Christian polemic along these lines, also in Iraq: Joel Thomas Walker, "Against the Eternity of the Stars: Disputation and Christian Philosophy in Late Sasanian Mesopotamia," in *La Persia e Bisanzio*, ed. Gherardo Gnoli and Antonio Panaino (Rome: Accademia Nazionale dei Lincei, 2001), 513–535.

In another passage of his *Book of Living* al-Jāḥiẓ names another cosmological force, rivaling the celestial sphere as a principle that determines everything in the world around us:

> Some exponents of the *dahriyya* maintain that this world of ours consists of four principles: heat, cold, dryness, and moisture, and everything existing is (their) product, is combined and generated by them. They regard [jaʿalū] these four as bodies. Others maintain that this world consists of four principles: earth, air, water, and fire. They regard heat, cold, dryness, and moisture as accidents in these substances. Everything else, they then say, follows from this. Thus they regard these four things mentioned as the principles, saying [wa-jaʿalū] that smells, colors, and sounds are products [thimār] of those four, according to the degree of mixture (of the four), little or much, loose or compact. Thus they give absolute priority to the tactile sense, leaving aside the other four senses.[8]

Still another rhetorical argument of the *dahriyya*, reported by al-Jāḥiẓ in his *K. al-Tarbīʿ wa-l-tadwīr*, is based on the fact that there is no anvil without a hammer and vice versa, and no egg without a chicken and vice versa: one always presupposes the other, and thus the world must have always existed. "Nothing can come to be that has not been preceded by its equivalent or its opposite to infinity."[9]

---

[8] Al-Jāḥiẓ, *K. al-Ḥayawān*, vol. 5, 40, reports here the account of his teacher al-Naẓẓām (died between 835 and 845). I follow the translation and textual emendation by Josef van Ess, *Theologie und Gesellschaft im 2. und 3. Jahrhundert Hidschra: Eine Geschichte des religiösen Denkens im frühen Islam*, 6 vols. (Berlin: de Gruyter, 1991–97), vol. 6, 66. For the theory of the "four principles" among the "naturalists" (*aṣḥāb al-ṭabāʾiʿ*) in eighth-century Iraq, see van Ess, *Theologie und Gesellschaft*, vol. 2, 36–41; see also Crone, "The Dahrīs According to al-Jāḥiẓ," 71, and Crone, "Ungodly Cosmologies," in *The Oxford Handbook of Islamic Theology*, ed. Sabine Schmidtke, in Oxford Handbooks Online, March 2014, 1–36, *aṣḥāb al-ṭabāʾiʿ*: 22–28. The epistemology of the *dahriyya*, which emphasizes the roles of sense perception and control by human reason, is not dealt with in my note; see Crone, "The Dahrīs According to al-Jāḥiẓ," 69, and "Ungodly Cosmologies," 10–13 ("Skepticism").

[9] Gimaret in Shaki and Gimaret, "Dahrī," 589, and Crone, "The Dahrīs According to al-Jāḥiẓ," 2, both with references. The passage is in Charles Pellat, trans., *Le "Kitāb at-Tarbīʿ wa-t-tadwīr" de Ǧāḥiẓ* (Damascus: Institut Français de Damas, 1955), sec. 46, pp. 29–30. Compare al-Ghazālī's argument, below.

Two generations after al-Jāḥiẓ, the groundbreaking and influential theologian al-Māturīdī (d. 944 in Samarkand), wrote the first extant summa of his field, the *K. al-Tawḥīd*. In his discussion of God's attribute of "speech" (*kalām*) and in particular in his chapter on "the refutation of the disbelievers," he deals extensively with the *dahriyya*. His adversaries remain as anonymous as they appear in al-Jāḥiẓ's accounts, but his presentation is considerably more scholastic: he differentiates between two groups. The first group (1) is convinced that the world has eternally maintained its present form, with one faction (1.1.) postulating a godless cosmos, regulated either by (1.1.1) unidentified eternal elements of nature or by (1.1.2) an eternal material substance (al-Jāḥiẓ's *ṭīna*) or by (1.1.3) the four elements about which al-Naẓẓām talks according to al-Jāḥiẓ, and another faction (1.2.) assuming primordial elements of nature together with a creator who has performed his act already in preeternity. The second group (2) combines an eternal material principle with a temporal origin to the world and is subdivided in four factions, respectively holding (2.1) that God created from a preexisting material substance, or (2.2) that the stars, once starting to move, caused the creation of the world, or (2.3) that eternal material substance at one point became differentiated by the emergence of "accidents," or (2.4) that the world owes its existence to the—temporally indeterminate— mixing of two primordial principles.[10] This is a rich offer, and later theologians and heresiographers have not added considerably to the multifaceted profile of the *dahriyya* as assembled by al-Jāḥiẓ and al-Māturīdī: the "Brethren of Purity," authors of a philosophical encyclopedia in the second half of the tenth century, describe the *azaliyya* (synonymous with *dahriyya*) as proponents of the eternity of the world (*bi-qidam al-ʿālam wa-azaliyyatih*) who ignore the

---

10 For an analysis of al-Māturīdī's views on the *dahriyya*, see Ulrich Rudolph, *Al-Māturīdī and the Development of Sunnī Theology in Samarqand* (Leiden: Brill, 2015) (German original: *Al-Māturīdī und die sunnitische Theologie in Samarkand* [Leiden: Brill, 1997]), 166–179, and "Index of Religious and Political Movements," s.v. "Dahrīya"; for al-Māturīdī's treatment of Dualism (no. 2.4 in his system), see "Index of Religious and Political Movements," s.v. "Dualists."

causa efficiens and focus entirely on the other three *causae*; the Jewish theologian Saʿādia Gaon (d. 942) refutes the doctrine of *dahr*, which regards "not only matter as eternal but the beings of the world which we see as invariable,"[11] and brandishes the sect as limiting knowledge to sense perception; the theologian al-Ghazālī (d. 1111), in his *Confessiones, al-Munqidh min al-ḍalāl* (*The Savior from Error*), declares in his chapter on the "stigma of unbelief common to all philosophers": "The first category, the Materialists, were a group of the most ancient philosophers who denied the existence of the omniscient and omnipotent Creator-Ruler. They alleged that the world has existed from eternity as it is, of itself and not by reason of a Maker. Animals have unceasingly come from seed, and seed from animals: thus it was, and thus it ever will be. These are the godless [al-zanādiqa][12] in the full sense of the term."[13] Later heresiographers, like Ibn Ḥazm al-Andalusī (d. 1064) and al-Shahrastānī (d. 1153) regroup the definitions of *dahriyya* assembled by their predecessors and associate it with various other deviant sects; al-Shahrastānī talks of "*dahriyya* philosophers," probably reacting to a development in the history of Islamic philosophy that would target al-Fārābī and Avicenna.[14]

Whereas al-Jāḥiẓ's various reports give the impression that in his time the *dahriyya* was still a living phenomenon, al-Māturīdī shows that by his time the *dahriyya* was already seen as a stagnant system of thought that was only discussed in a general and abstract manner. Al-Māturīdī himself names Aristotle as the great authority

---

11 Goldziher and Goichon, "Dahriyya," 95.

12 François C. de Blois, "Zindīḳ," in *Encyclopaedia of Islam*, new ed., vol. 11, 510–513; Crone, "The Dahrīs According to al-Jāḥiẓ," 74.

13 al-Ghazālī, *Freedom and Fulfillment: An Annotated Translation of al-Ghazālī's "al-Munqidh min al-Ḍalāl,"* trans. Richard Joseph McCarthy, S.J. (Boston: Twayne, 1980), 71.

14 Goldziher and Goichon, "Dahriyya," 96–97. The article ends with an interesting account of materialistic ideas in the Middle East and India under the influence of nineteenth-century European literature, and their refutation by Muslim authors. For *dahriyya* in the general context of Islamic heresiography, see Josef van Ess, *Der Eine und das Andere: Beobachtungen an islamischen häresiographischen Texten*, 2 vols. (Berlin: de Gruyter, 2011), Indices: Religionen, Sekten, Schulen, s.v. "Dahrīya."

behind the idea of the world's eternity, who "described an eternal elemental cycle, and assumed there to be a certain autonomy at work in nature, an autonomy that Islamic theology widely rejected."[15] A more specific target would be Proclus, whose *Propositiones* on the eternity of the world were better known in Arabic than in Greek and, more important, were read in al-Jāḥiẓ's time, together with their refutation by Johannes Philoponus.[16] Greek cosmology, in particular Galen's (d. c. 216) medical system, is clearly at the root of the concept, named by al-Naẓẓām, of the four bodies or qualities that constitute the world. The sources acknowledge the fact that the *dahriyya* continued a late antique trend. Some supposed that the Greeks and Romans were *dahrīs* before they became Christians, and Christian and Jewish orthodoxy located *dahriyya* in some surviving heterodoxies, quite as Muslim theology did.[17] In al-Jāḥiẓ's Baghdad, *dahriyya* probably was represented by individual scholars of the natural sciences, in particular astrology, medicine, and alchemy,[18] and perhaps it was a fashion among proclaimed skeptics, "the smart set,"[19] who attempted to apply the theological art of disputation, *kalām*, and the instruments of rationality to central domains of Islamic theology.

---

15 Rudolph, *Al-Māturīdī*, 169. For his second target, the Ismāʿīlīs, whose Plotinian concept of the Intellect as the origin of ideas in which the entire world was contained from the beginning was a major theological issue in tenth-century Transoxania, see Rudolph, *Al-Māturīdī*, 170.

16 Gerhard Endress, *Proclus arabus: Zwanzig Abschnitte aus der "Institutio theologica" in arabischer Übersetzung* (Beirut: Imprimerie Catholique, 1973). See also Elvira Wakelnig, "The Other Arabic Version of Proclus' *De aeternitate mundi*: The Surviving First Eight Arguments," *Oriens* 40 (2012): 51–95. For the reception of the works in the monograph on eternity by al-ʿĀmirī (d. 992), see Everett K. Rowson, *A Muslim Philosopher on the Soul and Its Fate: Al-ʿĀmirī's "Kitāb al-Amad ʿalā l-abad"* (New Haven, CT: American Oriental Society, 1988), Indices, s.vv. "Philoponus" and "Proclus"; see further van Ess's remarks in *Theologie und Gesellschaft*, vol. 4, 452, with n. 69.

17 For Zoroastrianism and its refutation of *dahrī* strands from the Sasanian period, see Shaki in Shaki and Gimaret, "Dahrī," 587.

18 This would correspond to the few extant works containing *dahriyya* elements: the writings of the alchemist Jābir b. Ḥayyān, Ps.-Apollonius's *Sirr al-khalīqa*, Nemesius of Emesa's *De natura hominis*.

19 Crone, "The Dahrīs According to al-Jāḥiẓ," 78.

Reflection

ETERNITY AND THE TRINITY

Christophe Erismann

An understanding of God as eternal is not peculiar to Christianity. It is intrinsic to any form of theism. A system that admits God usually tries to conceive of him in the most perfect way possible. Since eternity is a kind of perfection, it is common to predicate it of the divine entity that is admitted in one's system. However, the issue is more complex in the case of Christianity because God is a Trinity. The problem of the eternity of the Trinity stems from two claims:

1. God is eternal. (Eternity is a divine attribute.)
2. According to Christian dogma, God is a Trinity, that is, God is composed of three distinct persons—the Father, the Son, and the Holy Spirit—that are characterized by four relations: those of filiation/paternity for the relation between the Father and the Son and those of procession/spiration for the relation of the Holy Spirit to the other two persons.

This view constrains us to conceive of the eternity of God in terms of the eternity of the persons who constitute the Trinity. While the eternity of the Father is not questioned, the eternity of the other two persons has been debated, not as such, but in the context of the discussion of their divinity. For example, the Arian and Macedonian positions—which were both judged heretical—questioned, respectively, the divinity of the Son and of the Holy Spirit.[1] The response to these alternative theologies was a strong renewed statement of the consubstantiality of the persons, which implies that they are coeternal. The statement about the coeternity of the persons became a constant in medieval Trinitarian thought. An example is the formulation adopted by the Fourth Lateran Council in 1215, which demonstrates well the understanding of a close relationship between the consubstantiality and coeternity of the three persons of the Trinity. It reads as follows:

> We firmly believe and confess without reservation that there is only one true God, eternal [aeternus], infinite, and unchangeable, incomprehensible, almighty and ineffable, the Father, the Son, and the Holy Spirit: three Persons, indeed, but one essence, substance, or nature entirely simple. The Father is from no one, the Son from the Father only, and the Holy Spirit equally from both. Without beginning, always and without end [absque initio, semper ac sine fine], the Father begets, the Son is born, and the Holy Spirit proceeds. They are of the same substance and fully equal, equally almighty, and equally eternal [coaeterni].[2]

The conceptual difficulty lies in how to interpret the processes of filiation and procession (of the Spirit) in a way that does not express the change of an entity from one state to another but expresses the result. A process usually implies an initial condition that is then altered through the process to another different

---

[1] Arianism was a heresy upheld first at the beginning of the fourth century by the Alexandrian presbyter Arius, who claimed that Christ was not divine but a created being. Macedonianism (also called Pneumatomachian heresy) is a position named after the fourth-century bishop of Constantinople Macedonius; it denies the full divinity of the Holy Spirit.

[2] Translated in H. Denzinger, *Compendium of Creeds, Definitions and Declarations on Matters of Faith and Morals* (San Francisco: Ignatius Press, 2012) 43, section 800, p. 266.

condition. Let us take as an example the standard interpretation of individuation. Individuation describes the change of state from a universal to a particular entity. In medieval Aristotelianism, individuation is often explained through inherence in some given matter. However, the whole idea of individuation implies a previous state in which the entity was not individuated. So a philosopher who admits something like individuation—and not only individuality—must admit that in some way or another, there is a universal entity that becomes particular. In the case of the Trinity, there is no state that is previous to filiation and procession, given that they exist from all eternity. If there were such an anterior state, we would have to admit that there was a moment t when only the Father existed and therefore deny the coeternity of the three persons, and thus deny their consubstantiality and divine immutability. The Father always is the Father; he does not become so.

The problem becomes even bigger for the defenders of "Latin" Trinitarian thought who want to maintain an order: the filiation must take place "before" the procession of the Spirit so that the latter can proceed from both the Father and the Son (and not only from the Father).

Eternity must be shared by the three persons. Coeternity of the persons is required in order to conceive their consubstantiality and thus their equality. This clearly appears in the thought of Richard of St. Victor, a theologian from the second half of the twelfth century.[3] In his *De Trinitate*, Richard, who was a critical reader of Abelard, Gilbert of Poitiers, and Peter Lombard, the most renowned thinkers of the century, offers a subtle reflection on the divine being and eternity, notably on the relationship of eternity

---

[3] On Richard, see N. den Bok, *Communicating the Most High: A Systematic Study of Person and Trinity in the Theology of Richard of St. Victor* (Turnhout: Brepols, 1996); D. M. Coulter, *Per Visibilia ad Invisibilia: Theological Method in Richard of St. Victor (d. 1173)* (Turnhout: Brepols, 2006).

and aseity, i.e. the quality of being self-originated.[4] The textual basis of his reflection is the *Quicunque Vult* or so-called Athanasian Creed; this text, probably of Latin redaction but definitely not by Athanasius and highly influential during the twelfth century,[5] is remarkable in that it applies the divine attributes—including "being eternal"—not merely to the divine substance but to each of the three persons of the Trinity. This creed states that within the Trinity "nothing is before or after, neither greater nor less: but there are three whole persons coeternal [totae tres personae coeternae] and coequal to each other"[6] and insists on the unity in eternity: eternal Father, eternal Son, and eternal Holy Spirit are "not three eternals but one eternal" (*non tres aeterni, sed unus aeternus*).[7] Richard inherits this preoccupation with the three persons being eternal, which implies that the Son and the Holy Spirit are eternal in the same way the Father is eternal. According to Richard, eternity is duration without beginning or end that escapes mutability: ("aeternitas [est] diuturnitas sine initio et fine et carens omni mutabilitate").[8] This strong conceptual link between eternity and immutability is central to his Trinitarian thought. The relation to eternity and the concept of aseity allow him to define three modes of existence: eternal and from itself, noneternal and not from itself, and, finally, eternal but not from itself. God exists in the first mode; the second is that of the beings of the sensible world that are submitted to generation and corruption; the third seems to correspond to the Son and

---

4 For his contribution on eternity, see L.-A. Dyer, "Translating Eternity in the Twelfth-Century Renaissance," Ph.D. diss. (University of Notre-Dame, Notre Dame IN, 2011), 63–111.
5 See, for example, N. M. Häring, "A Commentary on the Pseudo-Athanasian Creed by Gilbert of Poitiers," *Mediaeval Studies* 27 (1965): 23–53.
6 Ed. C. H. Turner, "A Critical Text of the "Quicumque Vult"," *Journal of Theological Studies* 11, 43 (1910), [401-411], 408:25–26.
7 Ed. C. H. Turner, op. cit., 407:11.
8 Richard of St. Victor, *De Trinitate*, Ed. G. Salet (Paris: Cerf, 1999), Book 2, Chapter 4, p. 116.

the Holy Spirit, who are eternal but come forth from the Father. Eternity belongs to God, but also to each of the persons. This allows Richard to state that, obviously, in the Trinity, the different persons are absolutely equal and coeternal; for if they were not coeternal, they would not be equal[9]: "For one person to be eternal, it was necessary for the other to be co-eternal. It is impossible that the one came before and the other later. For, in this eternal and immutable divinity, there can be no passing old age, nothing new that comes to be. Thus, it is impossible that the divine persons are not coeternal".[10]

---

9 Richard of St. Victor, *De Trinitate, op. cit.* Book 3, Chapter 21, p. 214.
10 Richard of St. Victor, *De Trinitate, op. cit.* Book 3, Chapter 7, p. 180.

CHAPTER THREE

Eternity in Early Modern Philosophy

Yitzhak Y. Melamed

> God cannot be said to enjoy existence, for the existence of God is God himself.
> SPINOZA, *Cogitata Metaphysica* (1663)

It has frequently been argued that with the dawn of modernity, philosophers deserted the Boethian notion of eternity. Thus, Stump and Kretzmann write: "In the modern period, with the rejection of the medieval synthesis in theology, the notion of eternity, in the special sense at issue here, was largely abandoned. Hobbes is still aware of it in the Boethian sense '*as a permanent now*,' but Locke, for example, takes eternity to be just an infinity of temporal duration."[1] These claims, while broadly correct, are still quite misleading. Indeed, during the seventeenth century, an increasing number of philosophers questioned, and later deserted, the Boethian notion of eternity.[2] In place of it they adopted and advocated either the sempiternal (i.e., eternity as existence

---

1 E. Stump and N. Kretzmann, "Eternity," in *Routledge Encyclopedia of Philosophy*, ed. Edward Craig (London: Routledge, 1998), vol. 3, 422–427.
2 For a discussion of the Boethian and late-Platonic notion of eternity, see the end of chapter 1 above.

in all times)³ or the new Spinozist conception of eternity.⁴ Yet the Boethian notion was still strongly and broadly present in this period, and it could hardly be said that *any* seventeenth-century philosopher was not "aware" of it. Thus, Abraham Cohen de Herrera (c. 1570–1635), the major philosophical kabbalist of early seventeenth-century Amsterdam, celebrates the Plotinian/Boethian notion of eternity:

> And it should be noted... that eternity, as Boethius defines it in his *Consolation of Philosophy*, "is the whole perfect, and simultaneous possession of endless life," and as Plotinus proves in his profound treatise on eternity and time, it is infinite life or duration which, lacking beginning and end, remains entirely united and in oneness and is consequently so stable and fixed that it neither acquires nor loses anything.... And as the Prince of Tuscan Poets, Torquato Tasso, says in his dialogue the *Messenger*, eternity does not contain a first or last, a before or after, or elements in sequence or in motion: rather it is united and withdrawn into itself, like a very placid lake that has no ebb or flow, increase or loss of water.⁵

As I will soon show, many seventeenth-century figures hesitated and oscillated between asserting and questioning the very intelligibility of eternity qua *tota simul* life. Thus, a report on a 1648 conversation between Descartes (1596–1650) and a young theological student, Frans Burman (1628–1679), says that Burman suggested: "But eternity is all at once and once and for all" (*Sed aeternitas est simul et semel*).⁶ To this Descartes disparagingly replied: "That is impossible to conceive of

---

3 See Geoffrey Gorham, "Descartes on God's Relation to Time," *Religious Studies* 44 (2008): 414, and Luca Bianchi, "Abiding Then: Eternity of God and Eternity of the World from Hobbes to the Encyclopédie," in *The Medieval Concept of Time: The Scholastic Debate and Its Reception in Early Modern Philosophy*, ed. Pasquale Porro (Leiden: Brill, 2001), 453.
4 The epigraph to this chapter is from CM II 1 (G I/252/7).
5 Abraham Cohen de Herrera, *Puerta del Cielo*, bk. 5, chap. 4 (*Gate of Heaven*, trans. Kenneth Krabbenhoft [Leiden: Brill, 2002], 150–151). For the original text of Torquato Tasso (1544–1595), see Tasso, *Prose* (Milan: Riccardo Ricciardi, 1959), 44.
6 Suárez, too, cites approvingly the Boethian formula in Suárez, *Disputationes metaphysicae* 50.3.6.

[hoc concipi non potest]. It is all at once and once and for all insofar as nothing is never added to or taken away from the nature of God. But it is not all at once and once for all in the sense that it exists at once [simul existit]."[7] By the end of the century, Isaac Newton (1642–1727) pronounced that the Boethian notion was plainly unintelligible: "The human race is prone to mystery, and holds nothing quite so holy and perfect as what cannot be understood [quod intelligi non potest]. Yet in the conception of God this is dangerous, and conduces to the rejection of his existence. . . . Let them therefore consider whether it is more agreeable to reason that God's eternity should be all at once [totum simul] or that his duration is more correctly designated by the names Jehovah and 'He that was and is and is to come.'"[8] Newton's friend Samuel Clarke (1675–1729) quotes sympathetically the Boethian formula in his *Demonstration of the Being and Attributes of God* (1704),[9] yet in his sermons he describes it as unintelligible: "'Tis worthy of observation, as to the Manner of our conceiving the Eternity of God; that the Scholastic Writers have generally described it to be, not a Real Perpetual Duration, but One Point or Instant comprehending Eternity, and wherein all things are really co-existent at once. *But unintelligible Ways of Speaking have (I think) never done any Service to Religion.* The

---

[7] *Descartes' Conversation with Burman*, trans. John Cottingham (Oxford: Clarendon Press, 1976), 6 (AT V 149). For discussion of this passage, see Gorham, "Descartes on God's Relation to Time," 416–417, and Tad M. Schmaltz, *Radical Cartesianism: The French Reception of Descartes* (Cambridge: Cambridge University Press, 2002), 200–201. As I shall later show, on another occasion in the very same year, Descartes expressed agreement with the view that God's eternity is *tota simul*.

[8] J. E. McGuire, "Newton on Place, Time and God, An Unpublished Source," *British Journal for the History of Science* 11 (1978): 121.

[9] Samuel Clarke, *A Demonstration of the Being and Attributes of God (and Other Works)*, ed. Enzio Vailati (Cambridge: Cambridge University Press, 1998), sec. 5, p. 32: "Thus far we can speak intelligibly concerning the eternal duration of the self-existent being, and no atheist can say this is an impossible, absurd, or insufficient account. It is, in the most proper and intelligible sense of the words, to all the purposes of excellency and perfection *interminabilis vitae tota simul et perfecta possesio, the entire and perfect possession of an endless life.*" Clarke then quotes the claims of Gassendi and Tilloston, who harshly criticize the Boethian formula as unintelligible (33 n. 26). Enzio Vailati, *Leibniz and Clarke: A Study of Their Correspondence* (Oxford: Oxford University Press, 1997), 20, considers these quotes as intimating Clarke's rejection of Boethius. I tend to think that Clarke is genuinely ambivalent about the issue in the *Demonstration*.

true Notion of the Divine Eternity does not consist in making past things still present, and all things future to be already come."[10] Why did Descartes, Newton, and Clarke think the Boethian (or Plotinian) formula was an unintelligible mystification? The Boethian formula is committed to the simultaneity of God with all times, and to a notion of life, or duration, that is "all at once." But how can *duration* be all at once? Of course, various magnitudes, for example, infinite extension, can exist all at once. But if by duration one understands a *temporal* magnitude, the notion of infinite duration existing all at once seems to be just as paradoxical and contradictory as the notion of infinite extension existing completely at one and the same point.

In this chapter I will trace the history of eternity in the early modern period (roughly 1550–1750). Modernity seemed to be the autumn of eternity. The secularization of European culture provided little sustenance to the concept of eternity with its heavy theological baggage. Yet our hero will not leave the stage without an outstanding demonstration of its power and temptation. Indeed, in the two centuries of early modernity, the concept of eternity played important roles in the period's greatest philosophical systems.

The first part of the chapter concentrates on the debate about God's relation to time. While most of the great metaphysicians of the period—Suárez (1548–1617), Spinoza (1632–1677), Malebranche (1638–1715), and Leibniz (1646–1716)—ascribed nontemporal eternity to God, a growing number of philosophers conceived God as existing in time and eternity as having everlasting existence. For Newton, Gassendi (1592–1655), Henry More (1614–1687), Samuel Clarke, Isaac Barrow (1630–1677), and John Locke (1632–1704), God's eternity was simply the fact that he always was, is, and will forever be. In the second part of the chapter I will examine the concept of eternal truth (*aeternae veritates* or *vérités éternelles*). Though this concept has a long history, it became far more central in the early modern period with the emergence of the closely related notion of the "Law of Nature" (*lex naturae*).

---

10 Clarke, *The Works* (1738), 4 vols. (New York: Garland, 1978), vol. 1, 22. Italics added.

The third and most extensive part of the chapter will present Spinoza's original understanding of eternity as *self-necessitated existence*. Various elements of Spinoza's notion of eternity can be traced back to previous philosophers, but the core of his understanding of eternity is original and surprising. For Spinoza, eternity is primarily a *modal* notion, a unique kind of necessity and necessary existence. The final part of the chapter will address the reception of Spinoza's concept of eternity in the century following his death. Remarkably, Spinoza's notion of eternity was received positively by figures who would otherwise sharply criticize his philosophy. Oddly enough, the great "atheist" of the early modern period turned out to be *the* philosophical expert on eternity, namely, the very essence of God.

1 God, Time, and Eternity in Early Modern Philosophy

Attempting to point out communalities and differences between eternity and time, Abraham Cohen Herrera writes: "The infinite, unmoving present is thus like the origin or source of eternity, while the bounded, manifold, ongoing, and successive present is the cause of time. And in our way of understanding *both are measures, that is, eternity is the measure of infinite and abiding being, and time is the measure of limited and movable things*, except that time measures by diverse and successive repetitions many times over, as if it were counting or numbering, while eternity measures with a single, unmovable, permanence, unified and focused, in one unmoving moment or indivisible instant."[11] Herrera's dense language seems to assign eternity to only one being: "the infinite and abiding being," which he also calls "the First Cause," that is, God. Other early modern philosophers, however, appear to be somewhat less restrictive in terms of the kind of beings that enjoy eternity. Both Descartes and Spinoza consider the *essences*

---

11 Herrera, *Gate of Heaven*, bk. 5, chap. 4 (Krabbenhoft trans., p. 151). Italics added.

of things—even finite things—eternal.¹² In the next part of this chapter I will discuss the "eternal truths," and when I come to interrogate Spinoza, I will show that he believed that in some sense even finite things can be adequately conceived "sub species aeternitatis." In a very different manner George Berkeley (1685–1753) too would find an eternal abode for *all* things. Thus, in his *Three Dialogues*, Berkeley puts the following claims in the mouth of Philonous (the dialogues' character that represents Berkley's own views): "All objects are eternally known by God, or which is the same thing, *have an eternal existence in his mind*; but when things before imperceptible to creatures, are by a decree of God, made perceptible to them; then they are said to begin a relative existence, with respect to created minds."¹³ Another eternity-related question that engaged early modern philosophers was the issue of eternal punishment and reward.¹⁴ Finally, we should keep in mind that the threat of the unorthodox view of the universe as eternal *a parte ante* was clearly present, at least at the back of the mind, to most early modern thinkers. Yet, in spite of these reservations, it is clear that for the early modern philosophers, just as for their medieval predecessors, eternity was the kind of existence that belonged *primarily* to God. If other beings were said to be eternal, either their eternity would be considered as inferior to God's or they would be said to enjoy eternity by participating—in one manner or another—in God's eternity. For this reason, my main focus in this part of the chapter is the elucidation of the nature of God's eternity, that is, eternity in its full sense and complete colors.

Thomas Hobbes's attack on the Boethian notion of divine eternity was part of a more comprehensive assault he launched in the *Leviathan*

---

12 Descartes, *Meditations on First Philosophy* (Fifth Meditation) (AT VII 64). For Spinoza, see *Ethics*, part One, definition 8, explanation, and Part One, proposition 17, scholium (G II/63/18), and CM II 1 (G I/239/4).
13 George Berkeley, *Three Dialogues between Hylas and Pilonous*, ed. Jonathan Dancy (Oxford: Oxford University Press, 1998), dialogue 3, p. 133. Italics added.
14 See, for example, Nicolas Malebranche, *The Search after Truth*, trans. Thomas M. Lennon and Paul J. Olscamp (Cambridge: Cambridge University Press, 1997), 326–328 and 389.

(1651) against Scholasticism. In a crucial passage in chapter 27 of *Leviathan*—titled "Of Darkness from Vain Philosophy and Fabulous Traditions"—Hobbes writes:

> For the meaning of *eternity*, they will not have it to be an endless succession of time; for then they should not be able to render a reason how God's will and preordaining of things to come should not be before his prescience of the same (as the efficient cause before the effect, or agent before the action), nor of many other their bold opinions concerning the incomprehensible nature of God. But they will teach us, that eternity is the standing still of the present time, a *nunc-stans* (as the Schools call it), which neither they, nor any else understand, no more than they would a *hic-stans* for an infinite greatness of place.[15]

At the beginning of this passage Hobbes suggests an interesting diagnosis of the motive for the wide adoption of the Boethian formula among the Scholastics, claiming that the alternative view of divine eternity as everlasting existence faces the serious objection that if time existed before the creation of the world, God's choice to create the world at any specific instant of time would seem to be arbitrary. Before creation there seem to be no events, and thus all instants of precreation time should be uniform. But in a uniform time, there *cannot* be a reason to act at t_1 rather than t_2, since whatever is true about the one must be true about the other.[16] According to Hobbes, the Scholastics, attempting to avoid the danger of rendering God's actions arbitrary and incomprehensible, succumbed to another form of unintelligible talk by adopting the contradictory formula of "whole perfect, and simultaneous possession of endless life."

---

15 Hobbes, *Leviathan*, ed. Edwin Curley (Indianapolis: Hackett, 1994), 461. My discussion of Hobbes profited much from Luca Bianchi's excellent article "Abiding Then."
16 This kind of argument is most familiar from the Leibniz-Clarke correspondence, though it has a much longer history. See Bianchi, "Abiding Then," 544–545.

In his 1656 *Elements of Philosophy*, Hobbes addresses another argument that aims to vindicate the Boethian formula by pointing out a different problem with the understanding of eternity as everlastingness. Notice that the argument begins with a demonstration addressing the everlastingness of the *universe* and then shifts to the issue of the proper understanding of *God's* eternity.

> For who can commend him that demonstrates thus? "If the world be eternal, then an infinite number of days, or other measures of time, preceded the birth of Abraham. But the birth of Abraham preceded the birth of Isaac; and therefore one infinite is greater than another infinite, or one eternal than another eternal; which," he says, "is absurd." This demonstration is like his, who from this, that the number of even numbers is infinite, would conclude that there are as many even numbers as there are numbers simply, that is to say, the even numbers are as many as all the even and odd together. They, which in this manner take away eternity from the world, do they not by the same means take away eternity from the Creator of the world? From this absurdity therefore they run into another, being forced to call eternity *nunc stans*, a standing still of the present time, or an abiding now; and, which is much more absurd, to give to the infinite number of numbers the name of unity. But why should eternity be called an abiding now, rather than an abiding then? Wherefore there must either be many eternities, or *now* and *then* must signify the same. With such demonstrators as these, that speak in another language, it is impossible to enter into disputation.[17]

The argument Hobbes employs against the preeternity of the world goes back to medieval philosophy and ultimately to John Philoponus.[18] The core idea of the argument is that assuming the world had already

---

17 Hobbes, *Elements of Philosophy*, chap. 26, sec. 1, in Hobbes, *The English Works of Thomas Hobbes of Malmesbury*, 11 vols., ed. William Molesworth (London: J. Bohn, 1839–45), vol. 1, 413.
18 See Bianchi, "Abiding Then," 546–547, and Richard Sorabji, *Time, Creation, and the Continuum* (London: Duckworth, 1983), 210–231.

endured for an infinite duration of time, we are bound to get entangled in the paradoxes of infinity in which an infinite quantity turns out to be equal to its part (which is also infinite). That the whole is greater than its part was considered one of the most unshakeable truths for medieval and early modern philosophers.[19] Thus, if the preeternity of the world led necessarily to the violation of this unshakable truth, the assumption of preeternity would seem to be refuted. Hobbes's own contribution to this discussion was the note that this refutation of the preeternity of the *world* is just as much a refutation of the preeternity of *God*, that is, of the conception of divine eternity as everlastingness. Here again, Hobbes claims, attempting to avoid one absurdity, the Scholastics uncritically adopted another absurdity: the Boethian concept of divine eternity.

Hobbes's own position on the issue of divine eternity was quite delicate, as he openly admitted that we could not understand the notion of infinity and infinite time. "But the knowledge of what is infinite can never be attained by a finite inquirer. Whatsoever we know that are men, we learn it from our phantasms; and of infinite, *whether magnitude or time*, there is no phantasm at all; so that it is impossible either for a man or any other creature to have any conception of infinite."[20] Not being able to offer an intelligible alterative to the *nunc stans* conception of divine eternity, Hobbes's sophisticated polemics against the Scholastics and their adoption of the Boethian formula resulted in no more than an embarrassing stalemate.

I will turn now to Hobbes's great contemporary René Descartes and look closely at his understanding of divine eternity. I have already mentioned Frans Burman's report that in his 1648 conversation with Descartes, Descartes said that the Boethian formula is just "impossible

---

19 Early modern philosophers frequently refer to the proposition that "the whole is greater that its part" as the stock example of a necessary and evident truth. See Descartes to Mesland, May 2, 1644 (AT IV 110); Spinoza, *Ethics*, part Four, proposition 18, scholium (G II/222/22); and Leibniz, "Primary Truths," in Leibniz, *Philosophical Essays*, ed. and trans. R. Ariew and D. Garber (Indianapolis: Hackett, 1989), 31.

20 Hobbes, *Elements of Philosophy*, chap. 26, sec. 1, in Hobbes, *English Works*, vol. 1, 411–412. Italics added.

to conceive of."[21] In the very same conversation Descartes is also reported as saying: "We can divide God's duration into an infinite number of parts, even though God himself is not therefore divisible."[22] We might thus suspect that Descartes viewed divine eternity as mere everlastingness. Indeed, in the Fifth Meditation we find Descartes writing: "Apart from God, there is nothing else of which I am capable of thinking such that existence belongs to its essence ... and after supposing that one God exists, I plainly see that it is necessary that he *has existed from eternity and will abide for eternity* [ab aeterno extiterit, & in aeternum sit mansurus]."[23] The wording of the last sentence of this passage seems to indicate that Descartes indeed conceived divine eternity as everlastingness, but such a conclusion would be premature. In his 1644 *Principles of Philosophy* he addresses the nature of divine actions. "[God's] understanding and willing does not happen, as in our case, by means of operations that are in a certain sense distinct one from another; we must rather suppose that *there is always* a single identical and perfectly simple act by means of which he *simultaneously understands, wills and accomplishes everything.*"[24] The *Principles* passage presents God's *actions* as simultaneous, that is, as not having duration. This view still leaves open the possibility that God's *existence* is spread in time. I will turn then to a 1648 letter in which Antoine Arnauld (1612–1694) suggests to Descartes that human thought too is not successive. "The duration of a permanent and highly spiritual thing [rei permanentis & maxime spiritalis], such as the mind, is not successive but rather all at once [totam simul] (as is certainly the case with the duration of God)."[25] To this Descartes replies: "Even if no bodies existed, it could still not be said that that the duration of the human mind *was entirely simulatenous* [tota simul] *like the duration of God* [quemadmodum duration Dei]; because our thoughts display a

---

21 *Descartes' Conversation with Burman*, 6 (AT V 149).
22 *Descartes' Conversation with Burman*, 6 (AT V 149).
23 AT VII 68. Italics added.
24 Descartes, *Principles of Philosophy* 1.23 (AT IXB 14| CSM I 201). Italics added.
25 Arnauld to Descartes, June 3, 1648 (AT V 188).

succession which cannot be found in the divine thoughts."²⁶ We have thus two texts of Descartes, both dated 1648: in the letter to Arnauld he affirms that God endures "all at once"; in the conversation with Burman he claims that this very view is unintelligible.

We find a similar oscillation between affirmation and rejection of the Boethian formula in Nicolas Malebranche's *Dialogues on Metaphysics and on Religion* (1688). Attempting to explain divine immensity, Theodore, the Malebranchian spokesman in the dialogue, suggests the following analogy:

> Created extension is to the divine immensity what time is to eternity. All, bodies are extended in the immensity of God, as all times succeed one another in His eternity. *God is everything he is without succession in time. In His existence there is neither past nor future; everything is present, immutable, and eternal.* . . . God created the world, but the volition to create is not past. God will change the world, but the volition to change is not in the future. The will of God which was and will be is an eternal and immutable act whose effects change without there being a change in God. In a word, God was not, He will not be, but He is. We could say that God was in past time; but He was then everything He will be in future time. For His existence and duration, if it permitted to use that term, is completely in eternity, and completely in every passing moment of His eternity. Likewise, God is not partly in heaven and partly in earth. He is completely whole in his immensity, and completely in all the bodies which are locally extended in his immensity.²⁷

Theodore's explanation of divine eternity is very close to the Boethian formula, as he denies any succession in God and stresses repeatedly that

---

26 Descartes to Arnauld, June 4, 1648 (AT V 193| CSM III 355). Italics added. For a helpful discussion of this passage, see Schmaltz, *Radical Cartesianism*, 85–86 and 199–200.
27 Malebranche, *Dialogues on Metaphysics and on Religion*, trans. Nicholas Jolley and David Scott (Cambridge: Cambridge University Press 1997), dialogue 8, p. 132. Italics added.

God exists all at once. Since Theodore is the character representing Malebranche's view, it would seem that Malebranche fully endorsed the Boethian formula. Yet the response Malebranche puts in the mouth of Aristes—the skeptical disciple of Theodore—presses again the charge of unintelligibility or unclarity against the Boethian formula. "It seems to me, Theodore, that you are explaining an obscure thing by means of another which is not very clear."[28] The analogy Theodore draws between divine immensity and divine eternity is not helpful, claims Aristes, since divine eternity (which was supposed to help explain immensity) is unclear in itself. At this point Theodore responds by noting that Aristes granted that God is eternal earlier in the dialogue, and that it is for this reason that he (i.e., Theodore) brought the analogy with immensity.[29] However, since Aristes never explained what he understood by divine eternity, Theodore's response appears only partly satisfying.

Less we rashly infer that ambivalence was the only attitude toward the Boethian formula among seventeenth-century philosophers, I stress that one can find unhesitant partisans on both sides of the debate. Thus, Father Pierre Gassendi (1592–1655) rejected the Boethian formula in no unclear terms, claiming: "Eternity cannot be understood as anything else than perpetual duration . . . inasmuch as it lacks beginning and end," and placing the blame for the blunder of the Boethian formula in Boethius's reading of the *Timaeus*.[30] On the other hand, toward the end of the century, Anne Conway (1631–1679), defended the Boethian formula with equal decisive conviction:

> The eternity of creatures is nothing other than an infinity of times in which they were and always are and always will be without end.

---

28 Malebranche, *Dialogues*, dialogue 8, p. 132.
29 Malebranche, *Dialogues*, dialogue 8, p. 133.
30 Gassendi, *Opera Omnia in sex tomos divisa*, 6 vols. (Lyon: Laurent Anisson and Jean-Baptiste Devenet, 1658); reprinted in facsimile and with an introduction by Tullio Gergory (Stuttgart: Friedrich Frohmann, 1964), vol. 1, p. 225b. The English translation is quoted from Antonia Lolordo, *Pierre Gassendi and the Birth of Early Modern Philosophy* (Cambridge: Cambridge University Press, 2007), 127. See her helpful discussion there.

Nevertheless, this infinity of time is not equal to the infinite eternity of God since *the divine eternity has no times in it and nothing in it can be said to be past or future, but it is always and wholly present.* ... And the reason for this is obvious because time is nothing but the successive motions or operations of creatures, and if this motion or operation should cease, then time itself would cease and the creatures themselves would end with time. ... And since in God there is no successive motion or operation toward further perfection because he is absolutely perfect, there are no times in God or his eternity. Furthermore, because there are no parts in God, there are also no times in him since all times have parts and are divisible into infinity.[31]

Conway's two fine arguments in defense of Boethius appeared in her posthumously published *Principles of the Most Ancient and Modern Philosophy* (1690). At this point in time the tide was growing strongly against the Boethian formula. In his *Essay Concerning Human Understanding* (1690), John Locke attempted to explain the nature and origin of our idea of eternity. He thus writes: "By being able to repeat such *Idea* of any length of Time, as of a Minute, a Year, or an Age, as often as we will in our Thoughts, and adding them one to another, without ever coming to the end of such additions, any nearer than we can to the end of number, to which we can always add; we come by the Idea of *Eternity*, as the future eternal Duration of our Souls, as well as the eternity of that infinite Being which necessarily have always existed."[32] For Locke, eternity is nothing but infinite duration. Yet, Locke argues, we have no clear idea of eternity, just we have no clear idea of infinity of any other kind.

---

31 Anne Conway, *The Principles of the Most Ancient and Modern Philosophy*, ed Alison P. Coudert and Taylor Corse (Cambridge: Cambridge University Press, 1996), 13–14. For an illuminating discussion of Conway's understanding of eternity, see Christia Mercer, *Exploring the Philosophy of Anne Conway*, unpublished book manuscript, chap. 4.

32 Locke, *Essay concerning Human Understanding* 2.14.31 (*An Essay Concerning Human Understanding*, ed. Peter H. Nidditch [Oxford: Clarendon Press, 1975], 196).

Having frequently in our Mouths the Name *Eternity*, we are apt to think we have a positive comprehensive *Idea* of it, which is as much as to say, that there is no part of that Duration which is not clearly contained in our *Idea*. It is true that he that thinks so may have a very clear *Idea* of Duration; he may also have a clear *Idea* of a very great length of Duration; he may also have a clear idea of the comparison of that great one with still a greater: But it not being possible for him to include in his idea of any Duration, let it be as great as it will, the whole extent together of a Duration, where he supposes no end, that part of his *Idea*, which is still beyond the Bounds of that large Duration, he represents to his own thoughts, is very obscure and undetermined. And hence it is that in Disputes and Reasonings concerning Eternity, or any other *Infinite*, we are very apt to blunder, and involve ourselves in manifest absurdities.[33]

At the very end of the seventeenth century, Pierre Bayle (1647–1706) seemed to epitomize the growing embracement surrounding the notion of eternity by stating in his celebrated *Dictionnaire* (1697) that the common Boethian definition of divine eternity "is far more incomprehensible than the dogmas of transubstantiation [beaucoup plus incomprehensible que le dogme de la Transubstantiation]."[34]

2 Eternal Truths

In section 48 of the first part of his *Principles of Philosophy* (1644), Descartes writes: "All the objects of our perception we regard either as things, or affections of things, or else as eternal truths [aeternas veritates] *which have no existence outside our thought.*"[35] But what are precisely these "eternal truths" that have no existence outside our mind?

---

33 Locke, *Essay concerning Human Understanding* 2.29.15 (Nidditch ed., p. 369).
34 Pierre Bayle, *Dictionnaire Historique et Critique*, 5th ed. (Amsterdam: P. Brunel, 1740), vol. 4, 531 n. H. See Bianchi, "Abiding Then," 558.
35 Descartes, *Principles of Philosophy* 1.48 (AT VIIIA 22| CSM I 208). Italics added.

After listing the classes of things and their affections, Descartes turns to the category of eternal truths:

> *It is not possible—or indeed necessary—to give a similar list of eternal truths.*
>
> Everything in the preceding list we regard either as a thing or as a quality or mode of a thing. But when we recognize that it is impossible for anything to come from nothing, the proposition *Nothing comes from nothing* is regarded not as a really existing thing [res aliqua existens], or even as a mode of a thing, but as an eternal truth which resides within our mind. Such truths are termed common notions or axioms [communis notion sive axioma]. The following are examples of this class: *It is impossible for the same thing to be and not to be at the same time; What is done cannot be undone; He who thinks cannot but exist while he thinks*; and countless others. It would not be easy to draw up a list of all of them; but nonetheless we cannot fail to know them when the occasion for thinking about them arises, provided that we are not blinded by preconceived opinions.[36]

Earlier in the *Principles*, Descartes explained that the common notions are notions "from which the mind constructs various proofs; and for as long as it attends to them, it is completely convinced of their truth."[37] It seems thus that the eternal truths are just the most foundational principles of logic, mathematics, and metaphysics. Descartes's characterization of the eternal truths as "residing within our mind,"[38] and apparently not requiring the existence of anything (but God, as I will shortly show), places them in contrast to Descartes's understanding of laws of nature (*leges naturae*), the regularities that God established in

---

36 Descartes, *Principles of Philosophy* 1.49 (AT VIIIA 23–24| CSM I 209).
37 Descartes, *Principles of Philosophy* 1.13 (AT VIIIA 9| CSM I 197). See Descartes, *Rules for the Direction of the Mind*, rule 12 (AT X 419).
38 Descartes, *Principles of Philosophy* 1.49 (AT VIIIA 23| CSM I 209).

matter.³⁹ Still, what is most striking about Descartes's conception of both eternal truths and laws of nature is that he takes both to be *freely* chosen by God. Thus, in a letter of April 15, 1630, to his friend Father Marin Mersenne (1588–1648), Descartes writes:

> The mathematical truths which you call eternal have been laid down by God and depend on him entirely no less than the rest of his creatures. Indeed, to say that these truths are independent of God is to talk of him as if he were Jupiter or Saturn and to subject him to the Styx and Fates. Please do not hesitate to assert and proclaim everywhere that it is God who lays down laws in his kingdom.... It will be said that if God has established these truths he could change them as a king changes his laws. To this the answer is: Yes he can, if his will can change.⁴⁰

These claims must have surprised Mersenne, as the common Scholastic position, as formulated by Aquinas, restricted divine omnipotence to the realm of the logically possible.⁴¹ When Mersenne asks Descartes by what kind of causality God creates the eternal truth, Descartes does not seem to budge: "You ask me by what kind of causality God established the eternal truths. I reply: by the same kind of causality as he created all things, that is to say, as their efficient and total cause."⁴²

---

39 See Descartes's early work *The World* [Le monde], chap. 7 (AT XI 37).

40 AT I 145–146| CSM III 23. Just a few weeks later, on May 6, 1630, Descartes writes to Mersenne: "If men really understood the sense of their words they could never say without blasphemy that the truth of anything is prior to the knowledge that God has of it" (AT I 149| CSM III 24). The target of Descartes's criticism seems to be Francisco Suárez's claim that eternal truths "are not true because they are known by God, but rather they are thus known because they are true" (Suárez, *Metaphysical Disputation 31*, sec. 12.40. For an insightful study of Suárez's position that argues that Descartes's view is just an amended version of Suárez, see Amy Karofsky, "Suárez' Doctrine of Eternal Truths," *Journal of the History of Philosophy* 39 (2001): 46.

41 "God is unable to make opposites exist in the same subject at the same time and in the same respect.... Since the principles of certain sciences—of logic, geometry, and arithmetic for instance—are derived exclusively from the formal principle of things, upon which their essence depends, it follows that God cannot make the contraries of those principles." Aquinas, *Summa Contra Gentiles* 2.25.12 and 14. See Anthony Kenny, *The God of the Philosophers* (Oxford: Clarendon Press, 1979), 17.

42 Descartes to Mersenne, May 27, 1630 (AT I 151–152).

The very intelligibility of the claim that God can change the laws of logic was put under close scrutiny by both Descartes's contemporaries and modern scholars.[43] The issue comes up repeatedly in the *Objections and Replies*, which Descartes appended to his *Meditations* (1641). In the *Sixth Set of Replies*, more than a decade after the 1630 letter to Mersenne, Descartes still insists that God's will is *completely indifferent* in deciding which eternal truths to create.[44] Nothing can make God even incline toward one law rather than the other.

> If anyone attends to the immeasurable greatness of God he will find it manifestly clear that there can be nothing whatsoever [nihil omnino esse posse] which does not depend on him. *This applies not just to everything that subsists, but to all order, every law, and every reason for anything's being true or good* [nullamve rationem veri & boni]. If this were not so, then ... God would not have been completely indifferent with respect to the creation of what he did in fact create. If some reason for something's being good had existed prior to his preordination, this would have determined God to prefer those things which it was best to do.... Hence we should not suppose that eternal truths "depend on the human intellect or on other existing things"; they depend on God alone, who, *as the supreme legislator, has ordained* [instituit] *them from eternity*.[45]

---

43 For a fascinating and highly influential study of this issue, see Harry Frankfurt, "Descartes on the Creation of the Eternal Truths," *Philosophical Review* 86 (1977): 36–57. For a recent helpful overview of the current state of the debate, see David Cunning, "Descartes Modal Metaphysics," in *The Stanford Encyclopedia of Philosophy*, spring 2014 ed., ed. Edward N. Zalta, sec. 3.

44 There is a major tension between Descartes's insistence here on God's complete indifference as condition for divine freedom and his claims in the Fourth Meditation: "In order to be free, there is no need for me to be inclined both ways; on the contrary, the more I incline in one direction—either because I clearly understand that reasons of truth and goodness point that way, or because of a divinely produced disposition of my inmost thoughts—the freer is my choice. Neither divine grace nor natural knowledge ever diminishes freedom; on the contrary, they increase and strengthen it. *But the indifference I feel when there is no reason pushing me in one direction rather than another is the lowest grade of freedom*; it is evidence not of any perfection of freedom, but rather of a defect in knowledge or a kind of negation." *Fourth Meditation* (AT VII 57–58| CSM II 40). Italics added.

45 *Sixth Set of Replies* (AT VII 435-6). Italics added.

An interesting objection to these bold claims of Descartes was raised by Pierre Gassendi, who argued that Descartes's view of the eternal truths compromises the uniqueness of God's eternity by making *created beings*—the eternal truths—just as eternal as God.[46] At this point, one could expect Descartes to resort to the familiar strategy of making created things eternal only in a secondary and inferior sense.[47] Instead, Descartes seems to bite the bullet and insist that the eternal truths are just as eternal as God himself.

> You say that you think it is "very hard" to propose that there is anything immutable and eternal apart from God. You would be right to think this if I was talking about existing things, or if I was proposing something as immutable in the sense that its immutability was independent of God. But just as the poets suppose that the Fates were originally established by Jupiter, but that after they were established he bound himself to abide by them, so I do not think that the essences of things, and the mathematical truths which we can know concerning them, are independent of God. Nevertheless I do think that they are immutable and eternal, since the will and decree of God willed and decreed that they should be so. Whether you think this is hard or easy to accept, it is enough for me that it is true.[48]

Descartes's reply to Gassendi seems to admit that were he speaking of eternal truths as "existing things" [*de re existente*], Gassendi's objection would be in place. Indeed, at the beginning of my discussion I showed that in *Principles* 1.48, Descartes explicitly refers to eternal truths as

---

46 *Fifth Set of Objections* (AT VII 319).
47 There is some evidence that Descartes considered eternal truths as merely everlasting. See *Fifth Set of Replies* (AT VII 381): "Since eternal truths are always the same [eadem semper], it is right to call them immutable and eternal" (AT VII 381| CSM II 262). See Gorham, "Descartes on God's Relation to Time," 423, for a discussion of this passage.
48 *Fifth Set of Replies* (AT VII 380| CSM II 261).

a class distinct from both things and the affections of things. On the other hand in the First Set of Replies Descartes counters Johannes Caterus's claim that "an eternal truth does not require a cause," by insisting that an eternal truth, just like any other idea, "surely needs a cause enabling it to be conceived."[49] Thus, eternal truths are not just nothingness that does not require a cause. Just like the Cartesian God, (Cartesian) eternal truths are mental items, and just like God, eternal truths are eternal. For Gassendi, such a situation threatens the uniqueness of God. Of course, Descartes would claim that eternal truths do not exist outside the human mind, while God does. But since God exists outside the human mind as another (infinite) mind, Gassendi could easily reply to Descartes that eternal truths exist outside my mind, in the minds of other people. The main point of Gassendi is that unlike many of his medieval predecessors, Descartes does not locate the eternal truths in God's mind;[50] rather, he considers them as *created* beings. Granting full-fledged eternity to a created being may indeed be a small but crucial step on a path leading to idolatry.

Before I turn to discuss Spinoza's new conception of eternity in the next section, let me briefly point out Spinoza's radical reconceptualization of the notion of eternal truth. Descartes's view of God as legislating—like a king—the eternal truths and the laws of nature was primarily an object of ridicule for Spinoza, who considered such views as gross and childish instances of anthropomorphic thinking. Thus, in the scholium to the third proposition of Part Two of the *Ethics*, Spinoza writes: "By God's power ordinary people [Vulgus] understand God's free will and his right over all things which are, things which on that account are commonly considered to be contingent. For they say that God has the power of destroying all things and reducing them to nothing. Further, they very often compare God's power

---

49 First Set of Objections (AT VII 93| CSM II 70). The aim of Caterus's claim was to undermine the Cartesian proof of the existence of God in the Third Meditation.
50 See Kenny, *God of the Philosophers*, 15–17, who stresses this point.

with the power of Kings."⁵¹ In his *Theological Political Treatise* (1670) Spinoza draws a distinction between eternal truths and laws of nature that is very different from Descartes's (and indeed aims at eradicating Descartes's claim that God freely creates the eternal truths). In the following passage Spinoza explains the process by which a law of nature *turns into* an eternal truth: "As for the divine natural law whose highest precept we have said is to love God, I have called it a law in the sense in which philosophers apply the word 'law' [legem] to the common rules of nature according to which all things happen.... *Divine commandments seem to us like decrees or enactments only so long as we are ignorant of their cause. Once we know this, they immediately cease to be edicts, and we accept them as eternal truths, not as decrees.*"⁵² In describing natural regularities as *laws*, we conceive these regularities as expressions of the arbitrary will of the master of the universe. We are confident in the necessity of these regularities, yet being ignorant of their causes and not being able to explain their necessity, we conceive of them as decrees of a most powerful agent. One can see this analysis as Spinoza's attempt to understand the psychological processes that led people like Descartes to view God as freely legislating the laws of nature. In the above passage Spinoza refrains from judging this form of anthropomorphic thinking and only adds that the ignorance (of the causes of natural regularities) that is the ground of the conception of God as legislator is corrigible. Once we learn the causes of natural regularities they no longer appear arbitrary, and thus we no longer conceive of them as *laws*. Once we fully understand the causal and explanatory ancestries of these regularities we conceive of them as *eternal truth*. The more we learn the causes of

---

51 Unless otherwise marked, all references to the *Ethics*, the early works of Spinoza, and Letters 1-29 are to Curley's translation: *The Collected Works of Spinoza*. Vol. 1. Edited and translated by Edwin Curley. (Princeton: Princeton University Press, 1985). In references to the other letters of Spinoza I have used Shirley's translation: Spinoza, *Complete Works*, translated by Samuel Shirley (Indianapolis: Hackett, 2002). I have relied on Gebhardt's critical edition (*Spinoza Opera*, 4 volumes (Heidelberg: Carl Winter Verlag, 1925)) for the Latin text of Spinoza.
52 Spinoza, *Theological Political Treatise*, trans. Michael Silverthorne and Jonathan Israel (Cambridge: Cambridge University Press, 2007), chap. 16, n. 34 (G III/264). Italics added.

laws of nature, the closer we are to conceiving them as fully rational and transparent eternal truths.

Throughout the *Theological Political Treatise*, Spinoza brings several examples of eternal truths that were conceived by various biblical figures as laws or edicts.[53] I will conclude my brief discussion of eternal truths with a striking passage in which Spinoza points out that even the existence of God—an obvious eternal truth, for Spinoza—could be misconceived as an edict. The occasion for this misconception, according to Spinoza, was none other than God's revelation on Mount Sinai.

> It is for the same reason too, namely deficiency of knowledge [defectum cognitionis], that the Ten Commandments were *law* [lex] *only* for the Hebrews. Since they did not know [noverant] the existence of God as an eternal truth, i.e., that God exists and that God alone is to be adored, they had to understand it as decrees [legem]. If God had spoken to them as directly without the use of any physical means, they would have perceived this same thing not as an edict [legem] but as an eternal truth.[54]

4 Spinoza on Eternity as Self-Necessitated Existence

Spinoza's philosophy is as bold and original as it is difficult.[55] For this reason, I will precede my examination of Spinoza's novel conception of eternity with a brief exposition of the three building blocks of his ontology: substance (*substantia*), attribute (*attributum*), and mode

---

53 See, for example, *Theological Political Treatise*, chap. 4 (G III/63): "Adam perceived that revelation not as an eternal and necessary truth but rather as a ruling, that is, as a convention that gain or loss follows, not from the necessity and nature of the action done, but only from the pleasure and absolute command of the prince. Therefore that revelation was a law and God was a kind of legislator or prince exclusively with respect to Adam, and only because of the deficiency of his knowledge."

54 Spinoza, *Theological Political Treatise* chap. 4 (G III/63).

55 This section of the chapter relies partly on my article "Spinoza's Deification of Existence," *Oxford Studies in Early Modern Philosophy* 6 (2012): 75–104, and my *Spinoza's Metaphysics: Substance and Thought* (New York: Oxford University Press, 2013), 121–126.

(*modus*).⁵⁶ At the opening of part 1 of the *Ethics*, Spinoza provides the following definitions for substance, attribute, mode, and God (Spinoza's infinite and unique substance):

> Definition 3: By *substance* I understand what is in itself and is conceived through itself, i.e., that whose concept does not require the concept of another thing, from which it must be formed.
> Definition 4: By *attribute* I understand what the intellect perceives of a substance, as constituting its essence.
> Definition 5: By *mode* I understand the affections of a substance, *or* that which is in another through which it is also conceived.
> Definition 6: By *God* I understand a being absolutely infinite, i.e., a substance consisting of an infinity of attributes, of which each one expresses an eternal and infinite essence.
> Explanation: I say absolutely infinite, not infinite in its own kind; for if something is only infinite in its own kind, we can deny infinite attributes of it [NS:⁵⁷ (i.e., we can conceive infinite attributes which do not pertain to its nature)]; but if something is absolutely infinite, whatever expresses essence and involves no negation pertains to its essence.

Each of these definitions raises numerous interpretative questions and has been a source of many scholarly debates. Still, we can get the gist of Spinoza's understanding of substance and mode by noting that Spinoza defines substance as that which is *independent* both ontologically ("in itself" [*in se est*]) and conceptually ("is conceived through itself [*per se concipitur*]"), while mode is defined as that which is *dependent on another* both ontologically ("in another" [in

---

56 For a close study of Spinoza's understanding of these three concepts, see my article "The Building Blocks of Spinoza's Metaphysics: Substance, Attributes, and Modes," in Della Rocca *The Oxford Handbook of Spinoza*. For a study of the chronological development of Spinoza's understanding of substance and attribute, see my "A Glimpse into Spinoza's Metaphysical Laboratory."

57 "NS" is a reference to the 1677, *Nagelate Schriften*, the Dutch translation of Spinoza's works which was published simultaneously with Spinoza's *Opera Posthuma*, nine months after his death.

alio est]) and conceptually ("through which it is also conceived" [*per quod etiam concipitur*].

Both the attributes and the modes are qualities of the substance. However, the attributes are the *essential* qualities of the substance, while the modes are nonessential qualities of the substance (i.e., qualities that the substance can gain and lose). Since the attributes are the essential qualities of the substance, Spinoza argues that they must share the substance's defining characteristic of being *self-conceived* (i.e., no attribute can be conceived through another attribute).[58]

Relying on the asymmetric dependence of modes on the substance, Spinoza proves that two substances cannot share the same attribute.[59] Relying on the definition of substance as an independent being, Spinoza proves that one substance cannot be the cause of another.[60] These three crucial steps—the self-conceivability of attributes, the fact that two substances (if there are any) cannot share an attribute, and that substances are causally isolated—lead Spinoza to the proof that God, a substance of infinitely many attributes, must exist,[61] and shortly afterward to the demonstration that God is the only possible substance.[62]

Another crucial distinction we should have in mind is between what Spinoza calls *Natura naturans* and *Natura naturata*. Here are his explicit definitions of the two: "By *Natura naturans* we must understand what is in itself and is conceived through itself, *or* such attributes of substance as express an eternal and infinite essence, i.e. (by P14C1 and P17C2), God, insofar as he is considered as a free cause. But by *Natura naturata* I understand whatever follows from the necessity of God's nature, *or* from any of God's attributes, i.e., all the modes of God's attributes insofar as they are considered as things which are in

---

58 See *Ethics*, Part One, proposition 9, demonstration.
59 See *Ethics*, Part One, proposition 5, demonstration.
60 See *Ethics*, Part One, proposition 6, demonstration.
61 See *Ethics*, Part One, proposition 11, demonstration.
62 See *Ethics*, Part One, proposition 14, demonstration. For an excellent reconstruction of the entire argument (following the steps noted above), see Michael Della Rocca, "Spinoza's Substance Monism," in *Spinoza: Metaphysical Themes*, ed. Koistinen and Biro (Oxford: Oxford University Press, 2003), 11–37.

God, and can neither be nor be conceived without God."[63] Roughly speaking, *Natura naturans* is the realm of the essence of substance, and its infinitely many attributes, while *Natura naturata* is the realm of the modes, the nonessential qualities that follow from the essence of God, the unique and only substance.

With this cursory overview of the foundation of Spinoza's ontology at hand, I can now approach his understanding of eternity. My discussion will focus primarily on Spinoza's masterwork, the *Ethics*, though occasionally I will also refer to some of his earlier works and letters.

At the beginning of the climactic conclusion of the *Ethics*, Spinoza writes: "Eternity is the very essence of God insofar as this involves necessary existence [Aeternitas est ipsa Dei essentia, quatenus haec necessariam involvit existentiam] (by Definition 8 of Part I)."[64] In this passage Spinoza appeals to the definition of eternity that appears at the opening of the *Ethics*: "By *eternity* I understand existence itself, insofar as it is conceived to follow necessarily from the definition alone of the eternal thing [Per aeternitatem intelligo ipsam existentiam, quatenus ex sola rei aeternae definitione necessario sequi concipitur]."[65] To this enigmatic definition, which apparently sins in obvious circularity (by employing the definiendum, "eternity," in the definiens), Spinoza attaches the following explanation: "For such existence, like the essence of a thing, is conceived as an eternal truth, and on that account cannot be explicated by duration or time, even if the duration is conceived to be without beginning or end [Talis enim existentia ut aeterna veritas, sicut rei essentia, concipitur, propterequae per durationem aut tempus explicari non potest,[66] tametsi duratio principio

---

63 See *Ethics*, Part One, proposition 29, scholium.
64 *Ethics*, Part Five, proposition 30, demonstration. See Spinoza, *Cogitata Metaphysica* (=CM) II 1 (G I/251/11): "The reasons why Writers have attributed duration to God. The reason why these Writers have erred is threefold: first, because they have attempted to explain eternity without attending to God, as if eternity could be understood without contemplation of the divine essence—or *as if it were something beyond the divine essence.*" Italics added.
65 *Ethics*, Part One, Definition 8.
66 The recently discovered Vatican manuscript of Spinoza's *Ethics* has here "nequit" instead of "non potest." See Spinoza, *The Vatican Manuscript of Spinoza's "Ethics*," ed. Leen Spruit and Pina Totaro (Leiden: Brill, 2011), 84.

et fine carere concipiatur]."⁶⁷ In the existing literature there is some debate about Spinoza's understanding of eternity, and of the related issue of mind eternity. Some scholars interpret Spinoza's notion of eternity as mere sempiternity, or everlasting existence, while others consider it atemporal.⁶⁸ Although I have significant reservations about both readings, I find the atemporal interpretation more accurate. I will begin by registering that the primary meaning of *aeternitas* in Spinoza, as expressed in the official definition of the term in *Ethics*, part One, Definition 8, is *explicitly* contrasted with sempiternity or, in Spinoza's words, with "duration [that] is conceived to be without beginning or end."⁶⁹ Indeed, Spinoza stresses several times that "in eternity, there is neither *when*, nor *before*, nor *after*."⁷⁰ Similarly, he argues, "we cannot ascribe future existence to God, because existence is of his essence."⁷¹ Spinoza's claim in *Ethics*, part One, definition 8, explanation, that eternity is existence "conceived as an eternal truth" provides further support for the rejection of the sempiternal reading, since he clearly regards eternal truths as not enduring: "No one will ever say that the essence of a circle or a triangle, insofar as it is an eternal truth, has endured longer now than it had in the time of Adam."⁷²

---

67 *Ethics*, Part One, Definition 8, explanation. See CM II 1 (G I/252/17–18): "I call this infinite existence eternity, which is to be attributed to God alone, and not to created things, *even though its duration should be without beginning or end*." Italics added.

68 For the view of Spinoza's eternity as mere sempiternity, see Martha Kneale, "Eternity and Sempiternity," in *Spinoza: A Collection of Critical Essays*, ed. Marjorie Green (New York: Anchor Books, 1973), 227–224, and Alan Donagan, *Spinoza* (Chicago: University of Chicago Press, 1988), 107–113. The opposite, timeless reading of mind eternity, is advocated by Diane Steinberg, "Spinoza's Theory of the Eternity of the Mind," *Canadian Journal of Philosophy* 11 (1981): 55–65.

69 Even Kneale, who supports the sempiternal interpretation of mind eternity concedes that the definition of eternity at the opening of the *Ethics* is not consistent with the sempiternal reading. It is only "by the time he came to write Part V," claims Kneale, that Spinoza changed his view and "was thinking in a more Aristotelian way" ("Eternity and Sempiternity," 233). Against Kneale I would argue that it would be very odd for Spinoza not to revise such a key definition in the *Ethics*, had he abandoned his original nondurational understanding of eternity.

70 *Ethics*, Part One, proposition 33, second scholium. See CM I 3 (G I/243/12): "In Eternity there is no when, not before, or after, nor any other affection of time." See CM II 1 (G I/251/1). The last sentence is probably the closest Spinoza comes to the Boethian formula.

71 CM II 1 (G I/252/13).

72 CM II 1 (G I/250/29).

The definition of eternity at the opening of the *Ethics* (and its explanation) clearly rules out any conception of eternity as limitless duration,[73] but we should also pay close attention to the positive content of the definition. The definition not only tells us what eternity is—*existence*—but also tells what *kind* of existence it is—"existence itself, insofar as it is conceived to follow necessarily from the definition alone of the eternal thing"—that is, *the existence of a thing whose existence follows necessarily from its own essence.*[74] Indeed, in *Ethics*, part One, proposition 23, scholium, Spinoza relies on his definition of eternity in order to identify eternity with the "necessity of existence" (*necessitate existentiae*) that each attribute of God is conceived to express: "So if a mode is conceived to exist necessarily and be infinite, [its necessary existence and infinitude] must necessarily be inferred [concludi], *or* perceived through some attribute of God, insofar as that attribute is conceived to express infinity and *necessity of existence, or (what is the same, by D8) eternity* [quatenus idem concipitur infinitatem, et necessitatem existentiae, sive (quod per Defin. 8. idem est) aeternitatem exprimere], i.e. (by D6 and P19), insofar as it is considered absolutely."[75] On two other occasions in the *Ethics*, Spinoza uses the phrase "eternity or [sive] necessity." In *Ethics*, part One, proposition 10, scholium, Spinoza writes: "Nothing in nature is clearer than that each being must be conceived under some attribute, and the more reality, or being [realitatis, aut esse] it has, the more it has attributes which express necessity, *or* eternity [necessitatem, sive aeternitatem], and infinity." Similarly, *Ethics*, part Four, proposition 62, demonstration, reads: "Whatever the Mind conceives under the guidance of reason, it

---

73 See CM II 1 (G I/251/24) for Spinoza's detailed critique of those who consider eternity "a species of duration" and do not distinguish between God's eternity and the (infinite) duration of created things. I will shortly address the issue of the eternity of infinite modes.

74 The "conception of existence" in *Ethics*, Part One, Definition 8, must be *adequate*, for otherwise, Spinoza's proof in *Ethics*, Part one, proposition 19, demonstration—which relies on Definition 8—would be invalid. Thus, since eternity is existence, *adequately* conceived to follow from the definition of an eternal being, we may conclude that such existence indeed "follows from the definition alone of the eternal thing."

75 *Ethics*, Part One, proposition 23, scholium. Italics added.

conceives under the same species of eternity, *or* necessity [sub eadem aeternitatis, seu necessitatis specie concipit]." Almost always, Spinoza employs the Latin *sive* to designate equivalent terms, and the occurrences of the term in the above two passages are no exceptions. Thus, we have solid textual evidence showing that for Spinoza there is a very intimate relation between eternity and necessity. Some commentators have suggested that Spinoza identifies eternity with necessity, or necessary existence,[76] but this qualification is imprecise and insufficient, since for Spinoza the existence of *all* things is necessary.[77] What is truly unique to Spinoza's notion of eternity is its being *self*-necessitated (or necessitated by virtue of its mere essence), whereas all other things are necessary by virtue of causes that are not identical with their essences.[78] Here is Spinoza's presentation of this crucial distinction: "A thing is called necessary either by reason of its essence or by reason of its cause. For a thing's existence follows necessarily either from its essence and definition or from a given efficient cause [Rei enim alicujus existentia vel ex ipsius essentia, et definitione, vel ex data causa efficiente necessario sequitur]."[79] Let us look closely at the second sentence of this passage. The notion of existence necessarily following from the definition (and essence)[80] of a thing should be familiar by now: this is

---

76 See, for example, Kneale, "Eternity and Sempiternity," 235–238, and B. Leftow, *Time and Eternity* (Ithaca, NY: Cornell University Press, 1991), 63. Regrettably, these scholars are insensitive to Spinoza's important distinction between different kinds of necessary existence: necessary existence by virtue of one's essence as opposed to necessary existence by virtue of one's cause. See *Ethics*, Part One, proposition 33, scholium 1.

77 See *Ethics*, Part One, proposition 29, and *Ethics*, Part One, proposition 33. For an insightful discussion of Spinoza's necessitarianism, see Don Garrett, "Spinoza's Necessitarianism," in *God and Nature: Spinoza's Metaphysics*, ed. Yirmiyahu Yovel (Leiden: Brill, 1991), 97–118. On the intimate connection between necessity and eternity in al-Kindī and Avicenna, see chapter 2 here.

78 Herrera comes very close to identifying eternity with self-necessitated existence, claiming that what is "necessary by itself is, therefore, eternal." *Gate of Heaven*, bk. 5, chap. 4 (Krabbenhoft trans., p. 150). For Spinoza's adoption of the Maimonidean interpretation of "ego sum qui sum" (Exodus 3:14) as indicating God's self-necessitated existence, see my "Deification of Existence," 83–86.

79 *Ethics*, Part One, proposition 33, scholium 1.

80 For Spinoza, an adequate definition must capture the essence of the thing defined. See *Tractatus de Intellectus Emendatione* [Treatise on the Emendation of the Intellect], section 95. As a result, he frequently treats the two terms as interchangeable.

just the definition of eternity in *Ethics*, part One, definition 8. I am coming close to excavating Spinoza's understanding of eternity, but I am not yet there.

At this point it may be interesting to compare Spinoza's and Boethius's views of eternity. While both philosophers reject the conception of eternity as everlastingness, Spinoza, unlike Boethius, defines eternity without employing, or even referring to, the terminology associated with duration and time (such as Boethius's *tota simul* [all at once]). Only in the *explication* of his definition of eternity does Spinoza note that eternity cannot be equated with infinite duration, or time. This point is quite significant, since the role of a definition in Spinoza's system is to capture the essence of a thing.[81] Thus, it seems that for Spinoza, the essence of eternity is captured not by any relation—*not even negation*—to duration or time. Spinoza seems to be stressing this very point by noting that "eternity can neither be defined by time nor have *any relation* to time [nec aeternitas tempore definiri, nec ullam ad tempus relationem habere potest]."[82] Instead, he proposes a genuine definition of eternity as a unique kind of *modality: self-necessitated existence*. While quite a few of his predecessors associated eternity with necessity in one manner or another, he seems to go far beyond them in completely relocating this notion from the domain of temporality and defining it as a primarily a *modal* concept.[83] This new understanding of eternity as self-necessitated existence may also shed light on the circularity in Spinoza's definition of eternity (*Ethics*, part One, definition 8). Rather than a beginner's error, it seems to be a premeditated move attempting to capture the essential feature of eternity as *self*-necessitated.

---

81 See *Tractatus de Intellectus Emendatione*, section 95, and Ep. 9 (G IV/42/30).

82 *Ethics*, Part Five, proposition 23, scholium. Italics added.

83 Many of Spinoza's predecessors affirmed the identity of essence and existence in God, but they did not see it as constituting God's eternity. According to Gorham ("Descartes on God's Relation to Time," 422–423), Descartes considered this identity compatible with God's being in time, as he seamlessly moved, in the Fifth Meditation, from asserting the identity of God's essence and existence to ascribing temporality to God.

Apart from the definition of eternity at the opening of the *Ethics*, the other key text for understanding Spinoza's concept of eternity is Letter 12, famously known as the "Letter on the Infinite." Spinoza circulated copies of this letter also in his very late period,[84] and thus it seems to reflect his views during this time as well. In this difficult and intriguing text, Spinoza suggests a threefold distinction between eternity [*aeternitas*], duration [*duratio*] and time [*tempus*]. In explaining the distinction between the first two, Spinoza claims, "we conceive the existence of Substance to be entirely different from the existence of Modes. The difference between Eternity and Duration arises from this.[85] For it is only of Modes that we can explicate [explicare possumus][86] the existence by Duration. But [we can explicate the existence] of Substance by Eternity, i.e., the infinite enjoyment of existing, or (in bad Latin) of being [infinitam existendi, sive, invitâ latinitate, essendi fruitionem]."[87] Spinoza's use of the verb *explicare* in this passage may seem a bit odd. He is not looking here for an explanation of the causes of existence but is suggesting that existence can be explicated, or unfolded, as either duration or eternity. Eternity is the proper explication of the existence of substance, or the thing whose essence and existence are one and the same, while duration is the proper explication of the existence of modes, or things whose existence is distinct from their essence. A similar, though slightly different, distinction appears in a section titled "What Eternity Is; What Duration Is" in Spinoza's early work the *Cogitata Metaphysica*: "From our earlier division of being into being whose essence involves existence and being whose essence

---

[84] See Eps. 80 and 81.
[85] For a detailed discussion of the bifurcation between eternity and duration and its sources, see my *Spinoza's Metaphysics*, 105–112.
[86] I have altered Curley's translation here, translating *explicare* as "explicate" instead of "explains."
[87] Ep. 12 (G IV/54/16–55/3). Leibniz possessed a copy of this letter and annotated it. Commenting on the very last phrase in the foregoing passage, he writes: "This agrees well enough with Boethius's definition of eternity" (*Satis congruat cum definitions aeternitatis Boëtiana*). See Leibniz, *The Labyrinth of the Continuum. Writings on the Continuum Problem, 1672–1686*, trans. and ed. Richard T. W. Arthur (New Haven, CT: Yale University Press, 2001), 106–107.

involves only possible existence, there arises the distinction between eternity and duration."[88] In this early period (1663), Spinoza referred to the essence of modes as involving only *possible* existence, that is, being internally consistent. Later he will abandon the "involving only possible existence" terminology and instead claim that the essence of modes simply does not involve existence.

In another passage in the *Cogitata Metaphysica* Spinoza states the very same understanding of the nature of duration as the existence of modes in a critical note addressing certain unnamed opponents: "They have erred because they have ascribed duration to things only insofar as they judged them to be subject to continuous variation and not, *as we do, insofar as their essence is distinguished from their existence.*"[89] Whether modes are *in some sense* eternal is an important question that I will shortly address, but it is, I think, clear that God's existence is eternity, as defined in *Ethics*, part One, definition 8.

At this point we may wish to address a crucial problem. In propositions 21–23 of part One of the *Ethics*, Spinoza lays out an outline of his theory of certain infinite entities that follow from the attributes. Scholars commonly refer to these entities as "the infinite modes," though Spinoza never used the term. These are somewhat mysterious entities that appear in Spinoza's work from a very early stage, yet, as far I can see, he never fully developed this theory.[90] In the first of these three propositions he argues: "All things which follow from the absolute nature of any of God's attributes have always had to exist and be infinite, or are through the same attribute, eternal and infinite [Omnia, quae ex absoluta natura alicujus attributi Dei sequuntur, semper et infinita existere debuerunt, sive per idem attributum aeterna et infinita sunt]."[91] The immediate infinite modes, as these

---

88 CM I 4 (G I/244/13–15). See H. F. Hallett, *Aeternitas: A Spinozistic Study* (Oxford: Clarendon Press, 1930), 43–44, for a helpful discussion of this and the previous passages.
89 CM II 1 (G I/251/17–19). Italics added.
90 For a detailed discussion of the infinite modes, see chapter 4 of my book *Spinoza's Metaphysics*.
91 *Ethics*, Part One, proposition 21.

entities are commonly called,[92] are described here as "eternal," but this seems to conflict with Spinoza's claims in the Letter on the Infinite that the existence of modes is duration, while eternity is the existence of substance.

Though the last point may at first look like a blunt contradiction, we can sort it out if we pay attention to the following observations. First, the phrase in *Ethics*, part One, proposition 21, that asserts that infinite modes "have *always* had to exist" (*semper . . . existere debuerunt*) is more consistent with an everlasting, as opposed to atemporal, understanding of eternity.[93] Second, the demonstration of the aforementioned proposition does not show (or even attempt to show) that infinite modes are eternal in the strict sense of *Ethics*, part One, definition 8, but only shows that these modes "cannot have a *determinate* duration [non potest determinatam habere durationem]."[94] This obviously allows for the infinite modes to have *indeterminate* (or infinite) duration. Nowhere in this demonstration does Spinoza prove, or even *attempt* to prove, that infinite modes are atemporal. Since Spinoza was acutely aware of the distinction between endless duration and atemporality (recall my discussion of *Ethics*, part One. definition 8, explanation) it would be very odd for him to state one thesis and prove the other.

Third, the demonstration of *Ethics*, part One, proposition 21, does not at all mention Spinoza's official definition of eternity as atemporal self-necessitated existence (*Ethics*, part One, definition 8). Were Spinoza to argue that the immediate infinite modes are eternal in the strict sense of *Ethics*, part One, definition 8, the first thing he should

---

92 *Immediate* infinite modes are modes that follow directly from the nature of an attribute (see *Ethics*, Part One, proposition 21, demonstration). A *mediate* infinite mode is a mode that follows from the attribute only through the mediation of another infinite mode (see *Ethics*, Part One, propositions 22 and 23).

93 This point is also stressed by Harry Austryn Wolfson, *The Philosophy of Spinoza*, 2 vols. (New York: Schocken, 1969), vol. 1, 377. Another commentator who qualifies the eternity of infinite modes (in *Ethics*, Part One, proposition 21) unlike the eternity of substance, as mere everlastingness, is Martial Gueroult, *Spinoza: Dieu (Ethique 1)* (Paris: Aubier, 1968), 309.

94 *Ethics*, part One, proposition 21, demonstration (G II/66/13). Italics added.

do would be to appeal to the definition of eternity.[95] Spinoza does mention the definition of eternity in his discussion of the infinite modes in *Ethics*, part One, proposition 23, demonstration, and one might be tempted to consider this evidence that the infinite modes are eternal in the strict sense of *Ethics*, part One, definition 8. Yet, on closer examination, we should notice that in the demonstration of proposition 23, when Spinoza invokes definition 8, he does so in order to identify eternity with "an *attribute* of God . . . *insofar as it is considered absolutely*";[96] but this last characterization is clearly *not* true of the infinite modes, which "*follow from* the absolute nature of any of God's attributes" but are not this absolute nature itself.

Finally, we have clear evidence that Spinoza recognized a certain "second best" notion of eternity—eternity as everlastingness—as long as it is not applied to God (who is eternal in the strict sense of [atemporal] self-necessitated existence). Consider the following passage from the *Cogitata Metaphysica*:

> So we pass to the second question and ask whether what has been created could have been created from eternity.
> *What is denoted here by the words: from eternity*
> To understand the question rightly, we must attend to this manner of speaking: "*from eternity.*" For by this we wish to signify here something *altogether different from what we explained previously when we spoke of God's eternity*. Here we understand nothing *but a duration without any beginning of duration*, or a duration so great that, even if we wished to multiply it by many years, or tens of thousands of years, and this product in turn by tens of thousands, we could still never express it by any number, however large.[97]

---

95 Spinoza invokes *Ethics*, Part One, Definition 8, in almost all places where he proves the eternity of anything.
96 Italics added.
97 CM I 10 (G I/270/17–25).

In this passage Spinoza introduces a certain notion of eternity that is "altogether different" from God's eternity. Unlike God's eternity as self-necessitated existence, the "second best" eternity, which belongs only to created things, is identified with unlimited duration. Notice that Spinoza stresses that the "manner of speaking" that employs the expression "from eternity" [*ab aeterno*] indicates that what is at stake is the "second best" eternity of created things (i.e., everlastingness). In the following passage from the *Short Treatise,* Spinoza applies this very expression explicitly to the immediate infinite modes: "Turning now to universal *Natura naturata,* or those modes or creatures which immediately depend on, or have been created by God . . . we say, then, that these have been created *from all eternity* and will remain *to all eternity,* immutable, a work as great as the greatness of the workman."[98] We have, I believe, very strong evidence that the eternity of the infinite modes (even of the immediate infinite modes) is, unlike God's eternity, merely everlastingness. Spinoza stresses in several places that eternity truly belongs only to God.[99] Arguably, in all these places he is speaking of eternity in its strict sense (of *Ethics,* part One, definition 8), which completely excludes duration and time.

I have just clarified one exception to Spinoza's key claim in the Letter on the Infinite that the existence of God must be explicated through eternity, while the existence of modes should be explicated as duration, that is, we have seen that the infinite modes may be described as "eternal" but only in the inferior sense of the term as mere everlastingness. But Spinoza allows for another—and bolder—exception (or apparent exception) to his dichotomy between the existence of modes and existence of substance. In several places in the *Ethics,* Spinoza

---

98 *Korte Verhandeling van God de Mensch en deszelfs Welstand* [Short Treatise on God, Man, and his Well-Being] 1.9 (G I/48/3–9). Italics added. "From eternity to eternity" is also a translation of the Hebrew of Psalms 106:48 (*Min ha-Olam ve-ad ha-Olam*).

99 "And I call this infinite existence Eternity, which is to be attributed to God alone, and not to any created thing, even though its duration should be without beginning or enc" (CM II 1 [G I/252/17–19]). See Spinoza's critique of those who think that eternity is "something beyond the divine essence" (CM II 1 [G I/252/11]).

suggests that modes, even *finite* modes, may be adequately conceived as eternal, or *sub specie aeternitatis*.[100] I will turn now to examine briefly two important texts in which Spinoza develops this claim.

In proposition 45 of part Two of the *Ethics*, Spinoza argues: "Each idea of each body, or of each singular thing which actually exists, necessarily involves an eternal and infinite essence of God." To this proposition he adduces the following scholium:

> By existence here I do not understand duration, i.e., existence insofar as it is conceived abstractly, and as a certain species of quantity. *For I am speaking of the very nature of existence, which is attributed to singular things because infinitely many things follow from the eternal necessity of God's nature in infinitely many modes (see IP16). I am speaking, I say, of the very existence of singular things insofar as they are in God.* For even if each one is determined by another singular thing to exist in a certain way, still the force by which each one perseveres in existing follows from the eternal necessity of God's nature.[101]

For Spinoza, "singular things" [*res singulares*] are just the finite modes,[102] but strikingly Spinoza says here that if we conceive finite modes nonabstractly—that is, as completely imbedded in the substance—we can attribute to them "the very nature of existence," which is not duration but the very eternity of God. The reason for this bold claim is simple: when we conceive modes "insofar as they are in God," that is, as completely imbedded in God, we really conceive nothing but God, and God is eternal in the strict sense of self-necessitation.

The very same point is raised again toward the end of the *Ethics*, where Spinoza discusses the mind's eternity. When the human mind

---

100 See *Ethics*, Part Two, proposition 44, corollary, and Part Five, propositions 22, 30, and 31.
101 *Ethics*, Part Two, proposition 45, scholium. Italics added.
102 See *Ethics*, Part Two, Definition 7: "By singular things I understand things that are finite and have a determinate existence."

and body—two finite modes—are conceived as strictly flowing from God's essence, they take part in the very eternity of God. Thus, the 30th proposition of part Five of the *Ethics* and its demonstration read:

> Insofar as our Mind knows itself and the Body under a species of eternity, it necessarily has knowledge of God, and knows that it is in God and is conceived through God.
>
> Demonstration: Eternity is the very essence of God insofar as this involves necessary existence (by ID8). To conceive things under a species of eternity, therefore, is to conceive things insofar as they are conceived through God's essence, as real beings, *or insofar as through God's essence they involve existence.* Hence, insofar as our Mind conceives itself and the Body under a species of eternity, it necessarily has knowledge of God, and knows, etc., q.e.d.

Notice the italicized phrase in this passage. When the human mind and body are conceived through God's essence, they *thereby involve the existence which is God's essence, that is, eternity.* Here again, when the mind conceives itself (and its body) as completely imbedded in God, it really conceives God, the strictly eternal being.[103] Now, we can make sense of an intriguing note of Spinoza in the Letter on the Infinite. Note carefully the "insofar" clause:[104] "I call the Affections of Substance Modes. Their definition, *insofar as it is not the very definition of Substance,* cannot involve any existence."[105] When we conceive the modes as completely imbedded in the substance, they *are* defined through the substance, and to that extent their definition involves existence, that is, they are eternal.

---

103 For further discussion of Spinoza's understanding of mind eternity, see Hallett, *Aeternitas*, 72–98.
104 In a marginal note on the "insofar" clause, Leibniz asks: "So can the definition of a mode be the definition of substance in some manner?" (*Labyrinth of the Continuum*, 105). The demonstration of proposition 30 of Part Five of the *Ethics* indicates that the answer to this question is positive.
105 Spinoza, Ep. 12 (G IV/54/9–11). Italics added.

5 THE RECEPTION OF SPINOZA'S CONCEPT OF ETERNITY AS SELF-NECESSITATED EXISTENCE

Leibniz was no fan of Spinoza. He visited Spinoza once in November 1676 and was clearly impressed.[106] They also shared a close friend, Baron Ehrenfried Walter von Tschirnhaus (1651–1708). But Leibniz was a Christian philosopher, and Spinoza was not. The late Leibniz was also highly suspicious of Spinoza's attempt to speak in the language of mainstream theology. Thus, in a 1707 note he remarks that Spinoza's talk about "the intellectual love of God" (*amor dei intellectualis*) is "nothing but soap to the masses."[107] Earlier, in the 1686 *Discourse on Metaphysics*, Leibniz writes disparagingly about "the recent innovators who hold that the beauty of the universe and the goodness we attribute to the works of God are but the chimeras of those who conceive of God in terms of themselves."[108] Spinoza was clearly the direct target of this critique.

In light of this clear opposition to Spinoza, we might be surprised by Leibniz's unreserved adoption of Spinoza's understanding of eternity. The person whom many contemporary Europeans took to be the paragon of atheism tuned out to be the foremost expert on eternity, that is, God's essence. In 1678 Leibniz received a copy of Spinoza's *Opera Posthuma*, shortly after its publication. Leibniz's notes on the first few definitions of part One of the *Ethics* are quite critical. Yet when he comes to definition 7 ("That thing is called free which exists from the necessity of its nature alone"),[109] and what is more important for this discussion, definition 8 (i.e., Spinoza's definition of eternity), Leibniz notes: "I approve of both of these definitions."[110] Along the same lines, in his *New Essays on God on*

---

106 See my *Spinoza's Metaphysics*, 27–28, and 167–168. For illuminating discussions of Leibniz's friendship with Tschirnhaus, see Christia Mercer, *Leibniz's Metaphysics: Its Origins and Development* (Cambridge: Cambridge University Press, 2001), 408–409, and Mogens Laerke, *Leibniz lecteur de Spinoza: La genese d'une opposition complexe* (Paris: Honoré champion, 2008), 362–373.
107 Leibniz, *Philosophical Essays*, 281.
108 *Discourse on Metaphysics*, section 2, in Leibniz, *Philosophical Essays*, 36.
109 For a helpful discussion of Leibniz's reception of Spinoza's understanding of divine freedom, see Laerke, *Leibniz lecteur de Spinoza*, 834–843.
110 Leibniz, *Philosophical Papers and Letters*, ed. and trans. Leroy E. Loemker (Dordrecht: Reidel, 1969), 197.

Human Understanding, Theophilius, the Leibnizian interlocutor in the dialogue, rejects the view of eternity as infinite duration and identifies it instead with "the necessity of God's existence."[111]

A particularly interesting passage in which Leibniz allows—just like Spinoza—for eternity to have more than one sense yet insists that the more precise sense of the term is "necessity of existence" is the following: "Eternity, if it is conceived as something which is homogenous with time, will be unlimited time; . . . *But the true origin and the inmost nature of eternity is the very necessity of existing*, which does not of itself indicate any succession, even if it should happen that what is eternal coexists with everything. . . . Eternity *per se* does not indicate succession."[112] Samuel Clarke, with whom Leibniz had a celebrated exchange of letters, seems also to adopt Spinoza's notion of eternity as self-necessitated existence. Thus, in his 1704 *Demonstration of the Being and Attributes of God*, Clarke states the Principle of Sufficient Reason and then turns to equate eternity with existence that is necessitated by the nature of the thing. "Whatever exists has a cause, a reason, a ground for its existence, a foundation on which its existence relies, a ground or reason why it does exist rather than not exists, either *in the necessity of its own nature* (and then it must have been of itself eternal), or in the will of some other being."[113] Later, in the very same work, Clarke notes: "The ideas of eternity and self-existence are so closely connected, that because something must of necessity be eternal independently and without

---

111 Leibniz, *New Essays on Human Understanding*, translated and edited by Peter Remnant and Jonathan Bennett (Cambridge: Cambridge University Press, 1981), chap. 17, sec. 8. Two recent commentators who stress the influence of Spinoza on Leibniz's concept of eternity are Michael J. Futch (*Leibniz's Metaphysics of Time and Space* [Berlin: Springer: 2008], 191), and Vailati (*Leibniz and Clarke*, 19).

112 Leibniz, *De Summa Rerum*, translated by G.H.R. Parkinson (New Haven: Yale University Press, 1992), 41 (Ak. 6.3.484). Italics added. See Leibniz, *De Summa Rerum* 396 (Ak. 6.3.159): "But absolute existence is eternity or necessity. From this it immediately follows that such a being does not only exist, but also exists necessarily."

113 Clarke, *Demonstration*, 8. The curious phrase "necessity of its own nature" seems to be borrowed from proposition 16 of Part One of Spinoza's *Ethics*: "From the necessity of the divine nature there must follow infinitely many thing in infinitely many modes."

any outward cause, of its being, therefore it must necessarily be self-existing."[114] The view of eternity as mere everlastingness will become more and more popular throughout the eighteenth century.[115] In his highly influential *Metaphysics*, Alexander Baumgarten (1714–1762) will mostly adopt the conception of eternity as everlastingness, though his definition of eternity will change from one edition to another, and occasionally one could still trace echoes of Spinoza's definition.[116]

Close to the end of the century, Moses Mendelssohn (1729–1786) will write the following in his 1785 *Morgenstunden* (*Morning Hours*), a work that is partly directed against Spinoza but still echoes his definition of eternity as self-necessitated existence: "We must admit the sort of beginning of things that is in need of no further beginning, hence, a necessary being, whose existence does not depend upon efficient causes, whose duration however is not a temporal succession without beginning but instead a *timelessness, an immutable eternity* that can essentially have neither beginning, nor progression, nor end. . . . The necessary being has, like all the necessary truths of geometry, no past and no future time. . . . The immutable necessary substance is at once everything that can be thought of it, and its existence knows neither increase not decrease."[117] One of Mendelssohn's major literary (and political) projects was the translation of the Bible into German. Here, when he faced the question of how to translate the Tetragrammaton—the holiest divine name, which both Maimonides and Spinoza interpreted as indicating God's self-necessitated existence,[118] he settled on

---

114 Clarke, *Demonstration*, 31. It should be noted that for Clarke, unlike Spinoza, the self-necessitated, or eternal, being is everlasting (*Demonstration*, 32).

115 At the beginning of this chapter we encountered Newton's claim that it is more agreeable to reason to interpret God's eternity as the view condensed in the Tetragrammaton, i.e., "He that was and is and is to come." See note 8 here.

116 Thus, in sec. 303 of the first edition, we read: "a necessary being and substance, an infinite being, is eternal." See Alexander Baumgarten, *Metaphysics*, trans. Courtney D. Fugate and John Hymers (London: Bloomsbury, 2013), 424 n. 358.

117 Moses Mendelssohn, *Morning Hours: Lectures on God's Existence*, trans. Daniel O. Dahlstorm and Corey Dyck (Dordrecht: Springer, 2011), chap. 11, pp. 67–68.

118 See my "Deification of Existence," 81–86.

Der Ewige (the eternal).[119] Whether he had the claims of Maimonides and Spinoza in mind I will leave for the reader to decide.

Abbreviations

AT René Descartes, *Oeuvres*, 11 vols., ed. Charles Adam and Paul Tannery, new CNRS ed. (Paris: J. Vrin, 1974–86). Cited by volume and page number: "AT VII 23" stands for vol. 7, p. 23 of this edition.

CSM Cottingham, Stoothoff, and Murdoch (eds. and trans.), *The Philosophical Writings of Descartes*. 3 vols. (third volume edited also by A. Kenny). Cited by volume and page number: "CSM II 233" stands for vol. 2, p. 233 of this edition.

CM Benedict Spinoza, *Cogitata Metaphysica* [Metaphysical thoughts], an appendix to Spinoza, *Renati des Cartes Principiorum Philosophiae Pars I & II* [Descartes's *Principles of Philosophy*] (1663). Cited by part and chapter: "CM 1 10" stands for part 1, chapter 10 of this work.

EP Spinoza's Letters

G Benedict Spinoza, *Spinoza Opera*, ed. Carl Gebhardt, 4 vols. (Heidelberg: Carl Winter, 1925). Cited by volume, page, and line number: "G II/200/12" stands for vol. 2, p. 200, l. 12 of this edition.

Acknowledgment

I am indebted to Stephan Schmid and Zach Gartenberg for their most helpful comments on earlier versions of this chapter.

---

119 See Rivka Horwitz, "Mendelssohn's Commentary: The Eternal," *Jewish Studies* 37 (1997): 185–214.

Reflection

OUT OF TIME: DANTE AND ETERNITY

Akash Kumar

Writing about Dante and eternity in any sort of succinct fashion is like trying to fill a cup under a waterfall. There are eternal things and meditations on what it is to be eternal at almost every step of the way as we read the *Commedia*. But what sort of eternity are we talking about? Is it simply a matter of duration, lasting for all time, or something that is out of time entirely? As Jim Wilberding points out in chapter 4 here, there has long been a distinction between the idea of eternity as endless duration and timelessness, one that Boethius put forth as the separate terms *sempiternitas* and *aeternitas*. While many of the aspects of eternity that come out in Dante's *Commedia* have far more to do with the sempiternal, like duration of infernal punishment, we can see how the idea of the timeless also emerges once we step out of time and space in the extradimensional poetry of *Paradiso*.

There are eternal things throughout Dante's journey in the afterlife. In his programmatic statement in *Inferno* 1, where Virgil says that he will be Dante's guide, Virgil establishes the eternity of hell: "and I shall guide you, taking / you from this place through an eternal place" (e io sarò tua guida / e trarroti di qui per loco etterno) (*Inferno* 1.113–114).[1] We see here the opposition between the temporary, the place where the journey begins, and the eternal place through which Virgil will guide his charge. This sense of the eternity of hell is not exclusive to the endlessness of the torments but is something essential in the hierarchical order of creation. As we see in the famous inscription on the gate of hell: "Before me nothing but eternal things / were made, and I endure eternally" (Dinanzi a me non fuor cose create / se non etterne e io etterno duro) (*Inferno* 3.7–8). We have a sense here of both the primacy of the eternal, that hell was created before all noneternal things (even if not before all other eternal things), and of the sense of duration, measuring the eternal by how it outlasts all else. That it is hell itself that speaks to this essential quality of being eternal is quite notable, pointing to the paradox of putting the eternal into the temporal construct of language.

In spite of this sense of immutability and a scale that dwarfs any human construct, there is also a way in which Dante uses the eternal to describe how we seek to make ourselves last and even outlast our mortality. When Dante expresses his gratitude to his former teacher Brunetto Latini, he does so specifically for having taught him how man makes himself eternal: "Within my memory is fixed—and now / moves me—your dear, your kind paternal image / when, in the world above, from time to time / you taught me how man makes himself eternal" (che 'n la mente m'è fitta, e or m'accora / la cara e buona imagine paterna / di voi quando nel mondo ad ora ad ora / m'insegnavate come l'uom s'atterna) (*Inferno* 15.82–85). The image is fixed (*fitta*) yet it moves, past and present in one (*or m'accora*), as it strikes the

---

[1] Here and throughout, I cite Allen Mandelbaum's translation and the Petrocchi edition of the Italian text. See *The Divine Comedy of Dante Alighieri*, trans. Allen Mandelbaum (Berkeley: University of California Press, 1980–82) and *La Commedia secondo l'antica vulgata*, ed. Giorgio Petrocchi, 2nd ed. (Milan: Mondadori, 1994).

pilgrim's heart. The repetitive, almost habitual act of teaching (*ad ora ad ora*) is picked up by the imperfect verb *m'insegnavate* and shows the making of the eternal as a process bound in time that renews itself. This is contrasted to the fixed quality of the textual solution that is presented of eternity, or perhaps eternal memory, in Brunetto's writing of *Tresor*, the encyclopedic work he pleads to be remembered by—a text, as any human product, created in time yet potentially enduring and being eternal.

Purgatory is decidedly not eternal, lasting only until the last judgment, according to doctrine. Yet a sense of the eternal remains in Dante's *Purgatorio* and is even perhaps magnified. On the terrace of pride, when the painter Oderisi da Gubbio speaks of the impermanence of fame and the human voice (*Purgatorio* 11.103–108), he puts it in the context of different notions of time:

Before a thousand years have passed—a span
that, for eternity, is less space than
an eyeblink for the slowest sphere of heaven—

would you find greater glory if you left
your flesh when it was old than if your death
had come before your infant words were spent?

Che voce avrai tu più, se vecchia scindi
da te la carne, che se fossi morto
anzi che tu lasciassi il "pappo" e 'l "dindi,"

pria che passin mill'anni? ch'è più corto
spazio a l'etterno, ch'un muover di ciglia
al cerchio che più tardi in cielo è torto.

In the scope of a thousand years, Oderisi says, it will make no difference whether one dies still speaking like a baby or as an old man full of years and presumably capable of expressing himself eloquently. And yet those thousand years are nothing more than the blink of an eye from the perspective of the motion of the heavens. And still further, it is even less than that from the perspective of eternity. In this light, Dante's earlier words to Brunetto about man making himself eternal seem far too optimistic. The eternal has dwarfed any such human conception. The compelling combination of space-time here (*passin mill'anni* likened to *corto spazio*) presents a syncretic understanding of the nature of the universe and points us forward to the ever more philosophically dense language of the third canticle, *Paradiso*.

There, almost at the very end of Dante's vision of the divine, we find what is perhaps the most erudite and synthetic formulation of eternity in the *Commedia*. In *Paradiso* 29, Beatrice explains to Dante how God set about creating the angels. She does so because she has anticipated Dante's question, seeing both question and response "where, in one point, all whens and ubis end" (là ve s'appunta ogne *ubi* e ogne *quando*) (*Paradiso* 29.12). This terminal definition of the divine as the point where all time and space collapse is significant, especially for the notion of eternity that then comes out in Beatrice's subsequent discourse of creation (29.13–18):

Not to acquire new goodness for Himself—
which cannot be—but that his splendor might,
as it shines back to Him, declare, "Subsisto,"

in His eternity outside of time,
beyond all other borders, as pleased Him,
Eternal Love opened into new loves.

Non per aver a sé di bene acquisto,
ch'esser non può, ma perché suo splendore
potesse, risplendendo, dir *"Subsisto,"*

in sua etternità di tempo fore,
fuor d'ogne altro comprender, come i piacque,
s'aperse in nuovi amor l'etterno amore.

In describing the motive and implications of the creation of the angels, Beatrice offers a succinct qualification of the nature of God's eternity as being outside of time. We see first how she draws the line between God and created beings through a poetics of reflection and love. The motive of creation cannot be addition or change to the immutable creator and so turns to the reflection of splendor and the consciousness of existence in the temporal expression of language: "Subsisto" (I subsist) is an expression bound in time that nonetheless reflects on the eternal divine.

It is in the very act of creation, the introduction of time and change, that eternity is defined as being outside all such measures. In spite of numerous mentions of eternal things and even the eternal used as a subject throughout the *Commedia*, this nominal form of the word, *etternità*, is the only such occurrence in the entire poem. It constitutes a deliberate poetic distillation of a philosophical concept. Moreover, we find here not just a temporal specification but also an implication of God's eternity as both timeless and spaceless, beyond all possible standard of measure. With Dante's use of poetic *ripresa*—the technique of repeating the end rhyme word at the beginning of the following line—this sense of being outside time (*tempo fore*) is intimately linked to being beyond every possible limit (*fuor d'ogne altro comprender*).

But it is no more intimate a link than that between the creator and created. The act of creating angels is no abstract introduction of time and newness but is based on a principle of pleasure (*come i piacque*, l. 17) and the opening of eternal love to new loves (*s'aperse in nuovi amor l'etterno amore*, l. 18). What links the eternal and the temporal, the mutable, even the mortal, is love that opens and reflects, gives birth to language, and so makes the eternal, though out of time, so very timely.

Reflection

PERPETUUM MOBILES AND ETERNITY

Marius Stan

If the nineteenth century fathered energy physics, Leibniz was its distant grandfather, as he argued relentlessly that a certain physical entity—*vis viva*, an early form of mechanical energy—is conserved in all interactions. This conservation entails that a certain kind of mechanism, sometimes called *perpetuum mobile*, is physically impossible. So it is of great interest to learn what Leibniz himself thought of it and how that relates to some broader themes in his thought, for example, the world's eternity.

Early modern physics was of two minds about the possibility of perpetuum mobiles, theoretical and practical, and Leibniz is no exception in this regard. His first pronouncements on the issue show him confident that perpetual motion is feasible. In a draft from 1671, he outlines a mechanism with movable parts of different materials—wood and lead—alternately falling and rising in water without end, due to their different specific densities.[1] During his effervescent years of discovery in Paris (1672–1676), Leibniz considers again the possibility of perpetuum mobiles.

---

1 See his manuscript "Perpetuum mobile" of June 14, 1671, in Leibniz, *Sämtliche Schriften und Briefe. Achte Reihe: Naturwissenschaftliche, medizinische und technische Schriften; Erster Band: 1668–1676*, ed. E. Knobloch (Berlin: Berlin-Brandenburgische Akademie der Wissenschaften, 2009), 554–561.

In unpublished notes, he examines two proposed designs. One, originating perhaps with an unknown inventor, he dismisses as unworkable, on account of internal friction between the parts, which would bring the mechanism to a halt. Another, apparently proposed by himself, he endorses as technically possible.[2] But it is a perpetuum mobile improperly so-called: it uses a steady supply of external energy, for example, from wind or running water, to power a cyclical mechanism.

In Paris, Leibniz meets Huygens, who introduces him to the new mechanics, including Torricelli's principle and his own discovery that in a system of i colliding particles, the total $m_i v_i^2$ remains constant throughout the interaction. These two elements will put Leibniz on the path to his great discovery, in the 1680s, that a certain type of physical efficacy—which he baptized "live force" (vis viva) and on which he erected "dynamics," a new science—is conserved in all mechanical processes. Leibniz henceforth ties his new basic principle—Conservation of *Vis Viva*—to the claim that a perpetuum mobile is impossible. However, the conceptual connection between these two ideas shifts as Leibniz's thought evolves. In his 1686 *Brevis demonstratio erroris memorabilis Cartesii*, Leibniz takes the impossibility of perpetuum mobiles as a basic axiom and uses it as a yardstick for the right concept of force. Descartes's allegedly falls short, as it allows for mechanical processes that increase motive force and hence can be used to power a perpetual motion. In the 1690s, the impossibility of perpetuum mobiles becomes a corollary of Conservation of *Vis Viva*, now a fundamental principle, itself justified by deeper, metaphysical principles, for example the equality of the full cause and the whole effect, as in Leibniz's unpublished *Dynamica de potentia et legibus naturae* (1689–1690).

---

2 See V. Kirsanov, "Leibniz in Paris," in *The Global and the Local: The History of Science and the Cultural integration of Europe*, ed. M. Kokowski (Krakow: 2008, 353–358).

In any event, from 1686 on, Leibniz is resolute that perpetuum mobiles are physically impossible. Yet in 1715, a certain Karl Elias Bessler comes forward, under the nom de plume Orffyreus, claiming to have invented a self-sustaining mechanism. Instead of dismissing him as a fraud, Leibniz encourages further study of Orffyreus's machine and declares: "he is one of my friends." Leibniz's disciple Christian Wolff does no better: though an ardent believer in Conservation of Vis Viva, which rules out perpetuum mobiles, Wolff too extols the study of Orffyreus's device: allegedly "philosophers no doubt will receive from [its study] new light, so as to know by it other hidden things." To be sure, the machine turned out to be an ingenious hoax, secretly powered from the outside by Orffyreus's acolytes from a chamber adjacent to it.[3]

Now, a closer look at Leibniz's usage of "perpetuum mobile" reveals an ambiguity—which likewise plagues much early modern discussion of the idea. First, a "physical" perpetuum mobile is an artificial setup in which the total quantity of "force" increases over time. Leibniz declares this sort of mechanism to be impossible, on account of vis viva, the exemplar of force, being conserved in the world. Second, on the other hand Leibniz in his exchange with Denis Papin mentions a "mathematical" perpetuum mobile: an idealized mechanism in which friction has been abstracted away.[4] These devices are perpetuum mobiles in the sense that, once set in motion, they would go on forever.[5] Such idealized machines are

---

[3] Bessler described and advertised his invention in Bessler (1719). Leibniz discussed it briefly, endorsing Orffyreus, in Leibniz, *Nachgelassene Schriften physikalischen, mechanischen und technischen Inhalts*, ed. E. Gerland (Leipzig Teubner, 1906), 119–121. Wolff extolled the alleged philosophical merits of Orffyreus's mechanism in Wolff, *Mathematisches Lexicon* (Leipzig, 1716), 1042–1043. A brief account of the Orffyreus episode is in F. Klemm, "Vom Perpetuum mobile zum Energieprinzip," in F. Klemm and H. Schmank, *Julius Robert Mayer zum 150 Geburtstag* (Munich: R. Oldenbourg, 1965), 15–17.

[4] See Gideon Freudenthal, "*Perpetuum Mobile*: The Leibniz-Papin Controversy," *Studies in History and Philosophy of Science* 33 (2002), 609.

[5] An elementary example is a pendulum oscillating in a frictionless medium or in a vacuum. Another, closer to the spirit of Leibniz's *Brevis demonstratio*, is a perfectly elastic ball dropped from rest on a rigid surface in a vacuum.

fully in accord with Leibniz's metaphysical dynamics, insofar as they work either by perpetually converting actual *vis viva* into latent (i.e., kinetic energy into potential) and vice versa; or by circulating *vis viva* around the system, for example, as a Newton pendulum does. The more interesting question is, are such Leibnizian perpetuum mobiles ever instantiated in the natural world? Leibniz seems to think that at least one exists: the world itself. In his exchange with Clarke, he comes to reproach Newtonians for their theory of matter. That doctrine postulates hard atoms: *rigid, inelastic* bits of matter, which are peculiar in that they "destroy motion," as it were; speed—and so Leibnizian vis viva, a function of it—is lost in rigid-body collisions. The long-term yet predictable outcome of ever more Newtonian interactions is that as matter loses motion, the world will slow down or even grind to a halt. "God had not, it seems, sufficient foresight to make it a perpetual motion," Leibniz observes reproachfully of the universe according to Newton.[6] This consequence, however, Newton was glad to accept, thinking it showed God indispensable to a world that needs periodic rewinding. Leibniz, in contrast, thought his own cosmology—in which *all* matter is endowed with a *grand principe du ressort* and animated by a constant amount of vis viva—does better justice to God's craftsmanship. The Leibnizian cosmos allegedly needs no rewinding, as the relative motion of its parts are conserved, which makes it a physical analogue of Leibniz's mathematical perpetuum mobile. Thus, the world is an eternally self-moving machine.

Still, we have reasons to question this conclusion, from facts that even Leibniz knew. To see that, consider an insightful point Papin raised in his debate with Leibniz.[7] Leibniz had remarked, predictably, that the Cartesian measure of force, which Papin

---

6 Leibniz's First Letter to Clarke, sec. 4, in *The Leibniz-Clarke Correspondence*, ed. H. G. Alexander (Manchester: 1970), 11.

7 The definitive study of the exchange between Leibniz and Papin is Freudenthal, "*Perpetuum mobile.*"

defended, entails the possibility of physical perpetuum mobiles, or mechanisms that increase force. In response, Papin accepted that force cannot be increased but objected that, sometimes, force may be *lost*; for instance, in inelastic collisions. To deflect this difficulty, Leibniz privately granted that it "is not impossible" to think that some motion "in the bodies before the collision has been transferred to an insensible matter," such that, though the motion of the colliding bodies themselves may not be conserved, "it is nevertheless conserved in the total universe by an exceptional contrivance of nature."[8] That may be so, but this move fails to solve a grave problem for Leibniz: some of this transferred motion is *lost forever*, in that it cannot be retransmitted from the insensible parts of matter back to the macroscopic bodies. Some motion, and thus some Leibnizian *vis viva*, is dissipated, or irreversibly transferred to the smallest parts. Over time, these losses accumulate, draining the motion out of the large-scale matter in the world, threatening Leibniz's system in two ways. First, bodies will grind to a halt relative to each other, just as a manmade machine does, on account of friction and drag.[9] Leibniz in fact knows that friction is the greatest impediment to the practical feasibility of self-sustaining mechanisms: as he examined a proposed perpetuum mobile in his Paris years, he noted with skepticism that friction "will considerably hinder the motion" of the machine's moving parts.[10] But more generally, friction, drag, dampened oscillations, wear, material fatigue, and other dissipative factors threaten the Leibnizian perfection of this world—they turn

---

8 Leibniz, manuscript LH XXXV, 9, 7, pp. 24r–v, cited and translated in Freudenthal, "*Perpetuum mobile*," 623.
9 Drag forces become relevant at large scales in the Leibnizian cosmos, because he operates with a vortex theory of planetary motions, unlike Newton's gravitation theory of distant attraction across empty space. Though the Leibnizian ether in which planets swirl around is very thin, it is nevertheless a resisting medium, and motion through it generates drag forces, which dissipate speed.
10 Leibniz, manuscript LH XXXVII, 5, pp. 58–59r–v. For an account of the machine's design, see Kirsanov, "Leibniz in Paris," 355.

it from a perpetuum mobile into rattling machinery, with every turn jangling imperceptibly to a halt. Second, because of that, the variety of speeds—hence the individual values of *vis viva* in single bodies—diminishes in the world, thereby reducing the *varietates formarum* (diversity of forms), which for Leibniz is a mark of this world's perfection.[11] To put it in anachronistic language, Leibniz, as he discovered friction and other types of lost motion, came face to face with entropy but chose to look away, perhaps too dismayed to contemplate what it would do to his optimal, eternally, self-moving cosmos.

---

11 Leibniz claims that the "diversity of forms" is a criterion and mark of optimality for possible worlds in his 1697 *On the Ultimate Origination of Things*; see Leibniz, *Philosophical Essays*, ed. and trans. R. Ariew and D. Garber (Indianapolis: Hackett, 1989), 149–154.

CHAPTER FOUR

Eternity in Kant and Post-Kantian European Thought

Alistair Welchman

French philosopher Alain Badiou (1937–) is on a mission to rescue the concept of eternity. In modern European thought, he argues, the idea has fallen into near-oblivion as a result of the baleful influence of a philosophy of "finitude" tracing back to Immanuel Kant (1724–1804).[1] As a result, both of the twin pillars of European thought—the historicism of G. W. F. Hegel (1770–1831) and the phenomenology of Edmund Husserl (1859–1938)—are hostile environments for the eternal: for historicism, the eternal must be historicized, that is to say, temporalized; and while many strands of phenomenology are critical of everyday "clock" temporality, the contrast they develop to everyday

---

1 The very first page of Badiou's *Logics of Worlds: Being and Event, 2*, trans. Alberto Toscano (London: Continuum, 2009) excoriates the "dogma of our finitude," 1; similarly 'Kant is the inventor of the disastrous theme of our 'finitude,'" 535.

temporality is not eternity but a deeper, lived, ecstatic temporality of which the eternal is only a distorted, theoreticized image.[2]

Perhaps this outcome is not surprising. Traditionally the eternal is a primarily theological concept, and the equation of modernity and secularization would lead one to expect a declining interest: one commentator operating under such assumptions bemoans a paradoxical "death of eternity."[3] But news of this death has been proverbially exaggerated. Although it is indeed unusual for a modern European thinker to emphasize the eternal to the extent that Badiou does, many philosophers in the nineteenth century—among them F. W. J. Schelling (1775–1854), Arthur Schopenhauer (1788–1860), Søren Kierkegaard (1813–1855), and Friedrich Nietzsche (1844–1900)—were preoccupied with the eternal. Even Kant—and Badiou is right to say that he dominates the development of European philosophy in the nineteenth century—is less hostile to the idea than Badiou suggests.

The story of eternity is not as simple as a secularization narrative implies. Instead it follows something like the trajectory of reversal in Kant's practical proof for the existence of god. In that proof, god emerges not as an object of theoretical investigation but as a postulate required by our practical engagement with the world; so, similarly, the eternal is not just secularized out of existence but becomes understood as an entailment of, and somehow imbricated in, the conditions of our practical existence.

The sections that follow discuss some of those central figures in modern European philosophy whose views prominently feature some consideration of eternity. I start with Kant in section 1. Kant's critique of speculative theology is well known, and this hostility would appear to make it unlikely that the eternal, with all its theological baggage,

---

[2] As Heidegger remarks in his 1928 *Being and Time* (New York: Harper and Row, 1962), "even the 'non-temporal' and the 'supra-temporal' are 'temporal' with regard to their being," 40.
[3] Carlos Eire, *A Very Brief History of Eternity* (Princeton, NJ: Princeton University Press, 2010), 205.

would feature prominently in his critical philosophy. But in fact Kant's transcendental idealism endorses no fewer than three different concepts of the eternal, including what turns out to be the most historically influential idea: that practical reason involves a kind of eternal, nontemporal action. Kant shifts this notion of a nontemporal act from its original theological context of god's *actus purus* to a practical context, setting the stage for Schelling's and Kierkegaard's later development of this theme. Sections 3 to 9 detail these developments. But before that, section 2 is devoted to Hegel. Hegel's radical historicism is perhaps more than anything else responsible for making "the nineteenth century preeminently the historical century."[4] Hegel is not fertile soil for the concept of the eternal, but his historicism does turn out, at a crucial moment in the philosophy of nature, to presuppose a certain conception of eternity as an eternal present. Perhaps more important for the further development of eternity in nineteenth century thought, however, is that both Schelling and Kierkegaard situate their views of the eternal in the context of a collective rejection of Hegel. Section 3 discusses Schelling, who returns to Kant's conception of nontemporal choice, seeing human capacities for free eternal self-creation as rivaling god's. Such powers are required, Schelling argues, to resist the sublimation of the individual human person into the blankness of the Absolute. Section 4 briefly considers Schopenhauer's view that the in-itself of everything is an endlessly striving will. Section 5 concerns Kierkegaard, who is strongly committed to the eternal and indeed criticizes Hegel for compromising his conception of it by thinking it temporally; but Kierkegaard is obsessed by the paradoxical question of our practical "access" to the eternal within a particular temporal moment: the decisive moment, imbued with significance, that can turn life around and create a new person, pushing Schelling's concerns even further. The remaining, shorter sections present briefer

---

4 Peter Gay, *The Naked Heart: The Bourgeois Experience Victoria to Freud* (New York: Norton, 1996), 193.

accounts of more recent figures who make important use of some conception of the eternal: Nietzsche's eternal return (section 6), the theory of sovereignty elaborated by Agamben (1942–) (section 7) and finally Badiou's unapologetic attempt to resuscitate eternity as the condition of revolutionary political change (section 8). I end with a concluding meditation (section 9).

1 Kant

The critical aspect of Kant's critical philosophy is an attack on rationalism both in metaphysics and theology. The basic contours of his case are well known. Cognition in general is, for Kant, split into two basic components: concepts and intuitions (broadly: sensory perceptions of spatiotemporal particulars), and objective cognitive experience is possible only through the synthesis of the two: "thoughts without concepts are empty, intuitions without concepts are blind."[5] This view concedes that it is possible to *think*—in some empty way—what goes beyond the possibility of sensory perception (B146).[6] But rationalist metaphysics and theology falsely conflate thinking with genuine cognition, presupposing that one can gain objective (synthetic) knowledge of the way things are using procedures of pure reasoning a priori. For human beings, actual cognition is limited by the finite nature of human sensory apparatus.

More than half of the *Critique of Pure Reason* is devoted to this attack and attempts to demonstrate that rationalist metaphysics and theology must make use of illicit "dialectical" arguments. In two important cases this attempt bears on questions concerning the

---

5 *Critique of Pure Reason, Gesammelte Schriften*, 23 vols. (Berlin: Königlich Preußichen Akademie der Wissenschaften, 1910–), A51/B76; translation from Allan Wood and Paul Guyer (eds) *Critique of Pure Reason* (Cambridge: Cambridge University Press, 1999), 193–194. All references to Kant's texts hereafter refer to the English title (where appropriate) followed by the Akademie edition volume and page number; the *Critique of Pure Reason* will be referred to by the standard A/B formula for the first (1782, vol. 4) and second (1787, vol. 3) edition pagination.
6 Wood and Guyer translation, 254.

eternal: in the Paralogisms Kant takes on the immortality of the soul; and in the First Antinomy, the question of whether the world is eternal, that is, unlimited in temporal extent. Applying the distinction between mere thought and cognition, Kant does not argue *against* the immortality of the soul or the unlimited temporal extent of the world. Rather, he argues that although the requisite entities can be coherently thought, they can never be produced in intuition, hence cognition of them is impossible. So arguments that purport to demonstrate that we can know that anything is true of such objects must be unsound.[7]

Despite this apparent rejection of the eternal, however, Kant in fact appropriates for his own purposes three traditional conceptions of the eternality: first, the notion of an eternal past as a past that has never been present; second, the notion of the sempiternal, that is, something that exists at every moment of time; and third, the notion of eternal as nontemporal.

First, a priori knowledge represents the form of an eternal past. Kant's critique of the rationalist attempt to gain synthetic a priori cognition of objects that transcend the possibility of experience is not a rejection of a priori cognition in general. A priori knowledge is also possible both of concepts (in analytic judgments) and of synthetic judgments. If metaphysical claims are those synthetic propositions that can be known a priori, then Kant by no means rejects all metaphysics. Rather, the aim of the critical philosophy is to limit synthetic knowledge a priori to a priori knowledge of the conditions of possibility of experience, what Kant terms "transcendental" conditions. As many commentators have pointed out, most insistently Heidegger, the "prior" of a priori is itself a temporal determination. But it is not an empirical temporal determination. The priority with which we know something a priori is not the priority of a past that could once have

---

[7] In particular, Kant thinks a particular form of the fallacy of equivocation causes the illicit inferences, what he calls a "subreption" (e.g. A402, Wood and Guyer translation, 442), in which a term equivocates between a transcendental and an empirical meaning.

been a present. If it were, then the knowledge would be a posteriori. What one might call "transcendental" temporality—one connected in the first instance with the temporality of our epistemic access to the transcendental—appears therefore to be that of an *eternal* past: not a past that extends eternally back through the sequence of presents but a past that is *eternally* past. This eternal past resonates clearly with both ancient philosophy (it is the past that Socratic "recollection" refers to), and it anticipates some central problematics in the nineteenth century.

Second, Kant's endeavor to identify synthetic a priori knowledge with the conditions of possibility of experience arguably commits him to a kind of sempiternity, since any such conditions represent aspects of experience that must obtain in every experience, hence at every temporal moment. This creates an internal conceptual connection between sempiternity and transcendental conditions.

Still, the connection is loose because Kant is methodologically wary about hypostatizing structural conditions of experience as existent features of that experience: this is one of the lessons of the Paralogisms. So there is general reason for being skeptical about the move from transcendental conditions, as claims that are sempiternally true by virtue of their structural role in the constitution of experience, to the actual existence of correspondingly sempiternal entities. Nevertheless, it might be possible to make out a case for this move, at least in the case of substance, and perhaps space.

On the face of it the case of space looks rather unpromising. Kant explicitly denies that "pure space and pure time" as "forms of intuition" are to be thought of as "themselves objects that are intuited" (A291/B347).[8] So space appears to be a case in which forms—qua structural conditions of experience—are misinterpreted if viewed as objects within that very experience. But actually, at least in the case of space, the denial is likely motivated not by this structural consideration but

---

8 Wood and Guyer translation, 382.

by a more phenomenological argument that experiences are intrinsically (although only in part) conceptual. On this account, intuitions of space and time are unavailable experientially for the same reason that *all* intuitions are unavailable experientially, that is, because they are theoretical constructs: experience is always experience of intuitions qua "ingredients" in a conceptually mediated structure.⁹ The unavailability of "bare," unconceptualized intuitions, however, is clearly not intended to have the consequence that empty space is completely experientially unavailable. For if it were, it would be hard to see how Kant's account of the synthetic a priori nature of Euclidean geometry—so crucial, at least pedagogically, to his case—would be possible. Rather, we reason about the structure of space in an intuition that constitutes an experience only by being conceptually mediated but that should still presumably be understood as an experience of space itself. Indeed, in the Aesthetic, Kant maintains explicitly that "space is represented as an infinite given magnitude" (B39).¹⁰

Kant is more stringent in his denial that time can be "perceived," hence in his denial that it can be an object of experience (B225; B233),¹¹ because this premise plays an important role in the First and Second Analogies (which defend, respectively, a priori knowledge of substance and of the claim that every event has a cause) as well as in the Refutation of Idealism. But in the First Analogy it is the very fact that time *cannot* be perceived that necessitates a single substance within experience that *can* be. *Something* must persist in order for change to be possible, Kant argues. It is time itself that in fact so persists, it "lasts and does not change." Hence it is just because time itself cannot be perceived, that

---

9 John McDowell presents an extended version of this kind of view in his *Mind and World* (Cambridge, MA: Harvard University Press, 1996). Henry Allison interprets the "givenness" of space in a similar way in his *Kant's Transcendental Idealism*, rev. and enl. ed. (New Haven Conn.: Yale University Press, 2004), 112–116. I borrow the word "ingredient" from Aquila's phenomenological reading of Kant *Matter in Mind: A Study of Kant's Transcendental Deduction* (Bloomington: Indiana University Press, 1989).
10 Wood and Guyer translation, 175.
11 Wood and Guyer translation, 300, 304.

we can infer the necessary existence of something within perception that "represents time in general" (B225).¹² This something is substance, which exists "at every time" (*was jederzeit ist*) (A182/B225),¹³ exactly corresponding to the traditional notion of sempiternity.

Kant's transcendental reasoning can therefore be understood as providing a new ground for understanding sempiternity. The very notion of a transcendental condition implies the existence of propositions true at every moment of time and, arguably underwrites at least some claims for existents that are present at every moment of time (*jederzeit*).

More significantly, although somewhat more controversially, Kant's thought also bends the notion of nontemporality to its own ends. The very same argument by means of which Kant subjects the dialectical arguments of rationalist metaphysics and theology to critique also appears to authorize a strong conception of the nontemporal eternity of things. Transcendental conditions are conditions of experience, that is, of the way things must appear to us in order to be experienced as any kind of objects. We may *think* things as they may be in themselves independently of experience of them, but we cannot have any definite cognition of things in themselves: thoughts without intuitions are empty. But at the outset of the *Critique of Pure Reason*, Kant makes a famous argument from the fact that time is a condition of the possibility of experience (for human beings) to the widely accepted claim that time is therefore a formal property of experience and then to the highly controversial view—which he takes to be definitive of his doctrine of transcendental idealism—that time is not a feature of things as they are in themselves (A19/B33).¹⁴ If this argument is successful it appears to

---

12   Wood and Guyer translation, 300.
13   Wood and Guyer translation, 300. Translation modified.
14   Wood and Guyer translation, 155. Kant's actual arguments are about space *and* time (with the latter often being word-for-word identical with the former). Many people (e.g. contemporary cognitive scientists) accept the claim that we have a priori representations of time and space; Kant's further inference that things aside from our representation are not spatiotemporal was attacked most famously by Adolf Trendelenburg, who complained that Kant had "neglected" the "alternative" that time (and space) might be forms of human sensibility *and* a feature of things in themselves (even if we cannot know this). See Allison, *Kant's Transcendental Idealism*, 128–132, for one of many discussions.

legitimate some quite strong knowledge about things in themselves: that things in themselves are not temporal and hence are eternal.[15]

Whether this view can be reconciled with Kant's other claim that we can have no knowledge of things in themselves is debatable. But in a way Kant argues that it does not matter, for one of the most famous and influential conceptual moves associated with Kant is that even if eternal entities cannot be objects of cognition, we may still rationally believe in them because of their practical significance: the are conditions for the possibility of practical action (*Critique of Practical Reason*, 5:133).[16] This move underwrites a shift in the understanding of the concept of eternity in modern European thought that is broadly a kind of secularization. But by no means does this change suggest that the eternal is less important for modern thought. The opposite is true: by tying the eternal specifically to *practical* concerns, it has become a more important and urgent issue.

Such a transition of the concept of the eternal from the cosmological to the human scale is particularly visible in the Third Antinomy. The manifest content of the antinomy is the vindication of the possibility of human freedom: it answers the question of whether "causality in accordance with laws of nature" is the only kind of causality from which appearances can be derived, or whether "it is also necessary to assume another causality through freedom in order to explain them" (A444/B472).[17] As with the other antinomies, its resolution depends on an appropriate sensitivity

---

15 There is much discussion in the Kant literature about whether this picture of Kant as committed to the existence of "two worlds" (one of which is populated with nontemporal objects) is really accurate and even more about whether it is defensible. For the purposes of reconstructing the history of the concept of eternity in Kant, however, I will take Kant at his word and presuppose the "two world" reading, in part because this reading was standard during the nineteenth century. Gerold Prauss's *Kant und das Problem der Dinge an sich* (Bonn: Bouvier, 1974) stands at the head of a significant strand of sympathetic contemporary reconstructions of Kant that deny the "two world" reading. Henry Allison, Kant's most prominent defender in the English-speaking world, also gives such a reading, in particular in his notion of "epistemic condition" in *Kant's Transcendental Idealism*.

16 *Critique of Practical Reason*, trans. by Mary Gregor in *Immanuel Kant: Practical Philosophy*, trans. and ed. Mary Gregor (Cambridge: Cambridge University Press, 1996), 133–271, here 246–247.

17 Wood and Guyer translation, 484.

to the distinction between transcendental and empirical application of terms. Empirically, everything that happens must do so in accordance with natural law.[18] But this is not, Kant argues, logically inconsistent with the claim that an empirical event might have "grounds" in an "intelligible cause" (A537/B565)[19] or "transcendental cause" (A546/B574)[20] as well.

What is striking about the antinomy is how little the formulations of the conflicting thesis and antithesis seem to have to do with this eventual goal. The proof of the thesis argues that to explain the occurrence of an event causally is, in accordance with the result of the Second Analogy, to postulate the existence of a prior event from which it follows according to a necessary law. If there are only natural causes, then this prior event, qua event, must itself be the necessary effect of some further natural cause. But, Kant argues, this means that the explanation of the first event cannot be complete if it refers merely to the prior event. Completeness would require an explanation of that prior event too. But since there is no end to the sequence of causes, there is no complete or, one might say, *sufficient*, explanation. But if nature is to be governed by laws, Kant argues, then "nothing happens without a cause *sufficiently* determined a priori" (A446/B474, my italics).[21] As a result, nature considered as a generalization of natural laws violates the basic principle of natural law, that everything that happens is sufficiently determined.[22] This is a contradiction. Hence there

---

18 On Kant's first *Critique* view, this follows from the claim established in the Second Analogy that every event must (as a condition of the possibility of experience) have a cause. In the *Critique of the Power of Judgment*, Kant appears to be more skeptical of the inference from (individual) necessary connection to natural laws (20:208–211, trans. Paul Guyer and Eric Matthews (Cambridge: Cambridge University Press 2000), 13–15)

19 Wood and Guyer translation, 535.

20 Wood and Guyer translation, 539.

21 Wood and Guyer translation, 484.

22 Officially Kant regards the principle of sufficient reason as unprovable because transcendent (A782–94/B810–22, Wood and Guyer translation, 665–671). He nevertheless identifies causal explanation with this principle in the Second Analogy (A200–201/B246) because it has taken on a transcendental role as condition of possibility of experience and is hence restricted in its application to appearances. The argument of the Thesis of the Third Antinomy appears to turn precisely on an interpretation of the "sufficiency" of a reason or ground (*Grund*).

must be some other form of causality, and the only other option is causality through freedom.²³ The fact that this proof refers to the absence of any "completeness" in the series of causes, along with the definition of freedom as "an *absolute* causal *spontaneity* beginning *from itself*" (A446/B474),²⁴ suggests that Kant has in mind a divine type of freedom, that is, the freedom to initiate the whole causal series that makes up the entirety of the empirical world.

The cosmological reach of this argument renders it somewhat promiscuous. Since the distinction between appearances and things as they are in themselves is quite general, it therefore appears to follow that what can be said of a (possible) divine, intelligible, and nontemporal cause must in fact also be true of everything: every appearance is the appearance of something that possesses an intelligible—although cognitively inaccessible—character.²⁵ But what Kant is concerned about is the *special* case in which, as he makes clear in the Deduction (and the Paralogisms), the distinction between the way things appear to us and the way things are in themselves is reflexively applied to our own self-knowledge: thus inner sense "presents even ourselves to consciousness only as we appear to ourselves, not as we are in ourselves, since we intuit ourselves only as we are internally *affected*" (B152–153).²⁶ It is *this* application that forms the paradigm case of the crucial distinction between empirical and intelligible character: *my* empirical character constitutes the way *I* appear to myself, and *my* intelligible character is the way *I* am in myself (A5385–41/B566–569).²⁷ This distinction in turn is the basis for Kant's claim that human beings are both subject to exceptionless causal laws (in their empirical characters) *and* yet may be metaphysically free and able (in some sense) to initiate

---

23 See Allison, *Kant's Transcendental Idealism*, 378.
24 Wood and Guyer translation, 484.
25 Schopenhauer effectively endorses this claim. See below.
26 Wood and Guyer translation, 257.
27 Wood and Guyer translation, 535–537.

novel causal sequences by means of the freedom of their wills in a way analogous to the way god may be able to create the world.[28]

For Kant, the intelligible character of each human being must be understood nontemporally. He makes the following crucial inference: "Now this acting subject, in its intelligible character, would not stand under any conditions of time, for time is only the condition of appearances but not of things in themselves. In that subject no action would *arise* or *perish*" (A539–540/B567–568).[29] So, while in one way Kant is treating a special case (human beings) of *all* things, in another way the traditional understanding of god as nontemporal is, on the contrary, generalized to human beings.

The last clause of the foregoing quotation is, however, interestingly ambiguous: "in that subject no action would *arise* or *perish*." Taken by itself, it might mean that, qua temporally unconditioned, this subject is incapable of action, since action is plausibly interpreted as something intrinsically intratemporal, that is, involving arising and passing away. But of course this interpretation is belied both by the context (in which Kant is talking precisely about free actions) and by the description of the subject as "acting" in the first clause. The alternative interpretation is therefore to be preferred: the free intelligible subject is indeed capable of actions, but ones that do not arise or pass away: *eternal* (nontemporal) actions. And in this sense, Kant's conception of intelligible character appears to transfer another traditional attribute of god, this time the biblical god rather than the so-called god of the philosophers: namely that despite being nontemporal, god can still *act*, and is—at least in this—a *living* god.

---

28 Famously, Kant is not attempting to establish much more than the bare logical consistency of these two claims. Since he regards transcendental arguments as showing the truth of his view of empirical character, he does not strictly maintain much more than the bare logical possibility of the freedom of the intelligible character. I will often omit these caveats, partly for the sake of brevity and partly because commonly almost all the post-Kantians thought humans can gain substantive knowledge of things in themselves.

29 Wood and Guyer translation, 536.

Kant's clearest attempt at reconciling intelligible freedom with the apparent causal determinism of the Second Analogy involves a carefully thought-out interpretation of this notion of a nontemporal act. Kant argues that the nontemporality condition on intelligible actions makes it impossible for any such action to have a causal antecedent: "in regard to the intelligible character... no *before* or *after* applies." The effects (not causal in a phenomenal sense) of such an action therefore emerges "from itself" (i.e., is not determined by anything else), but it does not constitute a *beginning*, since "beginning" is a temporal designation. Rather, every phenomenal act is the "immediate" effect of intelligible character. These intratemporal phenomenal acts are capable of beginning, but not absolutely, since as intratemporal, they must have precedent events to which they are, by the reasoning of the Second Analogy, causally related (A553–554/B481–282).[30] Thus the mistake of the Thesis of the Third Antinomy is to assert that there must be a faculty of "absolute causal spontaneity beginning from itself" (A446/B474).[31] Appropriate sensitivity to the transcendental/empirical distinction requires separating the absolute spontaneity of an intelligible cause that acts "from itself" from the intratemporal notion of beginning: the former is transcendental, while the latter is empirical.

Kant develops this idea of an eternal action most systematically in his late text *Religion within the Boundaries of Mere Reason* (1793), where he gives an account of what he terms "radical evil." This term equivocates between two different senses. On the one hand evil is radical if it is so widely spread among human beings as to constitute something that appears to be "woven into human nature" (6:30)[32] On the other hand evil may be radical in the sense that it involves a *positive choice* of evil, rather than resulting from ignorance or being overcome by one's

---

30 Wood and Guyer translation, 543.
31 Wood and Guyer translation, 484.
32 *Religion within the Boundaries of Mere Reason and Other Writings*, trans. and ed. by Allen Wood and George Di Giovanni (Cambridge, Cambridge University Press, 1998), 54.

passions or something similar. Understood in this scond way, the text takes on the form of a defense against the objection that Kant's moral theory actually excludes the possibility of evil as a positive choice. Why might one make this objection? In the famous first section of the 1785 *Groundwork for the Metaphysics of Morals* (4:393–406)[33] Kant presents a distinction between actions motivated by consciousness of the moral law and actions motivated by pathological incentives: interests and inclinations. Only actions motivated by the moral law are properly morally worthy. But if my action is pathologically motivated, then it does not stem from my rational will, and it does not seem that I can be responsible for it. Thus, so the objection runs, I only turn out responsible if my act was morally worthy. But then "evil" actions are merely those in which my rational will was (unfortunately) overrun by my passions, and I would be responsible for none of them—every evil act would be able to plead *crime passionnelle* in mitigation. Thus there would be no "radical" evil: if an act is consistent with the moral law, then I am responsible for it; if not, then this in itself is evidence that I was not choosing at the time and so I am not responsible for it.[34]

In *Religion within the Boundaris of Mere Reason*, Kant tries to make it clear that we *can* choose evil: "only our own act is something that can be morally evil (that is, evil that can be imputed to us)" (6:31).[35] But this entails that it is not the mere fact that we are creatures with a sensibility—and hence capacity to be motivated by our own interests or inclinations—that lies at the bottom of radical evil. It may be that in immoral actions we act on the promptings of our sensibilities, our

---

33 *Groundwork of the Metaphysics of Morals*, trans. Mary Gregor and Jens Timmerman (Cambridge: Cambridge University Press, 1997), 7–18.

34 For instance, in the *Critique of Practical Reason*, Kant maintains that "recognition of the moral law is, however, consciousness of an activity of practical reason from objective grounds, which fails to express its effect in actions only because subjective (pathological) causes hinder it" (5:79, Mary Gregor translation, 204).

35 Wood and Di Giovanni translation, 35–36 (translation modified). Kant does not think we can choose evil *for evil's sake*; only devils can do that. Rather, radical evil turns on a subtler ordering of principles within our maxims for action: an evil act stems from a maxim that prioritizes pathological over rational incentives (6:36, Wood and Di Giovanni translation, 59).

interests and inclinations. But they cannot overcome us if the action is to be genuinely evil. Rather, we must *choose* to act on such pathological inclinations: people's power of choice "cannot be determined to action through any incentive *unless the incentive has been taken up into their maxim*" (6:23–24).[36]

There is an obvious tension between these two projects. If radical evil in the first sense really is "woven into human nature," then it is hard to see how it can be chosen. Kant argues that we have "propensities" to action, each of which constitutes "a subjective determining ground of the power of choice" and hence "*precedes all acts*" (6:31).[37] The question of whether evil is built into human nature therefore devolves onto the question whether we have a fundamental "propensity" to evil. But the tension is clear: if the propensity is "physical," then, even if it is part of human nature, we are not responsible for it, and it cannot ground any notion of *moral* evil; on the other hand if the supposed propensity originates in a free choice (of evil maxims in general), then it cannot actually be a propensity, since propensities are logically prior to acts, and a free choice must be an act. Kant grasps this dilemma and argues, following on from his discussion of the Third Antinomy, that we must distinguish "two meanings" of the word "act": an empirical sense and a transcendental sense. The latter sense is of an action that is "intelligible" and that "can be recognized only by means of reason, independently of any temporal conditioning" (6:31).[38]

The same tension is clearly evident in the idea of original sin: to the extent that it is "original," it cannot be our responsibility; to the extent

---

36 Wood and Di Giovanni translation, 49 (translation modified). Kant envisages a hierarchy of ever more general maxims culminating in an overarching maxim whose structure gives priority either to the moral law or to pathological incentives (6:21, Wood and Di Giovanni translation, 47).
37 Wood and Di Giovanni translation, 55 (translation modified).
38 Wood and Di Giovanni translation, 55 (translation modified). It is not obvious that this view is consistent with the *Critique of Pure Reason*. There an eternal action of intelligible spontaneity appeared to be a character of *every* free action (which had both an empirical and a transcendental aspect); whereas in the *Religion* text, intelligible (eternal, nontemporal) acts are distinguished from ordinary everyday empirical acts as different kinds.

that it is our responsibility, it cannot be "original." Kant's resolution of this paradox depends precisely on the nontemporal nature of the act constituting the choice of fundamental maxim: "To have one or the other [good or an evil] disposition as an inborn constitution does not here mean that it has not been acquired by the human being who harbors it, that this being is not the author of it, but rather, that it has not been acquired in time" (6:25).[39] An evil disposition is *acquired* all right—otherwise it wouldn't be *evil*, that is, something for which the holder could be held responsible; but it is not acquired *in time*. Consequently, its phenomenal manifestation can only be as of something already there, something *inborn*.[40]

This metaphor of "inborn-ness" or "innateness" of evil integrates all three conceptions of the eternal at play in Kant, while refocusing the theological concept of the eternal on the individual moral agent. The result of an innate disposition to evil is that evil is present at every moment in human life. This corresponds to the eternal as present at each moment. But the explanation for this eternal presence is that evil is innate, that is, prior to the actual experience of any individual human being, constituting, from the point of view of the individual human life, an eternal past that was never present as such to any moment of human experience. This corresponds to the understanding of the eternal as an eternal past. And Kant's explanation for the *moral* nature of this innate predisposition is that it is the result of an action taken by the human individual qua intelligible subject, hence an eternal nontemporal action. This nontemporal action corresponds to the understanding of the eternal as lacking any temporal predicates at all.

---

39 Wood and Di Giovanni translation, 50 (translation modified).
40 One might argue that we are responsible in a quite ordinary way for at least some of our innate dispositions: those that we could eliminate (or eliminate easily) if we chose to. On Kant's understanding of radical evil, however, dispositions only become morally relevant if we choose to act on them by incorporating them into our maxims. So the question is not the ancient (Aristotelian and rather empirical) question of the extent to which we can be responsible for our dispositions but, rather, the more perplexing question of how we can be said to have chosen something (our most basic maxim) that appears to be prior to any act and hence a part of our natural constitution.

Schelling and Kierkegaard both extend the basic impetus of Kant's thought, treating eternity not primarily as a way of describing the distinctive temporality of god but rather as a way of understanding certain key aspects of human existence. Despite the profound roots of both of Schelling and Kierkegaard in Christian thought, this displacement of the conception of eternity onto a notion of human freedom and self-definition marks a quite radical change. But before addressing these developments, it is important to situate them in the context of the dominant figure of philosophy in the first half of the nineteenth century, Hegel.

2 Hegel

Hegel does not talk much about eternity. Indeed the standard view is that Hegel is primarily responsible for making the nineteenth century the "historical" century, dipping everything, including philosophical thought itself, in the universal solvent of time. At the end of his *Phenomenology of Spirit* (1807), for instance, Hegel identifies philosophical conceptuality with time itself, claiming that "*time* is the *existence* of the *concept* itself."[41] While this comment is a little opaque, it certainly suggests a radical attempt to immerse philosophical thought in time, an attempt that looks highly inimical to the development of any concept of eternity at all. Such interpretations lend credence to the view that the famous developmental structures of Hegelian thought represent historicized versions of Kant's categories, plunging concepts into time and history. On this reading there would be no stable transcendental forms to nonhistorically (and hence nontemporally) condition the possibility of finite human experience so as to be in some

---

[41] "Die *Zeit* ist der *Begriff* selbst, der *da ist*." Georg Wilhelm Friedrich Hegel, *Sämtliche Werke*, Jubilee edition in 20 vols., ed. Hermann Glockner (Stuttgart: Bad Cannstatt: Friedrich Fromann Verlag, 1964), vol. 2 (*Phänomenologie des Geistes*), 612; trans. A. V. Miller as *Phenomenology of Spirit* (Oxford: Clarendon Press, 1977), sec. 801, p. 487. Translation modified. Further references to Hegel in the text will be by standard English title, section number (if appropriate), and volume and page number of the *Werke*.

sense present with every moment of it, and there would be no eternal past out of which the "a priori" forms could be recollected.

Hegel's dialectic is clearly more complicated than a simple temporalization of Kant's categories, but the most obvious direction in which this simple reading needs complicating ends up making the dialectic inconsistent with the nontemporality of the intelligible character of things. The sequences of "shapes" of consciousness that Hegel's works follow are not simply a sequence of temporalized forms of experience. Rather they constitute a sequence that develops so as to include among its forms the very notion of experience as something that divides into form and content: each shape of consciousness "posits" content appropriate for it. This form and content pair appropriate to a given shape come into conflict ("contradiction"), driving the search for a more satisfying shape. But what consciousness ultimately realizes is that in all these shapes it was itself that it was (mis)recognizing: consciousness attains absolute knowledge when it recognizes that its consciousness of an other is an alienated form of its own self-consciousness. But if this is correct, then the distinction between appearances and things in themselves, which grounds Kant's understanding of the nontemporal nature of things in themselves, cannot be supported: the very distinction between how things appear and how they might be in themselves is itself temporalized, dissolved in the universal acid of Hegel's historicization of the concept.

But perhaps this complete victory of time over eternity is not the end of the story, for the terms "historicization" and "temporalization" imply a process that starts in a nontemporal place and ends up in a temporal one. Perhaps Hegel does have some positive understanding of eternity. Indeed the slogan from the end of the *Phenomenology* quoted above suggests just this, for it could be more literally translated: "the *concept* is *time*, as it *is there* or *exists* [da ist]." Here "is" should be understood not as the "is" of logical identity but as Hegel's speculative identity. The (nontemporal) concept posits the temporal as its alienated other so as to achieve a properly mediated self-consciousness. This means Hegel assumes a nontemporal starting point. Indeed, such is the nub of Heidegger's critique of Hegel at the end of *Being and Time*, where he takes Hegel to task for claiming that "the development of history falls into time" on exactly the grounds that it thinks of spirit/mind (*Geist*) as something nontemporal, such that it can "fall into" or "happen" in time, rather than as a deeper and more "authentic" experience of time itself.[42]

Hegel does not exploit this idea of the fall of spirit/mind into time in the *History* text. But he returns to something similar in the *Philosophy of Nature*. Unlike the *Phenomenology*, which starts with, and presupposes, an elementary form of human consciousness, the *Philosophy of Nature* starts, according to Hegel, where the *Science of Logic* finished, with an internally articulated, speculatively developed concept, or what Hegel calls an "Idea," something that he sometimes identifies with god. Here the Ideas with which the *Logic* terminates posit the whole of (material) nature as their antithesis. Hegel's *Philosophy of Nature* then takes the form of a speculative recovery of material nature for the Idea: "alienated from the Idea, nature is the mere corpse of understanding. Nature is the Idea, but only in itself; this is why Schelling described it as a 'petrified intelligence' and others have even described it as 'frozen.' God, however, does not remain dead and petrified, and the stones cry out and raise themselves up to spirit" (sec. 247 Addition/9:50).[43]

---

42 Martin Heidegger, *Being and Time*, sec. 82, with the citation from Hegel's *Lectures on the Philosophy of World History* on p. 480.

43 *Philosophy of Nature*, ed. and trans. M. J. Petry, 3 vols. (London: Allen and Unwin, 1970), 1:206, translation modified. This kind of "metaphysical" reading of Hegel was standard in nineteenth-century Germany and indeed remained dominant up to the British Hegelian tradition. It is much less common now, with contemporary interpreters like Robert Pippin in his *Hegel's Idealism: Satisfactions of Self-Consciousness* (Cambridge: Cambridge University Press, 1989) emphasizing continuity with Kant and treating the dialectic as a conceptual affair, albeit one complicated by the fact that what concepts are concepts *of* is also itself taken up conceptually, so that Hegel's move beyond Kant is akin to Sellars's critique of the "myth of the given," (see Sellars's 1956 "Empiricism and the Philosophy of Mind," reprinted in *Science, Perception and Reality* [London: Routledge & Kegan Paul, 1963]). I take the metaphysical reading both because it is historically appropriate and because it gives a clear sense to Hegel's understanding of the eternal, which he elaborates in the context of a philosophy of nature that has been very resistant to interpretation in nonmetaphysical terms, although for an exception see Alison Stone, *Petrified Intelligence: Nature in Hegel's Philosophy* (Albany: State University of New York Press, 2005).

This speculative identity (and hence dialectical transition) is not quite the same as that between the concept and time, but it is closely related. In the *Nature* text, Hegel argues not just that space is the form of exteriority in general, but also the converse, that nature must be spatial in order for it to constitute the externality of the Idea to itself. Time in turn emerges as the speculative solution to the contradictions embodied by space. So one would expect that the Idea itself must be eternal in the sense of nontemporal, and indeed Hegel goes on to describe the "eternal unity of the Idea" (sec. 247 Addition/9:49).[44] But what kind of eternity does Hegel attribute to the Ideas? He answers this question by again following Kant quite closely: the early sections of the *Nature* text take their inspiration from Kant's Antinomies by raising the question of the "eternity of the world" (sec. 247 Addition/9:51, 52).[45] Concerning the eternity of the world, Hegel again follows Kant in thinking that the question is poorly posed, but he has a different aim in mind from Kant's. For Kant the question of the eternity of the world is poorly posed because the world is neither infinitely extended in time nor merely finitely extended but is endless: the totality of the series of conditions is not given (*gegeben*) as such but is set for us as a task (*uns aufgegeben*) (A497–498/B526).[46] For Hegel, the question of the eternity of the world is poorly posed because it inhibits the development of a proper conception of the eternal by identifying the eternal with the sempiternal. The world, on the view that Hegel rejects, would turn out to be eternal just if it had no temporal beginning, that is, if time were unending: "the question of the eternity of the world," Hegel writes, concerns "a representation of time, an eternity as it is called, an infinitely long period of time, such that the world has had no beginning in time." But mere unendingness of time fails to articulate a proper notion of eternity for

---

44 Petry translation, 1:205.
45 Petry translation, 1:206, 207.
46 Guyer and Wood translation, 514.

Hegel: "infinite time, if it is still represented as time and not as time sublated [aufgehoben], is still to be distinguished from eternity" (sec. 247 Addition/9:52).[47]

One would therefore expect that Hegel would endorse a nontemporal conception of eternity in which all temporal predicates are removed. And this appears to be the case: he describes eternity as "absolute timelessness" while he is in the process of delineating it clearly from any conception of duration (sec. 258 Addition/9:81–82).[48] He explicitly refuses both the notion of an eternal past and its correlate, a messianic eternal future: "eternity is not before or after time, it is not prior to the creation of the world, nor is it when the world disappears" (sec. 247 Addition/9:52).[49] To entertain such conceptions of eternity would be to "make eternity into a moment of time" (sec. 258 Remark/9:80).[50] A correct conception of eternity needs to be purified of all temporal determinations to achieve complete nontemporality. And this task is perhaps not as easy as it seems.

Hegel's interpretation is therefore similar to—but more sophisticated than—Kant's conception of the nontemporal eternal, with the crucial exception that for Kant nontemporal things are cognitively inaccessible. For Kant, nontemporality only achieves significance in relation to practical concerns, that is, in relation to freedom. For Hegel such a limitation does not appear. Hegel wants to heal the dualisms of Kant's thought precisely by using cognition to overstep the bounds of possible experience. Hegel broaches the eternal in a solidly theoretical context (the philosophy of nature), and even in his local references to Kant he chooses to position his views in relation to the first two, theoretical, antinomies rather than the third, which opens up the way to a practical significance for eternity. "Philosophy" itself, Hegel writes, "is the timeless conceptualization [Begreifen] of both time,

---

47 Petry translation, 1:207 (modified).
48 Petry translation, 1:231.
49 Petry translation, 1:207 (modified).
50 Petry translation, 1:213 (modified).

and everything else, according to its eternal determination" (sec. 247 Addition/9:52).[51]

This is important to bear in mind before bringing up a final aspect of Hegel's conception of the eternal, for this difference—between a conception of the eternal as of primarily cognitive or primarily practical significance—is crucial for understanding the way Schelling ends up departing from Hegel. The final aspect of Hegel's view (also similar to Kant's conception of the nontemporal eternal) is that the eternal is bound up with a kind of nontemporal act of creation. The persistent identification of the Ideas with god as well as the overall structure of the relation between the *Logic* and the *Philosophy of Nature* suggests that Hegel's account of the eternal at the beginning of the latter text is part of an account of the creation of the world. If it is true that Hegel is describing the creation, then the fact that the Ideas are eternal and that they create the world appears to entail an eternal act of creation. Indeed, Hegel infers directly from his account of nontemporal eternity that "the world is created, is now being created, and has always been created" (sec. 247 Addition/9:52).[52]

Two things are striking about Hegel's view. First, this nontemporal act is not directly connected with individual practical action as it is in Kant: although Hegel elsewhere devotes much attention to the problematic of freedom, he does not think of freedom in nontemporal terms, except possibly in this case of the creation of the world. But it is not obvious that this eternal act by means of which the Ideas posit nature is in fact free. These issues will be central to Schelling's account of the eternal and his eventual break with Hegel's thought. Second,

---

51 Petry translation, 1:207 (modified).

52 Petry translation, 1:207. Stephen Houlgate in "Schelling's Critique of Hegel's Science of Logic," *Review of Metaphysics* 53 (1999): 99–128, takes specific issue with this "metaphysical" reading of the transition from the *Logic* to the *Philosophy of Nature* (118). But in *Petrified Intelligence*, Alison Stone refers to this notion of a nontemporal act (66) as a way of rescuing Hegel's philosophy of nature from its metaphysical interpretation: since the creation of the world takes place nontemporally, there is no "linear" sequence from logic to nature, so that the developmental order of the dialectic does not have to be identified with the chronological order of the events of nature.

despite Hegel's rejection of the notion of an eternal past (future) on the basis that such conceptions have not achieved complete conceptual liberation from temporal determination, Hegel nevertheless thinks of the eternal as an eternal present: immediately after rejecting the eternal past and future, Hegel goes on to say the eternal "is absolute present, the now without before or after" (sec. 247 Addition/9:52).[53] Kierkegaard will take strong issue with this claim.

3 SCHELLING

Schelling's middle period works (from 1809 to about 1815) contain his most original accounts of the eternal: in particular the 1809 essay *A Philosophical Inquiry into the Nature of Human Freedom* (known as the *Freedom* essay) and the three fragments (from 1811, 1813, and 1815) of the *Weltalter* (*The Ages of the World*).[54]

In the *Freedom* essay Schelling identifies the "human essence" with Kant's intelligible cause;[55] he then goes on to describe it in the following terms, clearly echoing Kant and reinforcing the importance of a parallel between human freedom and divine creation:

> The essence of the human being is essentially *his own act*. ... In the original creation a human being is ... an undecided essence ... [and] only the being itself can decide itself. But this decision cannot occur in time; it occurs outside of all time and hence occurs together with the original creation (although as an act different from it). ... The

---

53 Petry translation, 1:207 (modified).
54 References to the *Freedom* essay will be to the pagination of vol. 7 of *Schellings Werke*, 12 vols., ed. M. Schröter (Munich: Beck, 1927–59), and then to the pagination of *Philosophical Investigations into the Essence of Human Freedom*, trans. Jeff Love and Johannes Schmidt (Albany: State University of New York Press, 2006). References to the *Ages of the World* will be to the pagination of the German text of the second draft in *Die Weltalter: Fragmente in den Urauffassungen von 1811 und 1813*, ed. M. Schröter (Munich: Biederstien, 1946), and then to the translation by Judith Norman in Žižek/Schelling, *The Abyss of Freedom/Ages of the World* (Ann Arbor: The University of Michigan Press, 1997). I use the abbreviations *Freedom* and *Ages*.
55 "Das intelligible Wesen" (383/49).

act by means of which the human being's life is determined in time does not itself belong to time but rather to eternity: nor does this act precede life temporally, rather it goes through time (unmoved by it) as an act that is by its nature eternal. (*Freedom*, 385–386/51)[56]

Schelling establishes this conclusion through an analysis of the phenomenon of agency, arguing that both metaphysical determinism *and* metaphysical indeterminism are inconsistent with our understanding of free action.[57] Indeterminism is not consistent with agency because if it were true, we would be like Buridan's ass, able to act only on the basis of randomness, not agency. But if determinism were true, we would be moved by external causes, which is also inconsistent with our agency (*Freedom*, 382–383/48–49).[58]

Schelling thinks that only one view is consistent with our understanding of our own agency: human beings determine their own essences or intelligible characters in a nontemporal act. In this case, the transcendental decision is groundless, but our individual acts flow from the choice of character made in that original decision and hence have grounds; conversely, although our individual acts are determined by the combination of character and circumstance, we are still responsible for them because our character is our choice. For Schelling, the act of character choice can only fulfill these requirements if it is nontemporal. And this puts it—us—on a par with god: our individual acts of character-formation take place "together with" the "original creation," although they are acts, "different from it". Equally, Schelling uses the same Kantian arguments in defense of a rationalized version

---

56 Translation modified.
57 Peter Van Inwagen's "How to Think about Free Will," *Journal of Ethics* 12 (2008): 327–341, makes a similar claim.
58 *Freedom*, 382-383/48-50). Indeed, in the *Ages of the World*, Schelling goes so far as to claim that there would no dimensions of time under determinism: "If, as a few supposed sages have claimed, the world were a chain of causes and effects which ran backwards and forwards to infinity, then there would in truth be neither past nor future. But this nonsensical thought should rightly have vanished along with the mechanistic system to which alone it belongs" (*Ages*, 24/120).

of "original sin" (*ursprüngliche Sünde*) and "radical evil" (*Freedom*, 388/53). Both are explained by a nontemporal choice whose intrusion into the temporal world can only be experienced in terms of chronological priority: original sin is radical because it appears as prior to our first temporal choice yet is also evil because it is something (bad) that we are responsible for.

But Schelling does depart in significant ways from Kant, ways that prompt him to give a much richer account of nontemporal eternity and its relation to the temporal. This departure goes back to Kant's *Religion with the Boundries of Mere Reason* and effectively drives a wedge between the two senses of "radical" evil: original sin and evil as a choice. Recall that this text is, in part, Kant's response to the complaint that immoral actions turn out not be actions at all, so that we can only choose the good. The hierarchy of maxims Kant envisages in the *Religion* text is intended to head this danger off: on this view, what makes my act wrong is not that I was in fact overrun by pathological incentives but that my action was governed by a maxim according to which I explicitly chose to act in accordance with such incentives. It is unclear that this solution will work, however, since, for Kant, a nontemporal decision would be made precisely by a person qua purely intelligible, and it is hard to see how this decision could issue in anything other than a rational maxim, which Kant identifies with moral action. This criticism was leveled at Kant at the time, for instance by Carl Schmid in his 1790 *Attempt at a Moral Philosophy*: Kant's theory rescues us from the frying pan of phenomenal causal determinism but only to land us in the fire of what he picturesquely termed an "intelligible fatalism."[59]

So both determinism and its denial are inconsistent with agency at the temporal and phenomenal level. Agency must therefore be understood nontemporally. But to conceive the nontemporal realm as

---

[59] C. C. E. Schmid, *Versuch einer Moralphilosophie* (Jena: Cröker, 1790), cited in Michelle Kosch, *Freedom and Reason in Kant, Schelling and Kierkegaard* (Oxford: Clarendon Press, 2006), 50–52.

intrinsically rational is equally inconsistent with agency, if freedom is understood properly as a "capacity for good and evil," as the *Freedom* essay claims (*Freedom*, 352/23). As a result, Schelling infers that the nontemporal realm cannot be strictly identified with pure rationality and complete intelligibility: it must possess an opacity, a resistance to the full light of the intellect. This conviction has several important repercussions. First and most obviously, it yields a quite different understanding of the eternal act that constitutes the character of each individual from Kant's act of rational self-determination. Second, Schelling thinks of the acts of individual (self)-creation as coeval with god's act of creation, so that his accounts of the two reflect each other. The same arguments that motivate him to deduce an original opacity in human (eternal) action therefore also compel him to suggest the same for god: god's freedom to create requires an "irreducible remainder that cannot be resolved into reason by the greatest exertion, but always remains in the depths" (*Freedom*, 360/29).[60] And, in the *Ages of the World* fragments at least, Schelling goes so far as to think of god as self-creating in the same way that human beings' nontemporal decisions are self-creating.

We can see what is distinctive about Schelling's notion of the eternal act of divine (and human self-creation) by looking at the details as Schelling presents them in the *Ages of the World* text.[61] Schelling begins his account in eternity, describing a set of forces constituting the dynamism of god's eternal nature. These forces of primal nature are in a dynamic tension that he presents as at a sort of impasse: a contractive force acts as a negative drawing-in while an expansive force acts positively, pushing things out, and their "spiritual" synthesis is itself undone by the initially contractive force in a cycle that, being eternal, can never come to an end or give rise to anything else.

---

60 Translation modified.
61 This analysis draws in part on Alistair Welchman and Judith Norman, "Creating the Past: Schelling's *Ages of the World*," *Journal of the Philosophy of History* 4 (2010): 23–43.

As Judith Norman and I have glossed it: "The forces want to be recognized as god's own nature, which is to say: collectively posited as the eternal ground of god. These different elements (primal nature and the godhead) are, strictly speaking, both aspects of god, so the longing is really god longing for himself, for his own existence."[62]

So god longs to exist—to create himself. But, as Schelling repeatedly insists toward the end of the *Ages of the World*: "how is a decision [whether or not to exist] possible here?" (*Ages*, 119, 120 [twice], 124/171, 172 [twice], 174). Any mode of decision-making is apparently going to be either necessary or arbitrary, and in neither case free. Here Schelling takes a quite standard definition of freedom as the "ability to be something along with the ability to not-be it" (*Ages*, 45/132),[63] that is, as what he describes as an "indifference" between two positions. But he gives a quite radical interpretation of this position: any decision will hamper god's freedom, because if god comes down definitively on one side or another, he will no longer possess freedom qua indifference between the two. So god cannot both decide to exist (or not exist) and retain his freedom.[64]

Therefore, in order to retain his freedom, god must decide both to exist *and* not exist. But this is clearly a contradiction.[65] Schelling adopts a big guns approach here, and rather than resolving the contradiction he appeals to the clause in the traditional formulation of the principle of noncontradiction that stipulates that a thing cannot

---

62 Welchman and Norman, "Creating the Past," 34.

63 In this passage Schelling argues that only a will can have this property of freedom, since a will is something that can or can refuse to actualize itself, whereas everything else is simply either actual or possible.

64 Sylvia Plath's novel *The Bell Jar* (London: Faber, 1963) has an image of freedom that expresses this Schellingian view: the protagonist dreams of her possible choices as ripe figs on a tree, but when she picks one of the figs, all the others die (chap. 7).

65 Schelling claims that existence (life) is contradiction but that the eternal cannot be contradictory. So there is a contradiction between the nature of the eternal and the nature of existence. But there is also a contradiction between the negation of being and eternity. Being has two forms in this text: *Seyn* (Norman trans.: "being") and *Seyende* (Norman trans.: "what-is"). If eternity cannot possess either of the two forms, then it must be either *Nichtseyende* or a contingent being. But both of these are false. So there is another contradiction here. Elsewhere in the *Ages of the World* (*Ages*, 33/125), Schelling identifies a contradiction between what-is and being themselves.

have contradictory predicates "at the same time" (*Ages*, 122/173). This is the kernel of Schelling's answer: the only way of resolving the contradiction is by spreading the prima facie contradictory predicates out *over time*. In eternity, there is no time. So the only way an eternal god can make a decision regarding his existence *and* preserve his freedom is by creating time,[66] and by projecting the division in his willing onto the dimensions of time. Specifically, the decision not to exist belongs to the past (is, in a sense, definitive of the past), and the decision to exist belongs to the present. God now (but only now) exists. The disjunction of the decision defines the dimensions of time.

However, the time that is thereby created cannot immediately be identified with empirical "clock" time. God's decisive act here is also the consummation of god's own act of self-creation (*Ages*, 122/172–173). So when god pushes his nature back out into the past, creating the dimensions of time in the process, the past into which god pushes his own nature does not preexist the act of pushing it back. This act did not take place in a present moment that then moved temporally into the past: there "was" no such present moment when the act took/takes place—the act is eternal. Nevertheless, the act is *posited* as past; god's nature is now his ground. This past is therefore a past with which no present moment has ever coincided—an eternal past. And this is just the kind of eternal past that is at issue both in the ancient Socratic doctrine of recollection and implicitly in the Kantian doctrine of the synthetic a priori.

But Schelling's account is quite distinct from these doctrines, and in it our relation to this eternal past is quite different from Socrates's. For Socrates, the eternal past is, as it were, already there. So all Meno's servant has to do, in the famous scene from the *Meno* (82b–85b), is immerse himself in it in order to recall what he has forgotten. But for

---

66 Schelling's theory is really more involved than this. In the passage cited, he actually rejects Aristotle's formulation, introducing his own notion of "potencies"; but then he goes on to argue that these potencies in fact constitute time.

Schelling we are ourselves akin to god in our need to *create* a past. At the end of the *Ages of the World*, Schelling applies the lessons learned from his account of the theology—indeed in this case theogony—of creation to the human individuals whose self-creative acts are coordinated with god's:

> The man who cannot separate himself from himself, who cannot break loose from everything that happens to him and actively oppose it—such a man has no past, or more likely he never emerges from it, but lives in it continually. It is advantageous and beneficial for a man to be conscious of having put something behind himself, as it were—that is, of having posited it as past.... Only the man with the strength to raise himself above himself is able to create a true past for himself; he alone can savor a true present just as he alone looks forward to a genuine future. (*Ages*, 23–4/120)

We see here the fulfillment of the suggestion Schelling made in the *Freedom* essay, that the human act of self-creation is a recapitulation of the divine act, including the creation of (a personal conception) of time. And there is another way Schelling draws out the consequences of the view that human beings create themselves in a way analogous to the way god does:

> When we speak about the character of a man, we have in mind his distinctiveness, the particularity of what he does and who he is, which is given to him through the expressing of his essence. Men who hesitate to be wholly one thing or another are called characterless; but men are said to have character if they reveal a determinate expressing of their whole essence. Nevertheless, it is a well-known fact that nobody can be given character, and that nobody has chosen for himself the particular character he bears. There is neither deliberation nor choice here, and yet everyone recognizes and judges character as an eternal (never-ceasing, constant) deed, and

attributes to a man both it as well as the action that follows from it. Universal moral judgment thus acknowledges that every man has a freedom in which there is neither (explicit) deliberation nor choice, a freedom which is itself fate and necessity. But most men shy away from this freedom which opens like an abyss before them, just as | they are frightened when faced with the necessity of being wholly one thing or another. They shy away from this as they shy away from everything coming from that inexpressible; and where they see a ray cast by it they turn away as if it were a flash of lightning that brings harm to everything in its way. They feel themselves crushed by this freedom, as by an appearance from an incomprehensible world, from eternity. (*Ages*, 127–128/175–176)

This passage claims that human beings not only may choose their ("intelligible") characters but also may remain "characterless," an act (or omission) apparently comparable to god's choice not to exist: they can remain in the state described in the *Freedom* essay as "undecided" but still have an empirical existence in time, but an existence without essence: "characterless," as the *Ages of the World* text has it, a paradoxical choice not to choose.

For Schelling, therefore, the eternal enters twice into human actions: to make a true beginning, to perform a free and unconditioned act, what has gone before must be pushed into an eternal past; on the other hand such an act, in connecting individuals with the eternal acts that constitute their characters, threatens to crush individuals under the weight of the eternal. But these claims are quite paradoxical, for if the act of beginning occurs outside time, then there is no sense in which my phenomenal actions are being addressed. The beginning has always already happened in eternity; I cannot now (at some point in time) make it. Similarly, there is no sense in which I can now be oppressed by the weight of eternity into failing to make an eternal choice of character, since that choice (or failure of choice) has always already been made.⁶⁷ It seems that Schelling raises a kind of existential question of eternity: how are we to orient ourselves to the nontemporal? But by locating free human action in the nontemporal realm, he makes it impossible to answer. This is just the point at which Kierkegaard intervenes.

4 SCHOPENHAUER

Before tackling Kierkegaard, however, a brief note on Schopenhauer is in order. Schopenhauer is probably the philosopher who most radically develops the idea that the in-itself of things is a nontemporal action. But he does so in a way that departs from the dominant practical orientation in Kant and Schelling, an orientation that is then taken up again in Kierkegaard. This is because Schopenhauer does not believe practical, that is, normative, philosophy to be possible at all: philosophy, he claims, "can never do more than interpret and explain what there is" (2:320/298).⁶⁸

Rather than radicalizing the (moral) depth of Kant's conception of an eternal act, Schopenhauer broadens its scope. He essentially accepts Kant's transcendental idealism in his 1819 masterwork *The World as Will and Representation*: Kant's phenomenal world is Schopenhauer's world as representation. But Schopenhauer then adds to the world as representation the claim that it is possible to gain some kind of cognitive and theoretical (as opposed to practical) access to the in-itself of things. This in-itself is what Schopenhauer terms *will*. It is a much

---

67 Schelling is more consistent in some ways than Kant here, since in the *Freedom* essay he insists that conversion (i.e. a radical case of making a new beginning) is impossible since it is merely the working-out of something that was "already in" the original act of self-creation (*Freedom*, 389/34) and emphasizes that phenomenal actions are completely determined (386/51).
68 References to Schopenhauer are to volume and page number of the *Sämtliche Werke*, ed. Arthur Hübscher, 4th ed. (Wiesbaden: Brockhaus, 1988), 7 vols., and then to *The World as Will and Representation*, trans. and ed. Judith Norman, Alistair Welchman, and Christopher Janaway (Cambridge: Cambridge University Press, 2010).

wider category than human will: Schopenhauer's will is an endless striving without goal or purpose that is manifest in every phenomenon from inanimate matter under the force of gravitational attraction to animals under the dominion of instinctual actions. Human beings are included, but what is special about them is merely that they are capable of being occasioned to action by abstract representations.

Schopenhauer's arguments are often snappier than Kant's, but they definitely have a Kantian inspiration. For instance, having identified the striving will as the in-itself, he quickly infers that it must be "endless" because, as the in-itself, the will is not subject to the transcendental form of time; this nontemporal "eternal becoming" can have no purpose, and thus it is with human life: we strive, but there is nothing ultimately to strive for (2:195–196/188–189). But Schopenhauer probably retains more of the machinery of Kant's system than he is entitled to. For one thing, he tries to take over Kant's distinction between intelligible and empirical character (2:127/131), which he uses to ground Kantian-sounding (and Schellingian-sounding) claims that ground moral responsibility not in empirical choices (for we are empirically determined and do not make any) but in a single nontemporal choice that results in our (empirically unalterable) character (2:188–189/183). But the term "intelligible" is hardly appropriate for Schopenhauer's view of the in-itself of things, which is shorn of rationality and even consciousness. And, more pressingly, he persistently identifies the combination of space and time as the conditions of individuation, calling them the *principium individuationis* (2:134/137), which evidently implies that the will as such cannot be individuated: neither unity nor multiplicity can belong to it (2:134/138). But then there is no room for the *personal* nature of responsibility: the will as such is free (from determination by the principle of sufficient reason); but in being so it is no longer my individual will.

It is not clear that Schopenhauer ever solves this difficulty, and it does seem that the way he deploys the notion of a nontemporal act, of the will, purchases breadth at the expense of moral depth. To return to that morally deeper thread, I will now consider Kierkegaard.

5 KIERKEGAARD

Kierkegaard is the first of the thinkers I am considering to take seriously the problem of how to make contact with eternity *in time*. He addresses this issue in *Philosophical Fragments* and in *The Concept of Anxiety*, where he notes the philosophical deficiencies of earlier attempts to address this problem, explicitly those of Hegel and Socrates; but Schelling, whose lectures he had heard in Berlin, remains in the background.[69]

The notion of a (free, pivotal, defining, moral) *decision* is as critical for Kierkegaard as it is for Kant and Schelling. But for Kierkegaard the decision is *not* eternal, it is a *temporal* decision that is distinguished by the fact that it makes contact with the eternal. Indeed, Kierkegaard argues that to fail to understand how the eternal is embedded in time is to fail to account for the fact that a temporal instant can be imbued with what he terms "decisive significance" as an individual moment (*Fragments*, 4:183/13), leading to a new person and a clear break from one's past.

The problem with the Socratic conception of eternity is that it fails to account for the significance of the moment. Kierkegaard's argument for this in the *Philosophical Fragments* has the form of a *modus tollens*: if one adopts a Socratic perspective on the attainment of truth, then no conception of the decisive moment can be formed. But there are decisive moments, so the Socratic perspective must be wrong (*Fragments*, 4:183/13). How is it wrong? On the Socratic view, Kierkegaard argues, using the theory of learning elaborated in the *Meno*, the "follower" (i.e., the student, Meno's servant in the dialogue)

---

[69] References will be to the volume and page of *Søren Kierkegaards Samlede Værker*, vols. 1–14, ed. A. B. Drachmann, J. L. Heiberg, and H. O. Lange (Copenhagen: Gyldendal, 1962), and then to the page numbers of the following English translations: *Johannes Climacus: Philosophical Fragments*, trans. Howard V. Hong and Edna H. Hong (Princeton, NJ: Princeton University Press, 1985); *The Concept of Anxiety*, ed. and trans. Reidar Thomte in collaboration with Albert B. Anderson (Princeton, NJ: Princeton University Press, 1980), and *Kierkegaard's Concluding Unscientific Postscript*, trans. David F. Swenson (Princeton, NJ: Princeton University Press, 1968). The titles are abbreviated to *Fragments, Anxiety,* and *Postscript*.

is shown by the teacher *to have already known* what the teacher purports to teach. It is for this reason that Socrates refers to himself as a "midwife" (*Fragments*, 4:180/10). If Socrates's theory is correct, then it will *always* be the case that the follower *already* knows, and, as already mentioned, Socrates's theory of recollection pioneers a conception of eternity as an eternal past in the sense that its contents can never have been present. As a result, Kierkegaard argues, the moment of midwifery, when the servant realizes that he already knew what Socrates was trying to teach him, is "indifferent". This moment cannot, for Kierkegaard, be significant, since the result is the realization that something had been the case all along. "Viewed Socratically," Kierkegaard writes, "any point of departure in time is *eo ipso* something accidental, a vanishing point, an occasion" (*Fragments*, 4:181/11), and "in the same moment I discover that I have known the truth from eternity without knowing it, in the same instant that moment is hidden in the eternal" (*Fragments*, 4:183/13).

As I have shown, the Socratic conception of an eternal past can be detected (although in a subjective form) in Kant's understanding of the a priori. And it is decisively modified by Schelling, for whom the eternal past is not a reservoir of content to be tapped but a positive achievement of creative willing, the condition for making a decisive beginning, both for god and for the human individual. Although in many ways Schelling is aiming to understand how a beginning—a decision—is possible, he botches the task, on Kierkegaard's analysis, by locating the decision itself in eternity, as the act of choice of "intelligible" character. As with Socrates, the "moment is hidden in the eternal." For the decision to have any significance, Kierkegaard believes, it must be temporal. In some ways, therefore, Schelling's theory is even worse than Socrates's.

If the Socratic/Schellingian notion of an eternal past "loses" the moment, Hegel's attempt to theorize eternity on the basis of the present fares no better, according to Kierkegaard. In *The Concept of Anxiety*, Kierkegaard claims that the instant as a durationless temporal point leads Hegel in particular to incorrectly identify the instant with eternity in the notion of the *eternal present* (*Anxiety*, 4:354 note/84 note). Kierkegaard argues that Hegel's durationless temporal point is just as "indifferent" as the Socratic instant, albeit for different reasons. Kierkegaard's objection to Hegel's speculative identification of the temporal moment with the eternal is an instance of a more general objection he has to Hegel's thought, that the individual or the particular cannot (as Hegel believes) be dialectically "preserved" in the absolute or universal but will rather be (undialectically) lost altogether in it. But where the indifference of Socrates's instant leads to a misconstrual of the eternal as something eternally *past*, Hegel's parallel misconstrual is of the eternal as eternally *present*.

It is tempting to see these two notions of the (Socratic) eternal past and the (Hegelian) eternal present as unstable transitional forms in the thinking of eternity: they are positioned precariously between a sempiternal understanding of eternity, which is unambiguously still a form of temporality, and a fully nontemporal conception of eternity. On this interpretation, the eternal past and eternal present are conceptions of the eternal that *partly* violate the normal conditions of the passage of time: what is eternally past has always been past and is therefore not something that was once present and has become past; similarly, the eternal present is always present and was never in the past, nor will it ever someday be in the future. Thus Kierkegaard's objection is, in part, that the modification of the eternal by temporal predicates ("past," "present") represents a failure to think the eternal properly, that is, completely nontemporally.[70] And in the absence of a proper understanding of eternity, it is not possible to see how to synthesize it with time and thereby account for the "moment" imbued with significance that represents a genuine transition or "leap" to something "new"—a conception he calls "repetition" (*Anxiety*, 4:354/85).

---

70 See Louis Dupré, "Of Time and Eternity in Kierkegaard's *Concept of Anxiety*," in *Faith and Philosophy* 1 (1984): 160–176, especially 169.

The synthesis of eternity with time requires a prior thought of eternity that is cleanly distinct from the temporal, otherwise it will be too easy to enter it from time—the moment in which eternity enters time will lack significance and be simply taken up again into eternity.

Sometimes, it is true, Kierkegaard presents his alternative conception of the eternal in futural terms. In a crucial distinction, for instance, Kierkegaard contrasts the fact that the Greek eternal past can "only be entered backwards" with his own vision: "Here the category that I maintain should be kept in mind, namely repetition, by which eternity is entered forwards."[71] This formulation is important because it brings the problem of eternity into explicit relation with one of Kierkegaard's most important technical terms, "repetition": the significant moment, which eluded both Socrates and Hegel, is the moment of repetition. But the "forwards" movement of repetition should not be understood as entailing an eternal future that can never be present, by analogy with the eternal past. Kierkegaard does not want to make an analogy between recollection and repetition but to argue that only the *substitution* of repetition for recollection can ensure a proper conception of eternity and hence a proper positioning of the temporal individual in relation to eternity.

But what is repetition if it is not a form of futurity? In *Anxiety*, Kierkegaard presents repetition (and the moment in which it occurs) as involving a decisive change in one's identity, such that there is no longer any "immanent continuity with the former existence" but a radical break, a "transcendence,"[72] so that the decisive significance

---

71 *Anxiety*, 4:359–360 and note/89–90 and note. See also *Anxiety*, 4:289 note/17 note. In her "Kierkegaard's *Repetition*: The Possibility of Motion," *British Journal for the History of Philosophy* 13 (2005): 521–541, Clare Carlisle makes the important connection to the prospect of immortal life in just these terms: "while the Greek philosopher finds the truth in an eternity that existed before his birth, the Christian looks forward to an eternal life to come after his death . . . so both recollection and repetition are movements of truth: the former moves towards a past eternity, and the latter moves towards a future eternity" (525–526).

72 As Deleuze has noted in *Difference and Repetition*, trans. Paul Patton (New York: Columbia University Press, 1994), 5–11, Kierkegaard's notion of repetition is not a "vulgar" one in which repetition is a relation between two instances of the "same" concept. Rather repetition extracts a difference from the two, giving "novelty" beyond the "generality" and "law" of the concept.

of a decision is that it *constitutes* the self, remakes the self in a new way. Schelling and Kant had both grappled with similar issues of moral reform, but neither succeeded in showing how a nontemporal act of choice of one's fundamental character could explain this phenomenon.[73]

Kierkegaard describes the subjective nature of the movement of repetition, the way it is bound up with the (re)constitution of the self, as "inwardness." In his polemic against Hegel in *Concluding Unscientific Postscript*, Kierkegaard clearly articulates the charge that the impersonal nature of Hegel's thought makes it impossible for him to understand the "subjective" side of human "existence."[74] This aspect of Kierkegaard's thought sometimes leads him to be regarded as a protoexistentialist and raises the reasonable question whether his account of the moment is essentially a contribution to the phenomenology of time. For instance, in *Anxiety* Kierkegaard argues that we should take seriously the fact that the term "moment" (*Øiblikket*) is a "figurative expression" (*Anxiety*, 4:357/87). The Danish term is cognate with the German *Augenblick*, literally the "glance of an eye" and comparable to the English expression "in the blink of an eye." Although the expression is intended to convey metaphorically the transience of the moment, its literal meaning perversely lends the moment at least minimal duration in comparison with the dimensionless punctual instant of Hegel's dialectic. This suggests that the nub of Kierkegaard's critique of the now, of the punctual present moment, is in the spirit of Husserl and Heidegger and that he wants

---

73 Kant is conceptually committed to this position since (1) we ought to reprioritize the good, and (2) ought implies can (*Religion within the Boundaries of Mere Reason*, 6:46, Wood and George Di Giovanni translation, 68). But this is hard to reconcile with the conjunction of (1) original sin (so that each of us has a propensity to evil), and (2) this propensity is explained by a nontemporal act in which each of us chooses evil. Kant has to appeal to god's grace, which creates its own problems. Schelling (see above) simply gives up and argues that reform is either impossible or is merely the temporal working-out of the nontemporal act of character choice (*Freedom* 389/53).

74 See Merold Westphal, "Kierkegaard and Hegel," in *The Cambridge Companion to Kierkegaard*, ed. Alastair Hannay and Gordon Daniel Marino (Cambridge: Cambridge University Press, 1997), 101–124.

to replace this "vulgar" notion of temporality with an ecstatically spread-out lived present.[75]

But there is a crucial difference between Kierkegaard's view and the various ways a distinction between everyday and "lived" time has been taken up phenomenologically in the twentieth century. For phenomenology the distinction between two conceptions of time is intended to *replace* the distinction between the eternal and the temporal: the nontemporal is merely an inadequate and clumsy way of grasping the finiteness of authentic temporality. But this is not at all the case for Kierkegaard: Kierkegaard's moment is the result of a *synthesis* of (everyday) time and the eternal.

For Kierkegaard the eternal is not a distorted way of expressing an authentic, lived form of temporality, as opposed to an inauthentic series of punctual nows. Rather, it is only on the basis of the integration of the eternal into a moment of inward renewal (repetition) that authentic temporality can be achieved.

Kierkegaard's distinction between the eternal and the temporal cannot be plausibly assimilated to the phenomenological distinction between authentic and inauthentic temporality because Kierkegaard persists in thinking the eternal nontemporally. On the other hand this persistence also makes it all the harder to understand how the temporal moment can integrate the eternal, and what exactly is therefore meant by repetition. Kierkegaard is of course a Christian thinker, and many of the terms discussed here are elaborated explicitly in Christian terms: the "leap" to something new in *The Concept of Anxiety* is a leap of faith; the renewal of the self is to be "born again," and the presence of the eternal in a historical moment is the incarnation. But to see the

---

75 Heidegger was himself deeply influenced by *The Concept of Anxiety* in his own *Being and Time*. Jean Nizet attempts to defend such a phenomenological reading of this part of *Anxiety*, identifying Kierkegaard's analysis closely with Heidegger's. See "La temporalité chez Soren Kierkegaard," in *Revue philosophique de Louvain* 71 (1973): 225–246, especially 237–242. Carlisle, "Kierkegaard's *Repetition*," reads Constantin Constantius's references to his "measured tread" as well as to clocks as evidence of a similar distinction between inauthentic and authentic temporality in Kierkegaard's *Repetition* (531).

role of eternity in Kierkegaard wholly through this explicitly theological lens is in part to refuse the originality of his problematic.[76]

6 Nietzsche

The intertwining of the concepts of eternity and repetition in Kierkegaard anticipates Nietzsche's doctrine of the eternal return. Nietzsche formulates the doctrine as the claim "that all things recur eternally and we ourselves along with them, and that we have already been here times eternal and all things along with us."[77] The structure of this notion of eternity is very traditional: it recalls Plato and is in effect a temporal thought of eternity as sempiternal. But Nietzsche displaces the notion of eternity from a noun to an adjective that modifies the conception of repetition or return. For Nietzsche this amounts to a revaluation of the value of eternity, which is no longer a property of transcendence but is linked to a joyful affirmation of immanence: "all joy wants eternity—, wants deep, deep eternity," Nietzsche writes.[78]

Nietzsche's conception also resembles Kierkegaard's in its focus on the moment: "if we affirm one single moment, we ... affirm not only ourselves but all existence ... and in this single moment of affirmation all eternity was called good, redeemed, justified, and affirmed."[79] But this apparent similarity serves to reinforce a profound difference from Kierkegaard. For Kierkegaard, it is not any moment that can

---

76 Michelle Kosch, in *Freedom and Reason*, chap. 5, argues that there is an independent reason for the non-Christian to attend to Kierkegaard's analyses. These analyses form in part an argument that the ability to choose evil (i.e. radical evil, in a robust sense) entails the existence of norms external to human reason. And this externality in turn implies that such norms must be revealed to us. Reason can tell us that some revelation must be necessary, but, of course, it cannot tell us what the revelation is or explain how it is itself possible. Such an interpretation, powerful as it is, does little however to motivate Kierkegaard's conception of eternity because what is important about the divine on this story is that it can ground norms, and the temporal status of the divinity is peripheral to this concern.

77 *Thus Spake Zarathustra*, trans. Adrian del Caro (Cambridge: Cambridge University Press, 2006), 178.

78 Or rather, Zarathustra. See *Thus Spake Zarathustra*, 264 (see also 184, 263).

79 *Will to Power*, trans. Walter Kaufmann (New York: Vintage, 1967), sec. 1032.

have significance but the decisive moment of conversion that splits one's personal history in two (just as the incarnation is supposed to split public history in two). By contrast, for Nietzsche, *every* moment is capable of affirmation and hence a kind of eternalization. Indeed, on one important interpretation of Nietzsche, the eternal return is a kind of test to see whether you are strong enough to will that immanence be eternalized—that the same return.[80]

There is yet a third interpretation of the eternal return, however, and this one converges again with Kierkegaard.[81] Taking his cue from Nietzsche's insistent critiques of the "same," Deleuze (1925–1995) denies that the eternal return should be thought of as the return of the same: "The eternal return is not the permanence of the same, the equilibrium state or the resting place of the identical. It is not the 'same' or the 'one' which comes back in the eternal return but return is itself the one which ought to belong to diversity and to that which differs."[82] What returns eternally is the different, precisely as different. And it is this that explains why the eternal return of the different is "the closest approximation of a world of becoming to one of being."[83] And here, just as in Kierkegaard (with whom Deleuze compares Nietzsche), it is the conjunction of the eternal and repetition (or return) that produces novelty. According to Deleuze, however, Nietzsche's superiority to Kierkegaard lies in the fact that Nietzsche thinks "the different" through more rigorously so that it is not, as with Kierkegaard, recaptured by the sameness of god.[84]

---

80 In sec. 341 of *The Gay Science*, trans. Walter Kaufmann (New York: Vintage, 1974), Nietzsche uses this formulation: "the question *in each and every thing*, 'do you desire this once more and innumerable times more?' would lie upon your actions as the greatest weight." The interpretation of the eternal return as an imperative of a structure similar to Kant's is presented in Karl K. Jaspers, *Nietzsche: An Introduction to the Understanding of His Philosophical Activity*, trans. C. F. Wallraff and F. J. Schmitz (Tucson: University of Arizona Press, 1966), 359–362.

81 This interpretation is peculiarly French, pioneered by Pierre Klossowski in his 1969 *Nietzsche and the Vicious Circle*, trans. Dan Smith (Chicago: University of Chicago Press, 1998), and Deleuze in his 1962 *Nietzsche and Philosophy*, trans. Hugh Tomlinson (London: Athlone Press, 1983).

82 *Nietzsche and Philosophy*, 46.

83 *Will to Power*, sec. 617.

84 *Difference and Repetition*, 95.

Nevertheless, there is some irony in this outcome, since Deleuze is one of those who, like the phenomenological tradition, think of eternity as a kind of false philosophical problem, inconsistent with the intellectual demands of modernity.[85] Deleuze lauds Bergson, for instance, because he "transformed philosophy by posing the question of the new in the process of self-construction rather than the question of eternity."[86] But in view of Deleuze's interpretations of Kierkegaard and Nietzsche, it appears that the (modern) question of novelty *requires* eternity.

7 Agamben

Contemporary Italian political theorist Giorgio Agamben makes a disquieting use of the central argumentative feature of the nineteenth-century account of eternity: the free act as nontemporal decision that cannot therefore be localized in empirical time. Agamben applies this idea to the fundamental political decision, the founding of the polis as "an event [that cannot be] achieved once and for all but is continually operative in the civil state in the form of the sovereign decision."[87] The decision to found a political unit can *never* be completed because it takes place outside time, like the decision that constitutes my intelligible character in Kant's analysis of radical evil. But, Agamben argues, this exteriority to time is not blankly paradoxical but involves a transcendental temporality distinct from and irreducible to empirical temporality, something more like the *intrusion* of timelessness into time.

This forms the basis of Agamben's understanding of Carl Schmitt's famous theory of sovereignty. Schmitt defines the sovereign simply as

---

85 Dan Smith "On the Becoming of Concepts," in *Essays on Deleuze* (Edinburgh: Edinburgh University Press, 2012), 122–145.
86 Gilles Deleuze, *Cinema 1: The Movement Image*, trans. Hugh Tomlinson and Barbara Habberjam (Minneapolis: University of Minnesota Press, 1986), 3.
87 *Homo Sacer: Sovereign Power and Bare Life*, trans. Daniel Heller-Roazen (Stanford: Stanford University Press, 1998), 109.

"the one who decides on the state of exception [Ausnahmezustand]."[88] Agamben argues that the act is "continually operative" in the permanent possibility of sovereign intervention in the form of a decision that determines a state of exception. The grounding of the polis appears both as an event that is always already over (never took place in the present) *and* as impossible to complete because still ongoing, so that the exercise of sovereign power is effectively required as the permanent possibility of regrounding the polis. But the act of formation of the polis is an inherently violent act, unconstrained by the law since it is the foundation *of* the law. And hence Agamben generates a deeply pessimistic reading in which the life of the state is constantly interrupted by episodes of state-driven sovereign violence whose ultimate source is the uncompletable (because nontemporal) act of foundation of the state itself.

8 BADIOU

Badiou thinks of contemporary (especially French) thought as mired in finitude because of its Kantian intellectual heritage and hence as uninterested in and unable to address the concept of eternity. Badiou wants to change that. And certainly Badiou has a muscular understanding of eternity: he identifies ontology with mathematics,[89] mathematics with set theory,[90] and hence being with sets; and sets are resolutely nontemporal.[91] This move leads to several problems, not least of which is how to account for time. Badiou has not yet developed an elaborated theory of time, but time is clearly

---

88 *Political Theology: Four Chapters on the Concept of Sovereignty*, trans. G. Schwab (Cambridge, MA: MIT Press, 1985), 5. *Ausnahme* is literally an exception, and the term *Ausnahmezustand* is often translated (including by Agamben) as "state of exception," although its corresponding technical sense in English is "state of emergency."
89 *Being and Event*, trans. Oliver Feltham (London: Continuum, 2006), 6.
90 *Being and Event*, 14.
91 *Deleuze: The Clamor of Being*, trans. Louise Burchill (Minneapolis: University of Minnesota Press, 1999), 47.

connected to the event, the second term of art in his central text *Being and Event*. Time, he writes there, "is the gap between two events."[92] For Badiou an event is a revolutionary change in a region of human affairs, paradigmatically a political change. This idea of an event raises two questions: how is any change possible, if being is eternal? How is a radical or revolutionary change possible? In answer to the first question, Badiou argues that events exploit ineradicable capacities for self-reference in natural human languages to violate or transcend (changeless, eternal) ontology: in a favorite example, the claim "this is the revolution" can, under some circumstances, partly constitute the revolution. To assure the radical nature of the change constituting an event, Badiou uses a kind of diagonalization to show that it is possible to construct an "indiscernible" set, that is, one whose membership is "unforeseeable" from the preevental situation. To constitute an event for Badiou is therefore to discern the membership of the indiscernible set. But how can you know you are doing so correctly? You cannot, at the time, even if it will seem retroactively comprehensible if you are right: you must have "faith." The result is that Badiou finds himself repeatedly drawing on Christian thinkers in order to understand this notion of "faith"—despite his conception of the event being radically secular. Over the span of his career he has given secularizing rereadings of Pascal, Saint Paul, and most recently Kierkegaard.[93] Perhaps Badiou is in a position to solve the problem that faced Kierkegaard, of giving content to the notion of the eternal that can appear in a decisive (for Badiou "eventual") moment. Badiou does this in a technical tour de force by showing how it is possible to guarantee the existence of a (novel) set that cannot even in principle be discerned by any available linguistic predicate.

---

92 *Being and Event*, 210.
93 See *Being and Event*, Meditation 21 (for Pascal), *St. Paul: The Foundation of Universalism*, trans. Ray Brassier (Stanford: Stanford University Press, 2003), and *Logics of Worlds*, 425–435 (for Kierkegaard).

9 Conclusion

On the face of it, the nineteenth century was a pretty hostile environment for the flourishing of any notion of eternity. The rise and spread of secularism challenged the often-theological background of the eternal; and the overwhelming importance of history to this "historical century" par excellence seems like a straight-out denial of the importance of the eternal.

And yet the eternal preoccupies a counter-canonical strand of European thought from Kant to the present day. Transcendental idealism opens up a renewed space both for the sempiternal and for the idea of an eternal past, and does so in several ways. To the extent that they leave traces within experience itself, transcendental conditions for the possibility of experience are necessarily present at every moment of that experience, and are hence sempiternal. And, when treated philosophically, as explanations for our synthetic a priori knowledge, the conditions of experience, rather than as potential components of experience, these same transcendental conditions represent a kind of eternal past, one that cannot have been itself present because it is always "prior to" every present experience.

These conceptions of eternity are important (and the eternal past returns in both Schelling and Kierkegaard). Still, what really preoccupied this counter-tradition was another kind of eternity: that human beings can perform, or are even constituted by, eternal nontemporal actions. In a way this idea is simply a specialization of the basic doctrine of transcendental idealism: that time cannot be predicated of things as they are in themselves. But it is not as a speculative, theoretical possibility that nontemporal actions interested either Kant or his followers. Rather it is overwhelmingly a *practical matter*.

For Kant this notion of a nontemporal act is first of all a way of safeguarding the freedom of human action: although our acts are determined at the phenomenal level, it is possible that those same acts are still free at the noumenal level, where temporal predicates cannot be applied. Thus we are free if and only if we are capable of performing nontemporal actions. We cannot know whether we are capable or not; but it is a condition of practical action that we assume it. In his later meditations on evil this practical postulate of nontemporal actions expands in scope, no longer applying only to individual actions but to a basic grounding action that nontemporally chooses the basic human personality.

Post-Kantian classical German idealists are often represented as going back on Kant's speculative modesty and creating systems predicated on what Kant constitutively denies: that we can have cognitive access to things as they are in themselves. Of the figures investigated here, this is most clearly true of Hegel (although he himself would claim to have dialectically sublimated the phenomenon/noumenon distinction) and Schopenhauer. Despite Hegel's commitment to a kind of universal temporalization or historicization, he nevertheless finds himself needing a conception of the eternal present as what is subject to such temporalization, what "falls into" time. And Schopenhauer, while extending the idea of a nontemporal act to cover the in-itself of everything, deprives himself of a normative outlook both by emphasizing the theoretical vocation of philosophy and by effectively denying the ultimate importance of individuals at all. Where the Hegelian individual is sublimated into the world spirit, the Schopenhauerian individual is dissolved into the universal will—or into nothingness.

But this speculative interest in the eternal is not what motivates Schelling or Kierkegaard. Schelling returns to the practical register and to Kant, arguing that the "essence" of a human being lies in a nontemporal act (*Freedom*, 385/51). However, he moves decisively beyond Kant first of all in driving a wedge between the nontemporal in-itself of things and their *intelligibility*. For Kant, at least to a first approximation, the in-itself of things just is their intelligible, noumenal aspect. But for Schelling, being determined by reasons is as much a denial of our eternal agency as being determined by causes. There is therefore an "irreducible remainder" of unintelligible opacity within the eternal itself (*Freedom*, 360/29). The second way Schelling moves beyond Kant is in explicitly associating eternal human self-creation with god's self-creation. The conjunction of these two views produces the characteristic Schellingian notion of a self-opaque god. Last, and building on the persistent analogy between human and divine freedom, Schelling also deduces the creation of time itself from god's eternal free choice. The problem that faces god, as Schelling sees it, is that of maintaining his freedom by choosing both to exist and not to exist. This is achieved by creating time itself in the nontemporal choice and sloughing nonexistence off into the past. But it must be a special kind of past: one that has never been present, and hence an eternal past (for at the moment it had been present, then god would have failed to exist). Human beings face a similar dilemma. But here Schelling interprets the failure to choose more existentially as a response to the crushing weight of an eternal decision that results in an absence of character.

Kierkegaard prolongs the existential tendency of Schelling's thought. For Kierkegaard the main problem of the eternal is how to make contact with it in our temporal existence: to do so there must be a particular moment of decision that has a decisive, life-changing significance that reflects the successful introjection of the eternal. Kierkegaard criticizes both Schelling's conception of an eternal past and Hegel's conception of an eternal present as failing to make it possible to distinguish such a decisive moment: to think the eternal through any temporal designation (past or present) is not to think the eternal in its nontemporal purity. Sometimes Kierkegaard seems to contemplate an eternal future (as in his technical concept of repetition), but this seems as much of a temporal designation as the past and present, and in the end it is not clear that Kierkegaard really can provide a satisfying understanding of the eternity that shapes a decisive moment.

After Kierkegaard the eternal does not play anything like the central role that it had. But it is still active, and where it is active it is still manifestly intertwined with practical concerns rather than being an object of merely theoretical interest. Nietzsche's eternal return, for instance, is often interpreted as a kind of moral test. More recent interpretations of Nietzsche emphasize instead the role of the eternal return in underwriting the possibility of novelty. But even these views come obliquely back to the relation of eternity to freedom, since to be free (in this Kantian tradition) is be able to initiate a new, that is, novel, causal series. Agamben, by contrast, sees an eternal nontemporal act at the heart of the violent creation of the polis and interprets the fact that a nontemporal act can never be over as the basis for the perpetual possibility of a resurgence of polis-founding violence in "states of emergency." Last, Badiou explicitly revives a number of Kirkegaardian themes, but like Agamben in a political rather than individually moral context, and claims to give a mathematical explanation for the intrusion of the eternal into the decisive, politically revolutionary moment of action. Whether this proof convinces only time will tell. But it is certainly clear that the eternal, in the form of a nontemporal act, has by no means been secularized or historicized into irrelevance but continues to exert pressure on our practical self-understandings.

Reflection

ETERNITY IN EARLY GERMAN ROMANTICISM

Judith Norman

The horrific aspect of eternity is eternal damnation; its hopeful aspect is eternal salvation. In the Faust legend, Mephistopheles is the figure of the first. But in Goethe's 1832 retelling, the figure of the second is something he refers to as the eternal feminine. The combination of eternity and femininity is striking and unexpected, an effect that is put into relief by its (first) occurrence in the final lines of the play: "the eternal feminine leads us upwards."[1] This suggests that the feminine plays the (familiar, theological) function of mediator into the eternal—although the mediation is now somehow eternal as well.

Women have sometimes been given leadership roles in the ascent to heaven—Mary, Beatrice, Gretchen; Novalis also refers to the eternal virgin. This can be seen as romantic in the archaic, chivalric, sense that alludes to the inspiring role of the ideal woman. But the romantic movement that first formed in Jena in the 1790s developed a different conception of the relation between women, mediation, and eternity, and it is this that I would like to briefly explore, using the notion of an eternal feminine as a point of departure.

---

1 Goethe, *Faust: Der Tragödie* zweiter Teil, lines 12110–12111. Project Gutenberg. Accessed on November 23, 2015. Translation is my own.

The Jena romantics were feminists by the standards of the time, with woman participating in the intellectual life of the group. But their writings also contain another traditional rhetorical use of the feminine namely an identification of women and wisdom and a playful analogy between philosophy and erotic pursuit—a heterosexualization of the themes of Plato's *Symposium*. The feminine is presented as the object of philosophical inquiry in Novalis's *Novices at Sais,* as well as in the first of Friedrich Schlegel's fragmentary *Ideas*, where the search for truth is portrayed as an attempt to unveil the goddess. (Novalis's real-life fiancée was coincidentally named Sophie.) In Schlegel's novel, *Lucinde*, the feminine is presented as the philosophical mentor, a Diotima figure who initiates the protagonist into the mysteries of love and Fichte.

But the romantics give a distinctive twist to these familiar themes. Diotima's attempt to initiate Socrates into the mysteries of love was a success (Socrates ultimately comes to know the true form), whereas Lucinde by contrast was a Fichtean, not a Platonist, and within the terms of Fichte's idealism, the highest truth is unattainable—it is the unattainable goal of an eternal striving. Similarly, it is unclear whether Novalis's novices manage to unveil the goddess.

This is in keeping with the traditional understanding of romanticism as centrally concerned with love, longing, and eternal striving. In perhaps the most famous statement of romantic self-description,[2] Schlegel describes romantic poetry as a "progressive, universal poetry" and writes that "its essential nature [is] that it is eternally becoming and can never be perfected." Moreover,

---

2 *Athenaeum Fragment* 116, in Schlegel, *Kritische Friedrich Schlegel Ausgabe*, ed. E. Behler, J.-J. Anstett, and H. Eichner (Paderborn: Schöningh Verlag, 1958–), vol. 2, 1967. Passages from Schlegel's *Athenaeum Fragments* and *Ideas* quoted in the text are from *Friedrich Schlegel's "Lucinde" and the Fragments*, trans. and ed. P. Firchow (Minneapolis: University of Minnesota Press, 1970).

he assigns romantic poetry the task of "leading us upwards"—raising "[poetic] reflection again and again to a higher power," asymptotically approaching its highest goal. Eternal striving constitutes the progressivity of romantic poetry. Its universality can be understood in terms of the goal that poetry is (eternally) striving toward: it is supposed to "reunite all the separate species of poetry and put poetry in touch with philosophy and rhetoric" and on and on, combining, uniting, organizing, synthesizing all of literature into one infinite, uncompletable system. And it is this task of romantic poetry that Schlegel compares to the role of love in life, feminizing it (without involving any actual women), since, as he writes, "[women's] very essence is poetry."[3]

If any author can be identified with this romantic project, it is Goethe. In his *Dialogue on Poetry* (a romantic refiguring of the *Symposium*) Schlegel writes of Goethe that he "has worked himself up ... to a height of art that for the first time encompasses the entire poetry of the ancients and the moderns and contains the seed of eternal progression."[4] Goethe might have thought that the eternal feminine leads us upward, but Schlegel is suggesting that it is Goethe himself who performs his task. The essence of women might be poetry, but men are the poets, the source of poetic production, and, apparently, fertility (Goethe's "seeds").[5] The themes of the *Symposium* no longer look quite so heterosexualized.

Returning to the Schlegel's definition of romantic poetry, we can see that it is universal precisely *because* it is progressive, that is, striving (upward) but, like a mathematical limit, eternally approaching without ever achieving its goal. Romantic poetry

---

[3] *Ideas* 127 in Schlegel, *Kritische Friedrich Schlegel Ausgabe*, vol. 2; *Friedrich Schlegel's "Lucinde" and the Fragments*.
[4] From *Gespräch über die Poesie*, in *Kritische Friedrich Schlegel Ausgabe*, vol. 2, 346–347. Translation my own.
[5] Fertility imagery is particularly apparent in Novalis, whose (assumed pen) name means "freshly plowed field" and whose collection of essays is entitled *Blüthenstaub* [Pollen].

universalizes all poetic genres, and then all literary genres, and then integrates philosophy. But elsewhere Schlegel extends the ambition of romanticism in the direction of religion as well, writing that "in a perfect literature all books should be only a single book, and in such an eternally developing book the gospel of humanity and culture will be revealed."[6] Schlegel calls this "eternally developing book" a bible and identifies it with the "new, eternal gospel that Lessing prophesied." There is a fidelity not just to the German cultural tradition in the reference to the Enlightenment author Gotthold Ephraim Lessing, but also to the German Lutheran tradition, with the notion of the book, the Bible, as mediator in the ascent to heaven, our eternal reward.

But Schlegel's conception breaks with both of these traditions. Schlegel's bible (i.e. literature) is eternal only in the sense that it is eternally developing, eternally becoming, it "can never be perfected." The revelation then will be of a peculiar sort, a revelation of a formative process; romantic poetry is poiesis in the traditional sense of production. But while this might distinguish it from Luther's Bible, it does not necessarily distinguish it from Lessing, who famously wrote: "If God were to hold all truth locked in his right hand, and in his left only the ever-rousing [immer regen] drive for Truth, albeit with the proviso that I would always and eternally [immer und ewig] err, and say to me 'choose!,' I would humbly fall on his left hand, and say: Father, give this to me! The pure truth is yours alone."[7] For Lessing, too, apparently, it is the path and not just the goal that is eternal. I had been arguing that this is a conception distinctive to German romanticism—but now we find it in one of the defining texts of the German

---

6 *Ideas* 95 in Schlegel, *Kritische Friedrich Schlegel Ausgabe*, vol. 2; Friedrich Schlegel's "Lucinde" and the Fragments.

7 *Anti-Goeze: Eine Duplik* (1778), in *Werke* ed. H. Göpfert (München: Hanser Verlag, 1979), vol. 8, 32–33.

Enlightenment. Is there anything distinctive about the romantic account?

I think there is. For one thing, we might note that Schlegel equates the eternal process not with epistemic humility but with sexual play. For another, Lessing takes God's left hand as an acknowledgment of and testament to our finitude and limitations. For Lessing, it is our erring and imperfections that are eternal; we are destined never to know the truth. The romantics, by contrast, do not suggest that there is anything deficient in the eternal becoming of the romantic poem; there is nothing it is failing to be, no Absolute perspective to make our own finite perspective seem small;[8] they would select God's left hand for the simple reason that God does not have a right hand; eternity is the name for the path, now not in addition to but instead of the goal.

There are several characteristic gestures of romanticism: deferral, incomprehensibility, chaos, irony, fragmentation: Schlegel joked that "almost" should be included in Kant's list of categories. All of these point not just to an incompletion of the projects of meaning-making or truth-finding,[9] but a valorization of the incompletion as such, an invitation to an eternal game.

---

[8] Manfred Frank gives a good explanation of this point, showing that for the romantics, the Absolute is a fictional construct. *The Philosophical Foundations of Early German Romanticism*, trans. Elizabeth Millán-Zaibert (Albany: State University of New York Press, 2004), 51, 174.

[9] I explore this theme in my article "The Work of Art in German Romanticism," in *Internationales Jahrbuch des Deutschen Idealismus* 6, ed. Karl Ameriks and Jürgen Stolzenberg (2009).

Reflection

ETERNITY IN HASIDISM: TIME AND PRESENCE

Ariel Evan Mayse

Hasidic teachings often refer to God as existing beyond time and outside all human notions of temporality. These homilies frequently interpret the sacred name Y-H-V-H as alluding to God's role as *mehaveh kol ha-havvayot*—the infinite One that brings all existence into being. In one formulation representative of many others, we read: "the Creator was [hayah], is [hoveh], and will be [yihiyeh] all at once. He is not within time, for He is beyond temporality. Time is a created thing, since the past, present and future are all the same [for the Divine]."[1] This understanding of God's eternity is not only declaration of divine transcendence. It also reflects Hasidic notions of panentheism, for Hasidic sermons claim that God is not the ultimate being but is the totality of Being itself. All aspects of time and existence are included within God, though an aspect of the Divine surely transcends the physical and temporal realms.

The name Y-H-V-H is also taken as a reference to God's eternal recreation of the world. R. Levi Isaac of Berdichev (c. 1740–1809), an early Hasidic master, declares the following: "The blessed

---

1 Menahem Nahum Twersky, *Yesamah Lev* (Benei Berak, Israel, 1997), *yoma*, 567–568. All translations, unless otherwise noted, are my own.

Creator made everything and is everything. In each moment, without ever ceasing, God bestows blessing upon His creatures and upon all the worlds above and below, onto the angels and onto all living beings. It is for this reason that we say in our morning prayers, 'Who *forms* light and creates darkness' and not 'Who *formed* light and created darkness.'[2] We use the present tense, because God is constantly forming, revitalizing all of life, moment to moment; all is from the blessed Holy One, who is perfect and all-inclusive."[3] God continuously infuses the cosmos with sacred vitality, without which the universe would instantaneously collapse. This sacred energy represents God's immutable and immanent presence in the physical world. The cosmos is a temporal embodiment of the infinite and eternal Divine, and it is constantly being revitalized with God's life-force.

Many Hasidic homilies apply similar notions of eternity to the Torah, claiming that the inner nature of Scripture exists beyond time. These sermons build on rabbinic traditions about the Torah predating Creation and combine them with mystical associations of Scripture with the *sefirot*, aspects of the Godhead that are themselves beyond space and time, though manifested through both.[4] R. Dov Baer, the famed Maggid (Preacher) of Mezritch (1704–1772), suggests that Scripture is invested in a textual "garment" of stories and laws but that its essence is atemporal and

---

2 See Isaiah 45:7.

3 *Kedushat Levi*, ed. M. Derbarmadiger (Monsey, NY, 1995), vol. 1, *bereshit*, 1. Based on the translation in Arthur Green, *Speaking Torah: Spiritual Teachings from Around the Maggid's Table*, with Ebn Leader, Ariel Evan Mayse, and Or. N. Rose (Woodstock, VT: Jewish Lights, 2013), vol. 1, 80.

4 See Gershom Scholem, "The Meaning of the Torah in Jewish Mysticism," in *On the Kabbalah and Its Symbolism*, trans. Ralph Manheim (New York: Schocken Books, 1996), 32–86; Moshe Idel, *Absorbing Perspectives: Kabbalah and Interpretation* (New Haven, CT: Yale University Press, 2002), esp. 26–136, and "Infinities of Torah in Kabbalah," in *Midrash and Literature*, ed. G. H. Hartman and S. Budick (New Haven, CT: Yale University Press, 1986), 141–157; and Elliot R. Wolfson, *Language, Eros, Being: Kabbalistic Hermeneutics and Poetic Imagination* (New York: Fordham University Press, 2005), esp. 190–260, 513–545.

infinite: "It is known that the Torah is pre-eternal [kedumah]; it is not a part of time. But how can a story [like the biblical narratives] be related to something that predates time [kadum le-zeman]? In truth this story is happening eternally [ha-hoveh tamid]. This [principle] also applies to a human being, who is called a microcosm ['olam katan]—this same story is taking place within him."[5] The Torah is not simply a historical record. Its stories are a textual embodiment of sacred events that are happening at all times; these take place within the world at large, within the individual, and even within the Godhead.

R. Menahem Nahum of Chernobyl (c. 1730–1797) suggests that the timeless Scripture enters language anew in every moment: "It is known that the Torah is made up of the blessed Holy One's names.[6] The Divine is the past, the present, and the future, the living and eternal One for all time. The Torah is also such. So why does Scripture recount what happened when the Tabernacle was made? How does it teach us the path [moreh derekh] today? Scripture is called *Torah* because it shows us the path upon which we should walk. Certainly in each and every moment the Torah is embodied [mitlabeshet] according to the need of that particular time."[7] This conception of Scripture is crucial to how the Hasidic masters interpret the Torah. All biblical tales, from patriarchs' deeds to the construction of the Tabernacle, are always relevant because the Torah is nothing less than divine eternity manifest within words and letters. Novel understandings of Scripture emerge in every generation because the infinite Torah is always taking on new forms that are appropriate for each time and place. We might say

---

5 *Likkutim Yekarim*, ed. A. Kahn (Jerusalem, 1974), no. 285, 105b. See also *Or Torah* (Brooklyn: Kehot, 2011), no. 245, *tehillim*, 298, in which the Maggid claims that "the blessed Holy One is preexistent [*kadmon*], and the Torah is preexistent [*mukdemet*]."

6 See Nahmanides's introduction to his commentary on the Torah.

7 *Me'or Einayyim* (Benei Berak, Israel, 1997), *pekkudei*, 213–214.

that the giving—and receiving—of Torah is an eternal process, for it continuously unfolds in infinitely new ways.[8]

The Hasidic masters do not restrict eternality to God or Scripture, and some explore the extent to which human beings may grasp atemporality as well. The Maggid describes worship as a particularly fecund opportunity for penetrating beyond time: "There is no passage of time above. The divine wellspring gushes forth in each instant. The flow is constant, and its nature is to do good and give blessing to God's creatures. But if you pray or study in this way [with love and awe], you may become a channel for that spring, bringing its blessing and goodness to the entire world."[9] Through prayer one may glimpse the realm of God's eternity, and the worshiper becomes a bridge spanning divine infinity and human temporality. The resulting flow of bountiful energy from one region to the other, says the Maggid, is the definition of an answered prayer. Elsewhere he refers to contemplation as a way of transcending time and detaching oneself from physical desires, thereby achieving unity with God (*ahdut 'imo yitbarah*).[10] Stepping into God's eternity can even empower the worshiper to perform miracles and effect change in the physical world.[11] In a daring teaching from his student R. Menahem Nahum, we read: "It is known that God is beyond time.

---

[8] R. Judah Aryeh Leib of Ger (d. 1905) describes this phenomenon as the "eternal life" (*hayye 'olam*) that God has imbued within the midst of his people; see Arthur Green, *The Language of Truth: Teachings from the Sefat Emet by Rabbi Judah Leib Alter of Ger* (Philadelphia: Jewish Publication Society, 1998), 71, 157, 202, 272, 291, 404. See also Gordon Tucker, "Taking in the Torah of the Timeless Present," in *Jewish Mysticism and the Spiritual Life: Classical Texts, Contemporary Reflections*, ed. Lawrence Fine, Eitan P. Fishbane, and Or. N. Rose (Woodstock, VT: Jewish Lights, 2011), 67–71; and Gershom Scholem, "Revelation and Tradition as Religious Categories in Judaism," in *The Messianic Idea in Judaism and Other Essays on Jewish Spirituality* (New York: Schocken Books, 1971), 282–303.

[9] *Or Torah* (Brooklyn: Kehot, 2011), no. 105, *ki tissa*, p. 146; based on our translation in Green, *Speaking Torah*, vol. 1, 231.

[10] See *Likkutim Yekarim*, no. 285 fol. 109a. See also Twersky, *Yisamah Lev, berakhot*, 507–508, where R. Menahem Nahum refers to rising above the seven days of the week, an embodiment of time also associated with the seven lower *sefirot*. One who accomplishes this can unite with the infinite Divine that both dwells within these days and yet transcends them all.

[11] See *Maggid Devarav le-Ya'akov*, ed. Rivka Schatz-Uffenheimer (Jerusalem: Magnes Press, 1976), no. 69, 116–117.

For Y-H-V-H, past [hayah], present [hoveh], and future [yihiyeh] are equal. God is, was, and will be in a single moment, since our blessed Creator is endless and has neither beginning nor end. Therefore whoever attains the brilliant light, which is the category of 'one,' becoming one with God, is also above time. Such a one is also able to see from one end of the world to the other, including past, future, and present."[12] One who journeys into the innermost recesses of the mind can cleave to the eternal unity that undergirds the temporal cosmos. Some Hasidic teachings situate this contemplative process in sacred rituals like prayer, study, or the other commandments,[13] but many sermons extend this paradigm to seemingly mundane actions performed with the correct mindfulness. This means that all human deeds are potential access points for attaining God's eternity.[14]

Some Hasidic homilies offer a more cautious perspective. In another sermon the Maggid declares:

> The prophet Elijah lives forever.[15] Although he is composed of the four elements that are part of time, these four elements are rooted in [divine] unity [ahdut]. When this unity is drawn into them, they grow and rise above time, arriving at a [realm of] oneness in which one lives forever.
>
> This is the explanation of [the sages' teaching]: "A single moment of pleasure [korat ruah] in the World to Come is greater than all of one's life in this world."[16] When one wishes to cross an expanse [that would take] five hundred years, he must walk for a very long time. But in a dream [or in his contemplative mind] the distance may be covered a single moment, since he is above temporality. The pleasure [ta'anug] he experiences in that moment above time

---

12 *Me'or Einayyim, shemot*, 147.
13 See *Kedushat Levi, eikhah*, 373.
14 See *Likkutei Moharan* 1:272.
15 See *Va-Yikra Rabbah* 27:4.
16 Quoted from m. Avot 4:17.

cannot be withstood in this world, which is within time. [Were he to draw the pleasure into this world], the existence of all time would be nullified [batel mi-metziut]. Time is a creation and cannot endure something from the eternal realm.[17]

The Maggid asserts that all physical, time-bound elements of the cosmos are grounded in God's unity. But he then reverses this thrust, claiming that the divine eternity in which the earthly realm is rooted cannot be directly revealed in the physical world. God's energy *must* be refracted through time and space, for otherwise it would overwhelm the cosmos and cause everything to sink back into the unlimited Divine.

We should note, however, that many Hasidic teachings emphasize God's manifestation in the temporal realm. They describe each of the various sacred days of the Jewish calendar as different expressions of the eternal Divine. In the introduction to his collection of homilies about the holiday cycle, R. Tsevi Elimelekh Shapira of Dynov (1785–1841) claims the following: "[Israel] has been sanctified with the [holy] times.[18] These are sacred vessels for receiving their [i.e., Israel's] holiness. Therefore, their souls truly know the beneficent changing of the times, and with affection, sweetness, and pleasantness, their souls sense the flow of holy vitality [shefa] according to the beloved and delightful [sacred] times. These are the wondrous gifts that the blessed Holy One has given unto Israel."[19] God constantly imparts to the world divine vitality, but this is a dynamic process. The sacred energy projected into the world manifests differently

---

17 *Likkutim Yekarim*, no. 290, fol. 108b. See R. Shne'ur Zalman of Liady's treatment of this issue in *Tefillot mi-Kol ha-Shanah* (Brooklyn: Kehot, 2008), 75b–76a. He links it with the notion of rising above time to the devotional practice of reciting the *Shema'*, described in Zohar 1:18b as an act of unification.

18 Quoted from b. Berakhot 49a.

19 *Benei Yissakhar* (Benei Berak, Israel 2005), vol. 1, *hakdamat ha-mehaber* (unpaginated).

on each of the holidays, and Israel has been bestowed with a heightened sensitivity to experience these fluctuations. To some extent they even control the passage of time, for the Jewish sages declare the holidays and appoint the sacred times.[20]

Many Hasidic texts actually prefer the image of God as manifest within the boundaries of time. These sermons describe God as having willingly limited Himself through temporality, compelled by His abundant love for Israel and the great pleasure He will receive from Israel's deeds. Their deeds are performed long after the cosmos was formed, but the eternal God delights in Israel's worship at the very moment of their inception. This explanation for Creation holds a subtle echo of the familiar Hasidic trope that God needs human action. The Divine restricted His eternity by accepting the temporal realm, and in return Israel's deeds sustain God.

Let us conclude with a homily emphasizing God's manifestation through temporality. The Maggid of Mezritch claimed that a worshiper transcends time through contemplative prayer. He concedes that God reaches toward humanity through a temporal matrix when we cannot rise above time, but the Maggid clearly prefers meeting God in the eternal. R. Levi Isaac, however, presents it differently:

> "And God looked at the children of Israel, and God knew" (Ex. 2:23). This may be interpreted in keeping with, "Y-H-V-H, Your works are in the midst of years. . . . In the midst of years, know them. In anger, remember Your compassion." (Hab. 3:2). Let us explain with a parable of a pauper who asks a rich man to fulfill his request. The rich man can do this, but he must connect his mind to the pauper's suffering in order to have compassion

---

20 This point, found explicitly in the Talmudic passage cited above, is frequently underscored in Hasidic texts.

upon him. The same is true when Israel is in pain. They cry out to God for compassion, since the blessed Holy One always fulfills the requests of His people Israel. One must attach his mind and prayer to the blessed One. "Prayer" [tefillah] also means a type of connection, as in "I struggled [naftulei] with God" (Gen. 30:8). The blessed Holy One connects Himself to them [i.e. the worshipers], in order to bestow compassion.

This is, "And God looked at the children of Israel, and God knew them." Knowing refers to a connection, as in "and Adam knew [his wife]" (Gen. 4:25). This is also, "and God knew."

Therefore, "In the midst of years, make them known." Suffering always happens within time, since there is no suffering, grief, or crying beyond time. "In the midst of years"—all suffering happens within the temporal—"know them"—connect to them. This bond brings about the end of the verse: "in anger, remember Your compassion."[21]

R. Levi Isaac stresses that the Divine can only be known through His engagement with the temporal world. Only within the realm of time can one accomplish prayer, an act here described as an intimate encounter between the worshiper and God in addition to a moment of supplication. There is an element of the Divine that remains elusively eternal, but it is precisely within the temporal that God and man may truly enter into a relationship.

---

21 *Kedushat Levi, shemot*, 135. The terminology and metaphor used by R. Levi Isaac are similar to that of the Maggid, and, although he does not cite his teacher explicitly, this homily seems to be offered as a world-affirming—or time-affirming—response to his master's more otherworldly teaching.

Reflection
ON WHITE ETERNITY IN THE POETRY OF MAHMOUD DARWISH

Abed Azzam

In the late works of the Palestinian poet Mahmoud Darwish (1941–2008), eternity becomes a central theme on the background of Darwish's own experience of death. At this time, Darwish's poetry expressed an intensive occupation with the Heideggerian idea of poetry. The following reflections concentrate on the being of eternity in Darwish's poetry. Although the present framework does not allow deepening into the whole of Darwish's poetic reflections on Heidegger's philosophy, it can nevertheless minimally enlighten the Heideggerian horizon it opens for itself.

After Heidegger, the theologian is "the legitimate expert on time [insofar as] time finds its meaning in eternity [But] if "the philosopher asks about time, then he has resolved *to understand time in terms of time.*"[1] We ask in continuity with Heidegger: How to proceed from time to eternity? In fact, the task of understanding eternity should first grant the *possibility* that eternity is able to reside somehow in the temporality of time. How can such a possibility be granted when eternity seems to escape, or even

---

[1] Martin Heidegger, *The Concept of Time* (Oxford: Blackwell, 1992), 1–2.

contradict, the domain of temporality, which is grounded in the certainty of death alone? Keeping with Heidegger, that which guarantees the possibility that the being of eternity can become intelligible in time is art. For Heidegger, *"All art*, as the letting happen of the advent of the truth of beings, is as such, in *essence*, *poetry*."[2] If so, what is eternity as poetry? How does (Darwish's) poetry let eternity unfold?

In 1998, Darwish survived a second heard attack, after which he wrote, besides other works, his epical poem *Mural* (جدارية), followed by *In the Presence of Absence* (في حضرة الغياب) in prose and poetry.[3] The occasion of these two works was the experience of that heart attack and the surgery following it. Darwish thought that *In the Presence of Absence* would be his last work. In this way, this experience became the sole subject of *Mural*, while *In the Presence of Absence* can be said to be an autobiography written under the influence of this same experience.

In *In the Presence of Absence*, Darwish divides himself into a narrating I who is heading "to a rendezvous . . . I have postponed more than once with a death,"[4] and an addressed I who is heading "to a second life promised . . . by language, in a reader who might survive the fall of a comet on earth."[5] In the chapter dedicated to this experience he narrates to his addressed I: "You were told that death had kidnapped you for a minute and a half and that an electric

---

[2] Martin Heidegger, "The Origin of the Work of Art," in *Martin Heidegger: Basic Writings*, ed. David Farrell Krell (San Francisco: HarperSanFrancisco, 1993), 197.

[3] Mahmoud Darwish, *Ǧidāriyah* [Mural] (Beirut: Riad El-Rayyes Books, 2001). Mahmoud Darwish, *Fī ḥaḍrat al-ġiyāb* [In the presence of absence] (Beirut: Riad El-Rayyes Books, 2006); English translation (quoted below): Mahmoud Darwish, *In the Presence of Absence*, trans. Sinan Antoon (New York: Archipelago Books, 2011). All English quotations (from the poems) hereafter are my translations. Darwish suffered from heart disease. The first heart attack, followed by a heart operation, took place in 1984. He died in 2008, three days after undergoing heart surgery in Houston.

[4] *In the Presence of Absence*, 16.

[5] *In the Presence of Absence*, 16. This division resembles in fact the division of the I in *Mural* into (1) an I that does not belong to the poet, and (2) a name whose past belongs to the poet.

shock had brought you back to life."⁶ This report seems to leave the debate about the certainty of death behind so as to be able to proceed precisely in accord with such certainty. Thus Darwish adds in *Mural*:

As if I had died before now ...
I know this vision, and I know that I am
heading to what I don't know. Perhaps
I'm still alive somewhere, and I know
what I want ...
someday I'll become what I want.⁷

On this ground, the poet Darwish follows Heidegger and René Char,⁸ while the vision of death lets him bring eternity into being:

And everything is white,
the sea hanging above the roof of a white cloud.
And the no-thing is white in
the absolute's white sky. I was, and I was not.
For I'm alone in the surroundings of this
white eternity. I came before my hour
so no angle arose to ask me:
what did you do, over there, in the world?
And I didn't hear the acclaim of the virtuous, not either
the moan of the sinners, I'm alone in the whiteness,
I'm alone⁹

---

6 *In the Presence of Absence* 100.
7 Darwish, *Ǧidārīyah*, 9.
8 In one part of his vision, Darwish reports seeing "René Char sitting with Heidegger ... drinking wine and not seeking out poetry"; see *Ǧidārīyah*, 18.
9 *Ǧidārīyah*, 7. In *In the Presence of Absence*, Darwish describes this same experience as follows: "White sleep. Dazzling sleep carries you like a feather on white clouds ... Light and transparent as if you were your soul, without past and present, emptied of time and feeling. You are neither something nor nothing. But you see as you have never seen before. You see white light, white clouds, white air." Darwish, *In the Presence of Absence*, 99–100.

Then, eternity "is-white." We should keep in mind that "eternity is-white" receives its significance from the poet's experience of death, as well as from its formulation in language as poetry. Within such a Heideggerian horizon of interpretation, the proposition "eternity is-white" brings us directly to inquire about the question "what is white?" in order to clear the essence of eternity. In other words, eternity may now reveal itself for the poet, who seeks its meaning in that which is white. But saying that "eternity is-white" seems to be like saying "eternity is eternity." How then does such a proposition remain meaningful outside the realm of metaphysics? Here we are reminded again that Darwish follows Heidegger. On this path, when Darwish says "eternity is-white," he allows for the location of (white) eternity behind the cupola "is." In continuity with Heidegger, this act of the relocation of eternity brings eternity into being. Eternity is transformed from the realm of metaphysics into the realm of being. Eternity gains through the mediation of the poetical saying "eternity is-white" its ultimate admission into the realm of being. When eternity reveals itself to be white, it becomes part of the poet's experience of time. Under these conditions, the poet may now seek after letting eternity show itself in that which is white, such as white flowers, white clouds, white sheets, and so on. However, the success of the poet in letting eternity reveal itself in the realm of experience is not always guaranteed. It happens sometimes that the poet may fail at describing the whiteness of some object. This is exactly what happens to Darwish when he attempts to describe the whiteness of the almond blossom:

To describe the almond blossom, neither the encyclopedia of flowers assists me, nor the dictionary
Speech shall kidnap me to the tricks of rhetoric
and rhetoric injures the meaning and praises its wound, like a male
   dictating on the female her feelings

So how can the almond blossom shine in my language
and I'm the echo? . . .
Which is the name of this thing in the poeticality of the
   no-thing? . . .
Words are neither homeland nor exile,
but the passion of whiteness to describe the almond blossom
neither snow nor cotton
so what is it
in its transcendence of things and names[10]

On the other hand eternity also happens to let itself reside in the most remote temporality of that which is white, as it does in the case of clouds,

like a white idea about the meaning of existence.
Perhaps some gods are revising the story of Genesis
(this universe has no final form . . .
the forms have no history . . .) . . .
Unknown Painters are still before you
playing, and painting the eternal absolute,
white like the clouds on the wall of the universe . . .
And the poets build houses in the clouds
and go . . .
Each feeling has an image
and each time a cloud
but the lives of the clouds are short in the wind
like temporal eternity in the poems
it neither fades nor lasts[11]

---
10 Mahmoud Darwish, *Kazahr al-lawz aw ab'ad* [Like almond flowers or further] (Beirut: Riad El-Rayyes Books, 2005), 47–48.
11 Mahmoud Darwish, *La ta'tathir 'emma fa'alt* [Don't apologize for what you've done] (Beirut: Riad El-Rayyes Books, 2004), 85–91.

The philosopher, keen to his path, handed to the poet the task of bringing into being whatsoever meaning of eternity, after he had called eternity to reside in the temporality of time. The poet accordingly, observes eternity in that which is white. Thus, insofar as eternity is white, the poet, in his description of the whiteness of the clouds, observes that eternity is that which is eternal. And eternity is, in fact, the eternal state of change. This is the eternal change of the ever-changing forms of the white clouds, which seem to be like the work of art: like the poem or, equally, like a mural that is being unceasingly painted on the wall of the universe.

Then, (white) eternity is the eternal flux of being that has no finality. In this way, the work of art, or equally the mural on the wall of the universe, defeats death: *Mural* itself has been a dialogue with death seeking after the possibility of the eternity of the work of art.[12] And Darwish, the poet describing the whiteness of eternity, proclaimed himself Troy's poet: if Homer narrated the victory of the Greeks, Darwish saw himself as the poet narrating the forgotten voices of the defeated Trojans. At one and the same time, and as quoted here above, this poet is not the dictating subject: he becomes "living and free" once he himself becomes forgotten as "person and . . . text," for he is part of the way, or "for the way," interpreting and interpreted, following and being followed, in a future past:

A past tomorrow precedes me. I'm the king of the echo.
I have no throne but the margins. And the way
is the how. My predecessors might have forgotten to describe
something, in which I may awaken a memory and a sense.[13]

---

[12] See Darwish, *Ǧidāriyah*, 29.
[13] Darwish, *La ta'tathir 'amma fa'alt*, 72; also see note 6 above.

CHAPTER FIVE

Eternity in Twentieth-Century Analytic Philosophy

Kris McDaniel

What parts of reality enjoy eternity? Let us provisionally understand eternity in such a way that anything that is eternal is *atemporal*, that is, neither in time (and hence not in space-time) nor subject to temporal (or spatiotemporal) properties or relations. This provisional conception of eternity is merely negative: it tells us what eternal things are not like. But it is a conception that is consistent with the possibility that anything eternal has positive aspects that are ultimately responsible for its atemporality. This provisional conception of eternity is also partly stipulative, but of course it does capture one long-standing traditional use of the word.

That at least some entities enjoy eternality has come to enjoy widespread acceptance from contemporary analytic philosophers, who for the most part are happy to traffic in mathematical entities such as numbers or pure sets, propositions, properties construed "platonically," and possible worlds construed either as complexes of properties or propositions or as sui generis abstracta in their own right, as opposed to complex concrete particulars. Given this ontological menagerie it is hard to know where to begin. I'll focus on propositions.

One philosopher whose work anticipated contemporary views about propositions is Bernard Bolzano, who called his abstract and atemporal bearers of truth-value *sentences-in-themselves*.[1] Later philosophers, such as Lotze, Frege, and Husserl, would endorse similar doctrines, although none of these philosophers would comfortably describe these entities as being part of reality.[2] Most contemporary metaphysicians do not distinguish between what is a part of reality and what there is, but the conflation of these concepts is a relatively recent development. And though I am happy to distinguish these concepts and understand reality as a specific mode of being that not all of what there is partakes in, I won't fight this fight here. Let's focus on the question of what entities among all of what there is enjoy eternity, setting aside issues pertaining to their specific modes of being.

As noted, Bolzano, Frege, and Husserl each recognized a class of entities that we can reasonably call *propositions*. Propositions as understood by these philosophers are eternal, that is, atemporal, bearers of truth and falsity that are nonetheless often made true by how things are in space and time. (Lotze seemed to recognize entities that have

---

[1] The source of Bolzano's doctrine is his *Wissenschaftleher*, published in 1837; an English translation of large portions of this book is *Theory of Science*, trans. Jan Berg (Dordrecht: D. Reidel, 1973). Bolzano called propositions "sentences-in-themselves." Bolzano is explicit that we must not ascribe being to propositions.

[2] Hermann Lotze, *Lotze's System of Philosophy, Part I: Logic*, trans. Bernard Bosanquet (Oxford: Clarendon Press, 1884), accepts (some) things that seem to have the structure of propositions, but they are ideal truths rather than parts of reality. Husserl, *Logical Investigations*, though should include first name "Edmund" too (1900–1901), vols. 1 and 2, trans. J. N. Findlay (London: Routledge, 2001), is an important of proponent of propositions, who also hesitates to ascribe the same kind of being to them as to "concrete" things. Things are less clear with Frege: although he embraces propositions, which he calls *thoughts*, he denies of them not being or reality but rather *actuality*. Incidentally, my understanding is that it is unlikely that Frege directly read Bolzano's work, although it appears that Lotze did, and Frege was a student of Lotze; thanks to Sandra Lapointe for discussion here.

the structure of propositions—for example, his ideal laws—but it is not clear to me whether it is appropriate to attribute to him the doctrine that there are propositions, since it is unclear to me whether he accepts timeless entities that are false. It is also not clear whether it is right to think of these entities as something more like truth-makers rather than truth-bearers.) In addition, Bolzano, Frege, and Husserl each recognized a larger domain of timeless entities, called "ideas in themselves," "concepts," and "meanings," respectfully.[3] Each of these philosophers offered a variety of interesting arguments for embracing these ideal entities, but I will focus on only one of them, which I will call the argument from antipsychologism about logic. The gist of this argument is that since we cannot understand the discipline of logic as being a subdiscipline of psychology, the best account of the subject matter of logic construes it as being about the domain of ideal entities. More on this argument momentarily.

The interest in the eternal stemming from the philosophy of logic and language was prominent in the late nineteenth and early twentieth centuries. In the middle of the twentieth century there was a revival of systematic, speculative natural theology, which was (surprisingly enough) practiced by self-described analytic philosophers. And with the return of systematic, speculative natural theology, questions concerning whether there is an eternal God were once again attended to.[4]

---

3 These figures were not the only advocates of propositions or meanings, so understood. G. E. Moore, "The Nature of Judgment," *Mind* 8 (1899): 176–193, is also an important and influential advocate of propositions, and Alexius Meinong, "The Theory of Objects" (1904), trans. Isaac Levi, D.B. Terrell, and Roderick Chisholm, in *Realism and the Background of Phenomenology*, ed. Roderick Chisholm (Glencoe, IL: Free Press, 1960), recognizes a species of proposition-like entities that he calls *objectives*. And of course, the importance of Russell cannot be overstated see David Goddin and Nicholas Griffin, "Psychologism and the Development of Russell's Account of Propositions," *History and Philosophy of Logic* 30 (2009): 171–186, for an overview of the development of Russell's views on propositions and how they connect up with issues concerning psychologism.

4 Recall that eternity is here is construed as atemporality—and the claim that a thing is eternal by itself leaves open other interesting metaphysical questions about the thing itself. So saying that, e.g., both propositions and a divine being are eternal does not commit one to saying that propositions and God have the same mode of being, or are both abstract entities, and so forth.

Perhaps the most exciting thesis is that *everything enjoys eternity*. The possibility that this is true has intrigued philosophers in every epoch, and the twentieth century is perhaps exceptional only in that during that time this conclusion was pursued more rigorously than ever before. At the end of the nineteenth century and through the beginning of the twentieth, many attempts were made to demonstrate by means of speculative metaphysics that time is a mere appearance. The most famous of these purported demonstrations is McTaggart's argument for the unreality of time. Although this argument is not widely viewed as successful, it did set the agenda for analytic philosophers pursuing the philosophy of time in the second half of the twentieth century.

Complementing the arguments of speculative metaphysics are the arguments of speculative philosophy of physics. The theory of special relativity appeared to many philosophers to show that temporality per se was not metaphysically fundamental but should instead be seen as an aspect of spatiotemporality. Of course the demotion of time to an aspect of space-time does not by itself imply that the things previously thought to be in time are in fact atemporal. But it does force us to rethink any ontological theory that trades on a sharp separation of spatial and temporal features. For example, certain "Cartesian" views of the nature of mental substances on which mental substances enjoy temporal properties but no spatial properties seem harder to sustain. On a standard interpretation of special relativity, nothing perfectly matches our ordinary conception of time, since there is no well-defined relation of simultaneity. But there is a well-defined relation of simultaneity relative to a reference frame, which can be used to define up the notion of a time relative to a reference frame: say that a time relative to F is a maximal class of space-time points that are pair-wise simultaneous to each other relative to F; being at a time relative to F consists of partially occupying one of these space-time points. And we can define up a notion of a region of space relative to a time relative to a frame as well: a region of space relative to a time t relative to a frame F is any subset of t relative to F; partially occupying a region of space (relative to a time relative to a frame) consists in occupying one of the members of the relevant subset in question. An upshot of these definitions is that anything in time must be in space. If being immaterial implies being nonspatial, then anything immaterial is thereby eternal.

Although this conclusion is interesting, it only takes us so far. Arguments from physics for the unreality of time were to come. Kurt Gödel argued for the unreality of time by appeal to considerations stemming from the theory of general relativity (rather than special relativity), although he was familiar with and probably influenced by the arguments of McTaggart. And more recently some physicists and philosophers of physics have entertained the hypothesis that spatiotemporality is itself a derivative feature that emerges from a more fundamental nonspatiotemporal framework.

My plan accordingly is as follows. In section 1 of this chapter, I will discuss in more detail arguments for the eternality of some entities, specifically focusing first on the case for ideal meanings, including propositions, and then turning to questions concerning the purported eternity of God. In section 2, I will first critically discuss some of the arguments of speculative metaphysics for the unreality of time, and I will follow this discussion by tracing some of the highlights of twentieth-century philosophy of time. I will then turn to a discussion of the hypothesis of speculative philosophy of physics that space-time derives from a more fundamental basis. This hypothesis has received comparatively little attention from metaphysicians, despite the tempting prospects for speculation it invites. Accordingly, I will discuss how the truth of this hypothesis would impact various other disputes in metaphysics, including disputes about what it is to be an abstract rather than a concrete object, the nature of material composition, and the relationship between necessity and eternity.

1 ARE SOME THINGS ETERNAL?
Antipsychologism in Logic

Although a number of philosophers fought against doctrines under the label of "psychologism," for the sake of space I will focus on just one of them, specifically, Edmund Husserl.[5] In 1900 and 1901, Husserl published volumes 1 and 2 of his *Logical Investigations*, which is one of the founding texts in the phenomenological tradition. It is also an excellent work of philosophy, in which the following topics (and many more) are explored in great detail: the metaphysics and epistemology of logic, the analytic/synthetic distinction, the nature of a priori knowledge, abstraction and the metaphysics of properties, the logic of parts and wholes, and puzzles and paradoxes of intentionality. But since my concern here is with that which is eternal in the Husserlian texts, I will focus on Husserl's antipsychologism in logic.

"Psychologism," like "empiricism," "rationalism," "neo-Kantianism," is really a name for a general class of doctrines that bear some family resemblances to each other. For this reason, it is not always clear which particular version of psychologism is being targeted when a historical figure argues against it (or, for that matter, in favor of it). In Husserl's case, the primary target seems to be the view that logic is properly construed as a subdiscipline of psychology, and more specifically as a subdiscipline devoted to normatively evaluating the mental states and processes, primarily judgments and inferences, that are the province of the other subdisciplines of psychology. Husserl's argumentative strategy is to provisionally concede that logic is a normative science concerned with evaluating mental states and processes but then to argue that (1) all normative sciences have as their foundation some nonnormative, that is, theoretical, science, and (2) the theoretical foundation

---

[5] For an extensive discussion of the many varieties of psychologism and antipsychologism, see Martin Kusch, *Psychologism: A Case Study in the Sociology of Philosophical Knowledge* (London: Routledge, 1995).

of the normative science of logic is not psychology but rather a discipline Husserl calls *pure logic*. Pure logic is the discipline that studies the necessary connections between an ideal realm of meanings, specifically propositions and their component parts. On Husserl's view, the normative science of logic does not rest on the findings of empirical psychology, for if it did, the claims of logic would be empirical rather than a priori, merely credibly believed rather than certainly known, and perhaps true only for human cognizers rather than just plain true. The normative science of logic rests instead on pure logic, which is an a priori science that is based on epistemically certain insight into a realm of meanings and that concerns itself with relations of consequence obtaining between these meanings. On Husserl's view, an inference understood as a concrete mental process that begins with judgments and ends with a judgment is a good inference just in case the proposition that is the content of the final judgment stands in appropriate atemporal relations of logical support by the propositions that are the contents of the initial judgments. In general, the evaluations of mental processes produced by the normative discipline of logic are parasitic on the relations between atemporal meanings.[6]

Husserl was not the first to find solace from psychologism in the realm of the eternal; Frege also argued that the province of logic is a third realm of entities he called thoughts and concepts. And, as noted earlier, both Frege and Husserl were anticipated by Bolzano and, to a lesser extent, Lotze.[7] But Husserl's critique of psychologism seemed

---

6 In addition to pure logic, Husserl also recognized a second a priori discipline called *pure ontology*, which studies the formal relations between categories of objects such as states of affairs, properties, relations, substances, and so forth. In fact, sometimes Husserl seems to conceive of pure logic as encompassing pure ontology and pure (propositional) logic.

7 One central text is *Lotze's System of Philosophy, Part I: Logic*. There is substantial controversy about the extent of Lotze's influence on Frege and the extent to which Frege's views on thoughts stem from positions of Lotze. Good starting points on these issues include Michael Dummett, *The Interpretation of Frege's Philosophy* (Cambridge, MA: Harvard University Press, 1981), and Hans Sluga, *Gottlob Frege* (London: Routledge, 1980).

to play a larger role in carrying the day, although of course there were sociological factors in play as well.[8]

Temporalism about Propositions

One might well accept the arguments against psychologism in logic and, as a result, embrace an ontology that includes propositions as mind-independent representational entities. But what is the motivation for holding that propositions are eternal entities as opposed to entities that are located in times? Is there work in the philosophy of logic that eternal propositions are more suited to perform than propositions in time? Or are there direct arguments for the timelessness of propositions?

It is difficult to distill clear answers to these questions from the works of these authors, which are in other respects remarkably clear. In Bolzano, it is clear that he does not wish to say that his sentences-in-themselves exist in the same way that ordinary concrete objects—in fact he goes further and denies that these sentences-in-themselves exist in any way whatsoever.[9] Perhaps for Bolzano being in time is sufficient for existing in some way. I detect in Husserl similar motivations; he does not wish to affirm the *reality* of the ideal, although he is committed to there being ideal entities. Moreover, Husserl is explicit that being temporal is a sufficient condition for being *real*, a mode of being that not everything shares.[10] As far as I can tell, then, their motivation for ascribing eternality to propositions, rather than omnitemporality, stems not ultimately from considerations having to do with psychologism in logic but rather from an inclination to affirm the being of propositions while denying that they have the mode of being that things like us and our surroundings enjoy.

---

8 For a lengthy discussion of the various philosophical and sociological factors involved in the controversy of psychologism, see Kusch, *Psychologism*.
9 Although of course on his view there are such entities as sentences-in-themselves, i.e., propositions.
10 See, for example, *Logical Investigations*, vol. 2, 351.

With respect to Frege, this motivation seems less prominent. Frege does deny that thoughts are real in the way that things are real, and he contrasts timeless thoughts with changeable things.[11] But what seems to be of central importance to Frege is not that thoughts are eternal but that their intrinsic representational properties are essential and unchangeable, and that thoughts cannot *cease* to exist. I read in Frege no deep inclination to deny that an omnitemporal being could have this kind of unchangeable and essential intrinsic nature.

Perhaps the reasons for ascribing eternity to propositions are relatively weak. Perhaps propositions needn't be timeless to be mind-independent bearers of truth and falsity. It is worth contemplating, however, whether propositions must be *necessary beings*, that is, existing in all possible worlds, in order to be suitable objects for a science of logic. And if so it is worth contemplating whether being in time is sufficient for being a contingently existing being. (Being at only some times rather than all times seems sufficient for being contingently existing beings, but it is far from clear that an omnitemporal being must be contingent.)

It is also worth contemplating whether there are positive arguments for the temporality of propositions. Some sentences are now true that once were false: "You are reading this paragraph" is now true, but it once was false. (Unless your reading habits are remarkably strange, this should strike you as a plausible example of a sentence whose truth-value changes over time.) What should we say about how propositions relate to sentences whose truth-value can change at different times? One might claim that a sentence changes truth-value from one time to another because that sentence expresses a different proposition from one time to another, while these propositions themselves do not change their truth-values. This is the position that Frege and Husserl endorse, and perhaps Bolzano as well.[12] And for what it is worth, it

---

11 Frege, "The Thought: A Logical Inquiry," *Mind* 65 (1956), 311.
12 See Frege, "The Thought," 309–310; Bolzano (1973), sec. 25; and Investigation 1 of vol. 2 of Husserl's *Logical Investigations*.

seems that this view is the dominant view among those who consider the question.

The alternative view is that a sentence changing truth-value from one time to another does so because the sentence expresses the same proposition at these different times and the truth-value of this proposition itself changes from one time to another. Would this view support the claim that propositions are temporally located? After all, they're truth-values, so mustn't they be in the times at which these changes in truth-value "occur"?

Not obviously. If a sentence like "Kris is eating a cookie right now" expresses a proposition that is true at some times and false at others, what should we say about "I am eating a cookie at 3:56 p.m. on July 10, 2012"? When uttered by some people, this sentence expresses a truth; when uttered by others it expresses a falsehood. Should we say that this sentence expresses a single proposition that is true at some persons but false at others? Should we then also say that this proposition is "in" persons or that it has some sort of spatial location? (Is it wherever it is true?) In general, even if we grant that truth-values of propositions can change at different indices, such as times, persons, locations, worlds, or whatnot, it is not obvious that we should grant the further step that the propositions in question are *located* at those indices, or even located at all.

Obviously, systematically examining the arguments for and against the eternity of propositions is a task too large to undertake here. I will now turn to another strand of eternity in twentieth-century analytic philosophy.

Eternity in Analytic Theology

Long thought dead and buried, rigorous systematic philosophy of religion enjoyed a resurrection (or perhaps a reincarnation) in twentieth-century analytic philosophy. Whether they deserve praise or blame is perhaps a matter of contention, but no one can argue with the claim that the bulk of responsibility for revitalizing the field falls on theistic philosophers such as William Alston and Alvin Plantinga. And with the return of rigorous, systematic philosophy of religion, a return to old concerns about the divine attributes was inevitable. What is germane to my purposes is the renewed attention to the question of how God relates to time.

God must not be limited by time in the way God's creatures are, that is by having temporal boundaries. I am an example of a temporally limited creature; I came into existence in 1976 and will go out of existence sometime around 2076. These years enclose my temporal boundaries. But is God unlimited by time by occupying every time, that is, by being omnitemporal? Or is God unlimited by time by virtue of being outside time altogether, that is, by being eternal in the sense used at the start of this chapter?

Or is there a third possibility for how God could be unlimited by time? One of the interesting conceptual developments is the idea of ET-simultaneity, which was developed by Eleanor Stump and Norman Kretzman.[13] The fundamental idea is that God is simultaneous with everything that is temporal even though not everything temporal is simultaneous with everything else that is temporal. On the face of it, the fundamental idea represents a distinctive way in which God might be unbounded by time; but on the face of it, the fundamental idea seems incoherent. If God is somehow simultaneous with me, and God is somehow simultaneous with Julius Caesar, how is it that I and Julius Caesar fail to be simultaneous with each other? In short, isn't simultaneity a transitive relation?

Stump and Kretzmann argue that, given what we know about simultaneity from modern physics, namely that it is not even a two-place relation but is relative to a frame of reference, we should be cautious in assuming the incoherence of the fundamental idea. Moreover, two things might be simultaneous at one frame of reference but not

---

13 Eleanor Stump and Norman Kretzman, "Eternity," *Journal of Philosophy* 58 (1981): 429–458

simultaneous at a different frame of reference. Stump and Kretzmann further develop it by introducing the notion of a divine "frame of reference" and the notion of ET-simultaneity; at the divine frame of reference, everything is simultaneous with God; this is ET-simultaneity. But at "ordinary" frames of reference, it is not the case that everything is simultaneous with everything else. Our intuition that I am not simultaneous with Julius Caesar is mollified on this view by making it clear that there is no ordinary reference frame at which Julius Caesar and I are copresent.

Whether Stump and Kretzmann's fundamental idea helps make sense of how God can be unlimited by time but still be a casually productive agent remains to be seen; but we shouldn't doubt that Stump and Kretzmann have articulated an interesting new way in which that which has been thought to be eternal can nonetheless relate to what is in time.

2 Is Everything Eternal?

I've examined some reasons to think that some objects are eternal. I will now turn to arguments for eternality of all things. As I mentioned in the introduction, I will examine two kinds of arguments: arguments from speculative metaphysics and arguments from speculative physics. I begin with the former.

Arguments from Speculative Metaphysics for the Unreality of Time

McTaggart was not the only philosopher of his period to argue for the unreality of time. Figures such as F. H. Bradley, who was perhaps the dominant mind of his generation, also argued against the reality of time, yet it is McTaggart's arguments that have commanded and continue to command the most attention. Before turning to the details of McTaggart's arguments, it is worth considering why this is the case.

I will briefly examine the case made by Bradley against the reality of time, which appears in chapter 4 of his masterwork *Appearance and Reality* and occupies a sum total of three and a half pages.

The first argument poses a dilemma: either time is a relation between durationless units, or it is not. If it is a relation between durationless units, then the whole of time is without duration—for how can something made wholly out of things with no duration have itself a duration? But if time is not a relation between durationless units, then it must be a relation between units with duration. But the notion of units with a duration is, according to Bradley, inconsistent. Reading between the lines here is tricky, but the idea behind this claim seems to be this: if the units themselves have duration, then something must unify them—and what could this thing be besides time itself? And so time itself "resolves" into nothing more than a relation between things that in turn require time to relate and so on without end. And this, according to Bradley, is impossible.[14]

Bradley seems to think that this first argument turns on conceiving of time as being analogous to space, and accordingly turns to an argument that purports to be independent of such a conception. So we should focus only on time as it is presented, which requires that we not consider more than the time that is now. Either the time that is now is simple and indivisible, or it is complex. It can't be simple, though, because time exists only if there are relations of before and after, so the time that is now must contain parts so related in order to even be time. But as soon as we concede that the time that is now has parts related by relations of before and after, the worries generated by the first argument arise here again.

Some things are worth noting about Bradley's arguments. First, like McTaggart's, they are ultimately detachable from the particularities of the metaphysical systems that their proponents defend. It

---

14 A similar argument can be found in a highly condensed form in Bradley, *Collected Works of F. H. Bradley*, 12 vols. (Bristol: Thoemmes Press, 1999), vol. 3, 109.

is actually surprising that Bradley's arguments for the unreality of relations and for the incoherence of the notion of inherence play so little of a role in chapter 4, given that these arguments first appear in the chapter immediately prior. (Bradley could have written off space and time understood as systems of relations or qualities as a mere corollary to chapter 3, and such a move is hinted at in chapter 3.)[15] So Bradley's arguments have interest independent of his other metaphysical commitments.

Now it is true that Bradley's other metaphysical commitments do seem to mollify the conclusion of these arguments, namely that the Absolute is nontemporal and that time is not real but rather mere appearance. For, nonetheless, Bradley says that time exists, although I take him to be saying that time exists *as an appearance*.[16] And since Bradley accepted degrees of truth, it was open for him to say that it is to some degree true that things are temporal. This muddies the waters; McTaggart on the other hand accepted neither degrees of truth nor degrees of reality, so the conclusion he offered is apt to seem clearer to contemporary analytic metaphysicians: on McTaggart's view, it is just flat-out false that things are temporal. McTaggart's view that objects apparently ordered by a temporal series really are ordered by some other series that in some way gives rise to the illusion that objects are temporally ordered in no way leads him to hold that it is true to any degree that things are temporal.

Perhaps this muddying of the waters is one reason why Bradley's arguments against the reality of time did not receive the scrutiny they deserved. An illustration of the tendency to spend more time on the conclusion rather than argument for it is Moore's lecture "Is Time Real?," which was delivered sometime during the winter session of

---

15 F. H. Bradley, *Appearance and Reality* (1893), 9th ed. (Oxford: Oxford University Press, 1930), 29.
16 Bradley, *Appearance and Reality*, 191.

1910–1911 and printed in *Some Main Problems of Philosophy* roughly four decades later. The vast majority of the lecture is spent on determining what Bradley is up to by saying that time is unreal yet exists, but nowhere in this lecture does Moore address the reasoning that led Bradley to this conclusion.

Bradley's arguments do not lack intrinsic interest either. I find the second argument to be pregnant with potential. Bradley, like McTaggart, accepted that time exists only if time passes. For example, Bradley complained that Russell's view of time as a (merely) objective ordering is not adequate because it does not account for the flow of time.[17] And if we accept that time passes only if there is a metaphysically distinguished part of time that is present, then something a lot like Bradley's second argument is compelling. Call *presentism* the view that that presently existing entities are the only entities that there are. On the presentist view, the present is metaphysically distinguished by way of being ontologically distinguished; note that it does seem to be a presupposition of Bradley's second argument that time is real only if presentism is true. Now either the time that is present is temporally extended or it is not. If it is not, then time is not real, since nothing bears the temporally before or temporally after relations to anything. If it is extended, then there are present parts of time that are not simultaneous with each other. But this seems impossible as well. So presentism must be false. Since time is real only if presentism is true, then time is unreal.

The argument I've just presented is definitely inspired by Bradley's second argument; in fact, I'd say it is basically his argument presented in a cleaned-up and streamlined way. And by my lights, regardless of whether it is ultimately successful, it is of as much intrinsic interest as McTaggart's more famous argument against the reality of time.

---

17 *Collected Works of F. H. Bradley*, vol. 4.

In this context, let me also note that Bradley anticipates the idea that the directionality of time is subjective.[18] Here is Bradley on the subject:

> For the direction, and the distinction between past and future, entirely depends on *our* experience.... But, if this is so, then direction is relative to *our* world.... For let us suppose, first, that there are beings who can come in contact in no way with that world which we experience.... And let us suppose next, that in the Absolute the direction of these lives runs opposite to our own. I ask again, is such an idea either meaningless or untenable? Of course, *if* in any way *I* could experience *their* world, I should fail to understand it. Death would come before birth, the blow would follow the wound, and all must be seem irrational. It would seem to me so, but its inconsistency would not exist except for my partial experience.[19]

This thesis of Bradley is also of intrinsic interest and is assessable even when one abstracts from the particularities of Bradley's larger metaphysical system. Bradley's interesting challenges to what were the dominant views about space and time have mostly been forgotten.

Let's turn now to a discussion of McTaggart's argument, which has many complex layers despite the apparent ease in which it can be initially summarized.[20] According to McTaggart, time is real only if there is genuine change in time as opposed to mere change of events in time. Events are located in time; most, perhaps all, have temporal extent, and so have temporal boundaries, roughly when they begin and end.

---

18 This view was later made prominent by Adolf Grunbaum, *Philosophical Problems of Space and Time* (New York: Knopf, 1963), 324–326; Grunbaum does not discuss Bradley, probably because Bradley's views on space and time were not seriously discussed by many philosophers at all during the time this book was published.

19 Bradley, *Appearance and Reality*, 189–190.

20 See my entry on McTaggart in the *Stanford Encyclopedia of Philosophy*, winter 2013 ed. Edward N. Zalta, http://plato.stanford.edu/archives/win2013/entries/mctaggart/, for more detailed background on McTaggart's views on time and eternity.

A change of events happens in an interval of time D just in case some event E either begins or ends at some part of D. But a mere change of events in time is not the same thing, according to McTaggart, as a change in time itself. Consider an event that begins in 1907 and ends in 1908. It will always be the case that this event begins in 1907 and ends in 1908, and it was always the case that this event has the temporal boundaries that it has. Moreover, this event bears temporal relations to other events; it is, for example, before McTaggart's death, and it is after McTaggart's birth, and moreover, it is always the case that these temporal relations obtain. It never will be the case that McTaggart's death comes before McTaggart's birth, and it will always be the case that E occurs between them. The existence of sequence of differentiated events in time does not suffice for real change in time.

There is real change in time just in case there is some feature F, had either by events in time, or by parts of time itself, such that although right now some time has this feature, it wasn't always the case that it had it, and it won't be the case that it has it. This condition is satisfied if there is a property of being present that is always had by one time and no others, but which time has the property of being present changes. Let's focus on this way of implementing the idea that there is real change in time.

According to McTaggart, real change in time is impossible. On McTaggart's view, were time to exist, it would have the following features. First, time, whatever its exact ontological constitution, would have something like parts—call them *times*—and each time would be as real as the others. (It might be that each time is identified with the sum total of what exists at that time, or it might be that times are sui generis entities. As far as I can tell, nothing in McTaggart's argument turns on this.) In McTaggart's argument against the reality of time, unlike in Bradley's, nothing like presentism is presupposed. These times are available to be quantified over, and we can attribute properties to them. The property of being present is supposed to be a property that some time simply has and other times simply lack. But there is one core commonality, namely that time is real only if time genuinely passes.

How then does McTaggart derive the claim that real change in time is impossible? It would be extraordinarily tedious to go through the myriad different interpretations of this part of his argument, so instead I will offer a quick gloss. First, put yourself in the position of someone in 1908 who thinks of her time as being present and our time as future. Surely 1908 is present, at least to her. Next, put yourself in the position of someone in 2208 who thinks of his time as being present and our time as past. Surely 2208 is present, at least to him. But if being present is a property that some times simply have—rather than, say a relation that a time can bear to an intelligent being who is located at that time—then it can't be that all of these times are present, for then they would each be past and future as well, and nothing can have all three of these incompatible temporal determinations. We can't deny that the determinations are incompatible without eliminating their use as agents of real change in time: if all three determinations are compatible, then every time has them—save perhaps the first and last time, if any of those exist—and so there is no genuine change in time that is marked by their exemplification.

Since time exists only if there is real change in time, and real change in time is impossible, time does not exist.

This is McTaggart's argument in a nutshell. Why was McTaggart's argument so important to the development of philosophy of time in twentieth-century analytic philosophy, while Bradley's argument had substantially less impact? There is no way to decisively answer this question, but the following factors strike me as playing a large role. First, McTaggart's argument for the unreality of time was first published as a stand-alone article and only later revised and incorporated into a substantially larger work (specifically, the second volume of *The Nature of Existence*), while Bradley's arguments for the unreality of time in *Appearance and Reality* appeared first in that work, where they occupy a substantially smaller quantity of pages. Moreover, other aspects of the book, such as the chapter immediately prior to it on relations and inherence themselves, were very influential. (A search on the *Philosophers Index* using the term "Bradley's regress" reveals a sizable and still growing secondary literature.)

Second, McTaggart enjoyed productive working relations throughout his career with many of the leading analytic philosophers of the early twentieth century, many of whom were affiliated with Cambridge University, where he worked to the end of his career. These philosophers of course included Russell and Moore, and Moore read through an entire draft of the first volume of *The Nature of Existence*, and so did well as C. D. Broad, who wrote a gigantic commentary on *The Nature of Existence*. It is fair to say that by the 1920s Bradley's shadow had already begun receding. On the other hand, by 1957, John Passmore would write that besides McTaggart no other contemporary philosopher had been commented on so extensively.[21] It also helped that McTaggart had as a champion a philosopher as influential as Peter Geach, whose father was a student of McTaggart and who exposed Geach to McTaggart at an early age.[22]

Finally, the project of dissecting McTaggart's argument proved fruitful. Through Russell, among others, it led to the development of an alternative theory of time as a manifold related by relations of before, simultaneous with, and after but in which the notions of past, present, and future were merely relative notions.[23] Sometimes this view is called "eternalism," since on it there is no change in time, and location of time is analogous to location in space: just as Syracuse, New York, is real although it is not here, so too Julius Caesar is real even though he is not now. On this view, the reality of time does not require the reality of passage or temporal becoming. Such a view of the nature of time immediately leads to interesting questions about how it is to be reconciled with our apparent perception of temporal

---

21 See John Passmore, *A Hundred Years of Philosophy* (London: Duckworth, 1957), 75.
22 See the preface to Peter Geach, *Truth, Love and Immortality: An Introduction to McTaggart's Philosophy* (Los Angeles: University of California Press), 1979).
23 Russell time and experience.

passage, our apparent possession of free agency, and our apparent possession of rational time-asymmetric preferences, such as our preferring that our pains be in the past rather than in the future. These projects of reconciliation are apt to generate a large secondary literature.

Various defenders of "absolute becoming," "real tense," and "genuine change in existence" collectively reacted to McTaggart's argument by developing a variety of distinctive and interesting ways in which these slogans could be cashed out. In addition to presentism, new views enjoyed prominence, such as C. D. Broad's growing block view, according to which (1) the present and the past are equally real while the future is unreal, (2) the universe can be conceptualized as a four-dimensional block in which the present corresponds to an outer surface, and (3) change in time consists of new layers being added to this block, embedding the moment that once was present in successively stacked slices of further reality.[24]

And one promising response to McTaggart's argument led to the development of tense logic. One might worry about McTaggart's treatment of phrases like "Queen Anne's death is past" as subject-predicate sentences in which the property of being past is attributed to the event of Queen Anne's death. But an alternative logical treatment is available, one that makes use of the idea that tenses are best represented by special sentential operators. Roughly, a sentential operator is a linguistic expression for which the operation of prefacing a grammatically complete sentence with it yields a more complex complete sentence. "It is not the case that" and "it is possible that" are sentence operators. On the alternative picture, we have (at least) three special operators, "it was the case that," "it is now that case that," and "it will be the case that." On this alternative view, "Queen Anne's death is past" is better represented as "It was the case that Queen Anne died."

---

24 See C. D. Broad, *Scientific Thought* (Paterson, NJ: Littlefield and Adams, 1959), chap. 2; Broad's book was originally published in 1923. It should be noted that, although Broad's formulation of the growing block view is clearer and more precise, an earlier formulation of it seems to appear in Arthur Lovejoy, "The Obsolescence of the Eternal," *Philosophical Review* 18 (1909), 482.

The idea that past, present, and future are incompatible determinations simply drops out from this picture, since there are no such determinations. Furthermore, on this kind of picture, it is metaphysically misleading to think of times as a manifold of entities related by relations of before and after: "it was the case that" is meant to be a primitive expression. We shouldn't, on this view, understand "It was the case that P" along the lines of "P is true at some time that is before now," where perhaps "now" simply indicates the time at which the utterance occurs. Rather, on this view, time is more like possibility than space—at least, time is more like possibility on a view in which talk of possible worlds is a mere figure of speech and modal claims are best expressed using primitive modal operators.[25]

3 Arguments from Speculative Physics

In 1949, a very interesting and very short article by Kurt Gödel was published in a volume in the *Library of Living Philosophers* series that focused on Albert Einstein. In this article, Gödel outlines an argument for the unreality of time that stems from considerations of the theory of general relativity. In this article, Gödel explicitly links his project to the speculative attempts of earlier philosophers and even cites McTaggart's 1908 article. For better or for worse, Gödel's argument against the reality of time does not seem to have captured the attention of philosophers in the same way McTaggart's argument has, a fact Yourgrau has noted with much regret, although in recent times interest in it has been revived.[26]

---

25 For a nice in-depth discussion of tense logic both as a semantic theory of tensed claims and as a metaphysical theory, see Ulrich Meyer "Time and Modality," in *The Oxford Handbook of the Philosophy of Time,* ed. Craig Callender (Oxford: Oxford University Press, 2011), 91–121.

26 Palle Yourgrau, *A World without Time: The Forgotten Legacy of Gödel and Einstein* (New York: Basic Books, 2005). In addition to Yourgrau's work, see also Mauro Dorato, "On Becoming, Cosmic Time, and Rotating Universes," in *Time, Reality and Experience,* ed. Craig Callender (Cambridge: Cambridge University Press, 2002), 253–276, and Steven Savitt, "The Replacement of Time," *Australasian Journal of Philosophy* 72 (1994): 463–474.

Before formulating Gödel's argument, two preliminary remarks are in order. First, Gödel's way of presenting the argument suggests that he assumes that time exists only if presentism is true. (Recall that a similar assumption seemed to have been made by Bradley.) Gödel explicitly writes that time exists (at least in the ordinary sense of the word "time") if and only if there is what he calls "an objective lapse of time," and he also says that the essence of an objective lapse of time is "that only the present really exists."[27] Furthermore, there is an objective lapse of time only if reality consists in an infinite sequence of nows that come into existence successively.[28] But whether Gödel wants to firmly commit himself to the view that time exists only if presentism is true, he definitely commits himself to the logically weaker thesis that time exists only if some version of the A-theory of time is true.

Second, Gödel grants that one might employ an argument based only on considerations stemming from special relativity: if special relativity is true, there is no well-defined relation of absolute simultaneity. But if there is no well-defined relation of absolute simultaneity, then there is no well-defined notion of "the now," hence presentism must be false. This argument or ones essentially similar to it have been well discussed in the literature. But it is not one that Gödel here rests on; Gödel claims that the existence of matter and the curved spatiotemporal structure it induces could allow for the privileging of certain ways of partitioning space-time into "local times,"[29] although he also indicates difficulties with certain attempts to privilege specific partitions.

Here is a concise summary of Gödel's argument. First, there are possible worlds in which general relativity is true and in which space-time

---

[27] Kurt Gödel, "A Remark About the Relationship Between Relativity Theory and Idealistic Philosophy," in *Albert Einstein: Philosopher-Scientist*, edited by Paul Schlipp, Open Court Publishing, 1970/1949, 562. Dorato, "On Becoming," 601–602, proposes an anodyne reading of Gödel according to which "the objective lapse of time ... referred to by Goedel amounts to the rather nonmetaphysical, almost self-evident claim that if "event E occurs (or, equivalently, tensely exists) at time t', at a later or earlier time", other events occur (or exist)." It strikes me as very implausible that Gödel intended a claim as weak as this. Interestingly, Dorato does argue that even a claim as weak as this faces Gödel's argument.

[28] Gödel, 558.
[29] Gödel, 559.

contains a timelike path from a region in space-time that terminates back at that region. If one traveled from such a region in space-time along this path in one direction one would emerge at the same region where one began. (Initially, a world like this might seem to be a world in which time travel is possible, but whether time exists in these worlds is one of the things that is at issue.) Second, these worlds are in fact physically possible worlds, since these worlds have the same laws as the actual world. But, third, in such worlds, no dimension of space-time is a temporal dimension, since time exists only if there is genuine passage of time, and there cannot be genuine passage in such worlds. For time can genuinely pass only if there is a global partitioning of slices of space-time into times. But in space-times of the sort considered by Gödel, there is no such partition. In short, there is no time in these worlds. Fourth, if there is no time in these worlds, then there is no time in the actual world either. Conclusion: there is no time in the actual world.

Although the first conclusion was originally contested—see Yourgrau for a brief discussion[30]—my understanding is that it is now conceded that Gödel's first premise is correct. The second premise, however, should be more contentious, since it is not actually clear that these worlds have the same laws as the actual world. Gödel focuses only on solutions to equations relevant to general relativity, but general relativity is arguably inconsistent with quantum mechanics. And so the conjecture that general relativity is but a mere approximation to correct laws of the actual world is not implausible.[31] Nonetheless, I won't focus on the second premise; frankly, I'd be speaking out of school if I did. The third premise turns on the idea, stemming at least as far back as McTaggart, that time is real only if time genuinely passes. Since I've already examined that premise earlier, I won't revisit it here.

---

30 *A World without Time*, 119–121.
31 Jill North has suggested to me a second consideration against the second premise. Suppose that a possible world is physically possible only if it contains time, and that these "Gödel worlds" do not contain time. Then these Gödel worlds are not physically possible. That is, although the possibility of Gödel universes is in some sense a mathematical consequence of the laws of general relativity, this possibility is nonetheless not a physical possibility.

So I will turn to the fourth premise. Why should we accept it? Gödel is explicit that there is no contradiction to denying it, so why think we can move from claims about what is possible to what is actual? First note that there are similar arguments as precedents both in earlier debates about the nature of space and in current debates in metaphysics. Recall Newton's famous rotating bucket argument for the existence of absolute space. Consider the difference between a bucket of water at rest and one undergoing rotation. As the bucket is rotated, the surface of the water is disturbed and creeps up the bucket's interior. Only sometime after the bucket comes to a rest does the surface of the water became a flat plane once more. Imagine a possible world with the same laws as ours (a putatively physically possible world) in which only a bucket of water rigidly rotating exists. In such a world, there is no external system of material objects for the bucket to be rotating *relative to*, yet the effects of rotation will (allegedly) still be present. So what explains them? In that possible world, the entity responsible for the effects of rotation must be absolute space itself: the bucket rotates relative to some fixed parts of absolute space. So absolute space exists in that other world.[32]

What should we conclude about the actual world? If we rely on the following general principle, we conclude that absolute space exists in the actual world: the facts about the nature and existence of the structured entity or entities that are occupied by material objects do not vary across worlds that are physically possible relative to the actual world. Some principle of this sort seems to be playing a role in both Newton's and Gödel's respective arguments.[33]

It's not clear to me that we should accept such a principle. Perhaps the following consideration tells against it. I see no necessity in thinking that space-time must be topologically unified, so consider a possible world

---

[32] For a somewhat different take on Newton's bucket, see Robin Le Poidevin, *Travels in Four Dimensions: The Enigmas of Space and Time* (Oxford: Oxford University Press, 2003), 46–50.

[33] I am confident that the analogy between Newton and Gödel's arguments has been made elsewhere and hence that this discussion is probably not original to me, but I have been unable to locate a source in which this analogy is discussed.

that consists in two completely disconnected space-times.³⁴ For simplicity's sake, let's consider a possible world that consists of two disconnected duplicates of the actual universe. Offhand, it is not clear to me why we would be forced to say that this world is not physically possible: perhaps general relativity governs the interactions of objects in both universes in this other world, so it is correct to say that the laws are the same in both worlds. *If* this is correct, then the general principle is incorrect. And it might be obvious, but it is worth pointing out that Gödel does allow that different "spatiotemporal" structures are physically possible, so some variation in the nature of "the arena in which objects find themselves" is allowed across physically possible worlds.³⁵ If we are trying to formulate a principle that bridges physical possibility and actuality, we need to formulate it in such a way that the above considerations do not serve as counter-examples to it. This seems to me to be a difficult task.³⁶

But perhaps this not the principle that motivates Gödel. He writes:

> The mere compatibility with the laws of nature of worlds in which there is no distinguished absolute time, and, therefore, no objective lapse of time can exist, throws some light on the meaning of time also in those worlds in which an absolute time *can* be defined. For if someone asserts that absolute time is lapsing, he accepts as a consequence that whether or not an objective lapse of time exists (i.e., whether or not a time in the ordinary sense of the word exists) depends on the particular way in which matter and its motion are arranged in the world.... a philosophical view leading to such consequences can hardly be considered satisfactory.³⁷

---

34 See Phillip Bricker, "The Fabric of Space: Intrinsic vs. Extrinsic Distance Relations," *Midwest Studies in Philosophy* 18 (1993): 271–294.
35 Compare with Chris Smeenk Earman and Christian Wüthrich, "Time Travel and Time Machines," in Callender, *The Oxford Handbook of the Philosophy of Time*, 597.
36 Compare with John Earman *Bangs, Crunches, Whimpers, and Shrieks: Singularities and Acausalities in Relativistic Spacetime* (Oxford: Oxford University Press, 1995), 198).
37 Gödel, 562.

In light of what Gödel says above, perhaps he accepts the following rationale for why the absence of time in some physically possible worlds entails the absence of time in the actual world. First, if time exists in one world rather than the other, it is only by virtue of the difference of the distribution of matter across the respective spatiotemporal manifolds at those worlds. (One consequence of general relativity is that there is a correlation between spatiotemporal curvature and the distribution of mass across space-time.) But, second, wouldn't one have thought that whether a dimension of space-time is a temporal dimension is determined entirely by the intrinsic nature of space-time rather than by the way it is occupied by material objects? So the different distributions of matter in the merely possible space-time and the actual space-time can't account for whether time exists in the actual world. Since time is absent in the merely possible and nothing extra in the actual world could account for the presence of time, time is absent in the actual world as well.

This does not strike me as a compelling argument. First, there are very tricky questions we must ask ourselves about the nature of the dependence between space-time and its occupants: there is a correlation between physically possible space-time structures and physically possible distributions of matter. But does space-time have the structure it has *in virtue of* material objects possessing certain properties and standing in certain relations to each other? (I am asking a question here not about causal dependence but about a strong kind of modal or essential dependence, just as one is not concerned with causal dependence but rather a stronger form of dependence when one asks whether something is good in virtue of God's loving it or whether God loves something in virtue of its being good.) If space-time does have the structure it has in virtue of facts about material objects, the second claim in the above rationale looks shaky.

But if spatiotemporal structure is not metaphysically determined by facts about the occupants of space-time, the first claim in the above rationale is dubious as well. If spatiotemporal structure is not metaphysically determined by facts about the occupants of space-time but is merely nomically (and so merely contingently) connected, then one could endorse the following package: there is a set S1 of all the metaphysically possible worlds in which space-time has the structure that it has in one of Gödel's physically possible worlds and is completely empty of material objects; there is a set S2 of metaphysically possible worlds in which space-time has the structure that it has in the actual world yet is completely empty of material objects; time exists in none of the worlds in S1 in virtue of the intrinsic structure of space-time found in those worlds; time exists in all of the worlds in S2 in virtue of the intrinsic structure of space-time found in those worlds. If this package of claims is correct, then the first claim in the reconstruction of Gödel's rationale is false.

There are many other ways to try to defend Gödel's claim that the lack of time in one of these physically possible worlds implies a lack of time in our own world. For example, Yourgrau and Savitt argue that inhabitants of a rotating universe world would have the same "experiences of temporal passage" that we have, yet time would not pass in those worlds, so we lack sufficient evidence to think that time passes in our world.[38] I confess to having inchoate worries about the style of argument this seems to exemplify: in an evil demon world we would have the same experiences of material objects that we in fact have, but this doesn't mean that we lack evidence for the existence of material objects. Does the fact that the relevant scenario is physically possible rather than merely metaphysically possible make a difference? Maybe. Perhaps a belief that P counts as knowledge only if the believer can rule out any *relevant* proposition that is logically inconsistent with it; perhaps being true in some physically possible world suffices for being a relevant proposition; but a proposition's merely being true in some metaphysically possible world does not suffice for

---

38 Palle Yourgrau, *A World without Time: the Forgotten Legacy of Goedel and Einstein*, (New York: Basic Book, 2005). 53; Savitt, "The Replacement of Time," 467–472.

it to be a relevant proposition. On this kind of epistemological view, we might be able to know that we have hands without being able to know that time is real.

One final attempt, and then I will move on. Some philosophers think that questions concerning the nature of time belong to a family of fundamental ontological questions such that answers to these questions are metaphysically necessary if true at all. Included in this family of questions are questions about the nature of possibility—are there really possible worlds or is talk of possible worlds merely a useful fiction?; questions about the structure of objects—are ordinary objects composites of form and matter, or complexes of substratum and attribute, or bundles of properties?; and questions about the nature of mathematics—are there Platonic mathematical objects or is some sort of "nominalism" correct? It is no surprise that philosophers who think that questions about the nature of time are like this have tended to be philosophers who believe in "objective lapsing of time." Perhaps Gödel thought it was part of the ordinary conception of time that questions about the nature of time cannot be merely contingent, so the question whether time passes cannot be merely contingent.[39] If so, then a proof of the physical possibility of worlds without time suffices for a proof for the nonexistence of time, since physical possibility suffices for metaphysical possibility.[40] But the *physical* possibility of these worlds is not an idle wheel, since one could quite reasonably hold that, in general, it is easier to acquire evidence for what is physically possible than for what is merely metaphysically possible but physically impossible. If Gödel had simply described a putatively merely metaphysically possible world in which space-time had an unusual structure, the dialectical and epistemological forces of his argument would have been much weaker.

---

39 This line of thought is suggested by Earman, *Bangs, Crunches, Whimpers, and Shrieks*, 197–199.
40 Steven Savitt, "The Replacement of Time," *Australasian Journal of Philosophy* 72 (1994): 466, suggests a related line of reasoning.

Gödel attempted to show that time was unreal by way of considerations from fundamental physics. More recently philosophers of physics have worried about the reality of time under the guise of what is called "the problem of time" in quantum gravity. This problem, as I dimly comprehend it, concerns the cancellations of any variable for temporal measurement in certain fundamental theories that attempt to reconcile quantum mechanics and general relativity.[41] If time is real, its reality is not explicitly expressed by these fundamental theories, which seems to suggest that time is merely an emergent feature of a manifold rather than a fundamental one. For example, Kiefer writes that "both the familiar time and its arrow can thus be understood from quantum gravity, which is fundamentally timeless."[42] But perhaps the nonfundamentality of time is not a radically new view, since it has an ancestor in the view that time is a mere aspect of space-time. So it is worth pointing out that some philosophers of physics have argued that spatiotemporality itself is nonfundamental. For example, Witten writes, "Contemporary developments in theoretical physics suggest that another revolution may be in progress, through which a new source of 'fuzziness' may enter physics, and space-time itself may be interpreted as an approximate, derived concept."[43]

Given that whether spatiotemporality is fundamental is currently a live issue, it's worth pausing to ask whether Gödel's argument could be modified. The basic idea is that a manifold counts as a spatiotemporal manifold only if certain constraints are met by it, one of which is that

---

[41] A number of papers in Craig Callender and Nick Huggett, *Physics Meets Philosophy at the Planck Scale: Contemporary Theories in Quantum Gravity* (Cambridge: Cambridge University Press, 2001), discuss this issue. See also Claus Kiefer, "Time in Quantum Gravity," in Callender, *The Oxford Handbook of the Philosophy of Time*, 663–678.

[42] Kiefer, 2011), 678.

[43] Edward Witten, "Reflections on the Fate of Spacetime," in Callender and Huggett, *Physics Meets Philosophy at the Planck Scale*, 125. The idea that space and time might be real without being fundamental is of course not a new idea; for example, Bernardino Bosenquet, 'The Impossibility of Creation from Eternity According to St Bonaventure," *Proceedings of the American Catholic Philosophical Association* 48 (1974): 121–135, endorses it. But that this possibility is now seriously entertained by philosophers of physics is part of an exciting new phase of the exploration of this idea.

there is some aspect to that manifold that can be at least roughly in the neighborhood of what we conceptualize as being temporal. But it is a contingent matter whether this constraint is met, and possible worlds containing manifolds with strange enough structures are not worlds in which the constraint is met. So being spatiotemporal is an emergent, contingent feature of a manifold and not a fundamental one.

If it is a central part of our concepts of space and time that they are metaphysically fundamental phenomena, then the conclusion that space-time is an emergent, nonfundamental aspect of the manifold suffices to show that there is no time. But frankly I doubt that it is a part of our concepts of space and time that they are fundamental; there might nonetheless be interesting philosophical conclusions to learn about them. (It is obviously a central part of the A-theory of time, for example, that time is not a mere aspect of the spatiotemporal manifold.)

There are a number ways space-time could fail to be fundamental. But I'll focus on the following possibility, which I will put in more metaphysical terms. It might be that being spatiotemporal is a complex and *extrinsic* property of things that are "smaller" than the universe. In order for me, for example, to enjoy spatiotemporality, I must be a part of something that contains me that has the emergent property of being spatiotemporal. Only the universe as a whole is a candidate for having any kind of intrinsic (yet nonfundamental) spatiotemporality. Arguments that then turn on the intuition that spatiotemporal features such as shape are intrinsic would be in trouble. For example, David Lewis objects to one popular view about how material objects persist through space-time by appealing to the idea that shape properties are intrinsic; this is the kind of argument that would be deeply problematized.[44] Briefly, Lewis's argument is as follows: things either persist through time by having temporal parts at each time they exist

---

44 David Lewis, *On the Plurality of Worlds*, (Basil: Blackwell Ltd., 1986), 204.

at or by being wholly present at each time they exist at. If something is wholly present at times t1 and t2 (or at space-time regions R1 and R2) and changes its shape from t1 to t2, then strictly the shapes it successively enjoys are not intrinsic properties but are something like relations to times (or space-time regions). But, according to Lewis, shape properties are intrinsic properties rather than relations to times (or space-time regions). So things that persist by being wholly present cannot undergo intrinsic change. But ordinary things do undergo intrinsic change as they persist through time. So ordinary objects have temporal parts. This interesting argument can be challenged in many ways, but for my purposes it suffices to say that the argument is completely undercut if shapes are extrinsic, nonfundamental features.

A second example: theories about the nature of parts and wholes that tie the possession of certain mereological features to the possession of certain spatiotemporal features would also be problematized. For example, Ned Markosian raises what he calls "The Simple Question," which asks for the necessary and sufficient conditions a thing must satisfy to be without proper parts and defends the claim that, at least for physical objects, this necessary and sufficient condition is being a maximally spatially continuous object.[45] But being maximally spatially continuous is a contingently possessed extrinsic property (if spatiotemporality is nonfundamental in the respect just mentioned) of the things that have it, whereas being without parts is an intrinsic property. So the latter cannot be necessarily equivalent with the former.[46]

A third example: if to be physical is to be spatiotemporal, then perhaps physical objects are only contingently physical. And if this is so, then, if to be abstract is to be nonspatiotemporal, physical objects

---

45 Ned Markosian, "Simples," *Australasian Journal of Philosophy* 76 (1998): 213–226.
46 The view that I prefer, according to which, roughly, there is no informative necessary and sufficient conditions for being a simple is not refuted by this possibility. See Kris McDaniel, "Brutal Simples," *Oxford Studies in Metaphysics* 3 (2007): 233–266, for an exposition and defense of this view.

could have been abstract objects. There is a worry that we will lose whatever grip we might have had on the concrete/abstract distinction if one consequence of spatiotemporality's being a nonfundamental and extrinsic feature is that nonspatiotemporality is also an accidental feature.

In general, we must be cautious, for any intuitive connections between spatiotemporality and modality are in danger of being severed if spatiotemporality is nonfundamental. We might have been inclined to explain or tie together the necessary existence of abstract objects such as numbers with their status as atemporal beings. Recall, for example, my earlier discussion about whether being a necessary being required being eternal. But on the hypothesis I am considering, some of the beings that are in fact spatiotemporal could have existed without being spatiotemporal, and some of these beings might still be good candidates for being contingently existing objects. Eternity on this view must not be taken as either a sufficient condition for or an indication of the enjoyment of necessary existence.

In short, the required changes to our worldview might be quite extreme. We should begin to consider the ramifications of space-time's being nonfundamental for our metaphysical inquiries. We might find them to be just as interesting as the consequences of older arguments for the unreality of time.

Acknowledgments

I thank Adam Elga, Sandra Lapointe, Jill North, Ted Sider, Christian Tapp, and the audience at the Eternity Conference at the University of Bochum, for helpful comments on earlier drafts of this chapter.

Reflection
BORGES ON ETERNITY
William Egginton

Throughout his writing Borges gravitates again and again to the same paradox: we are ineluctably temporal, which is also the reason why we inevitably desire eternity; and dreams of eternity can only fail, condemned as we are to temporality. While this paradox is expressed in the very title of one of his collections, *The History of Eternity*, Borges's thoughts on both time and its negation are scattered throughout his work, even as they reflect a remarkable consistency

"The New Refutation of Time," published in 1952 as part of Borges's *Other Inquisitions*, consists of two versions of the essay: one published in 1944 and a more concise revision of that essay republished in 1946. In the second section of the original 1944 essay Borges recounts an anecdote conveying a mystical sense of pure repetition he experienced on returning to a village he had visiting during his childhood: "I write it, now, like this: That pure representation of homogeneous facts—night in serenity, limpid little wall, provincial odor of the deep jungle, fundamental earth—is not merely identical to that of this corner from so many years ago; it is, without similarities or repetitions, the same. Time, if we can intuit that identity, is a delusion: the indifference and inseparability of a moment from its apparent yesterday and another from its apparent today, is enough to disintegrate it."[1] In the revision he dispenses with the anecdote, inserting in its place a discussion of the Confucian parable of Chuang Tzu and the butterfly. Imagining the possibility that at least one of the millions who have read Chuang Tzu's dream in the twenty-four centuries since he wrote it may have also dreamed it in identical form, Borges asks: "Is it not enough for one sole term to repeat itself in order to ruin and confuse the history of the world, to denounce that there is such a history?"[2]

The essays thus performs the very impossibility of its argument, which Borges himself mentions in the preface written for the 1952 combined edition: "It is not hidden from me that this is an example of the monster that the logicians have called *contradictio in adjectio*, because to say of a refutation of time that it is new (or old) is to attribute to it a predicate of a temporal nature, which installs the very notion that the subject wants to destroy."[3] In the final passages of the text he then explicitly affirms the futility of the refutation when he writes:

> *And yet, and yet* . . . [italicized phrase in English in the original] to negate temporal succession, to negate the I, to negate the astronomic universe, are obvious desperations and secret consolations. Our destiny . . . is not frightening because unreal, it is frightening because irreversible and ironclad. Time is the substance of which I am made. Time is a river that tears me down, but I am the river; it is a tiger that rips me apart, but I am the tiger; it is a fire that consumes me, but I am the fire. The world, unfortunately, is real; I, unfortunately, am Borges.[4]

---

1 Borges, *Obras completas*, 4 vols. (Buenos Aires: Emecé, 1996), vol. 2, 143. All translations in this chapter are my own.
2 *Obras completas*, vol. 2, 147.
3 *Obras completas*, vol. 2, 135.
4 *Obras completas*, vol. 2, 149.

Thus a text that begins by declaring its intention to refute time succumbs by its very arguments to time, and ends by affirming time's dominion over all and the author's ultimate identity with time.

This assertion of the dominance of the ephemeral over our intuition of the eternal appears in a variety of forms. In 1941 Borges wrote of a secret society that set itself the task of creating a world in exhaustive detail. The planet of Tlön, ostensibly the mythological creation of the people of the land of Uqbar, had invented an idealism so systematic, so realistic, that the mere imagination of something usurped the resistance of reality and that thing would emerge, as imagined. In a postscript to the story postdated to 1947, six years and a world war away from his own moment in time, Borges writes eerily, enigmatically of the incursion of Tlön into the real world:

> Almost immediately reality gave way in more than one regard. The truth was it desired to give way. Ten years ago whatever symmetry with the appearance of order—dialectic materialism, anti-Semitism, Nazism—was enough to bewitch men. How not to submit to Tlön, to the minute and vast evidence of an ordered world? It is useless to reply that reality is also ordered. Perhaps it is, but in accordance with divine laws—I translate: inhuman laws—that we will never perceive. Tlön may be a labyrinth, but it is a labyrinth woven by men, a labyrinth destined to be deciphered by men.
>
> The contact with and habits of Tlön have disintegrated this world. Enchanted by its rigor, humanity forgets and forgets again that it is a rigor of chess masters, not of angels.[5]

The prophetic nature of the postscript is strangely disguised by its author's pretending to have penned it long after the time he actually did; nevertheless, it reads with a somewhat

---

5 *Obras completas*, vol. 1, 442.

postapocalyptic tone. Its author speaks to us from a time when it is apparently too late to avoid the catastrophe we should have seen coming.

And what is the catastrophe? Just as the power of the idea, of the word, holds sway over material reality in Tlön, so too does the mere idea of a perfectly organized world hold sway over the imagination of men. The problem, Borges says, is our inability to distinguish properly between a rigor of chess masters and a rigor of angels. We invent worlds, invent systems, and then see in those aspects of the world that are not of our invention the traces of the same organizing rigor we have put into our inventions; in confusing the rigor of chess masters with a rigor of angels, we confuse the ephemeral order of our ideas with the eternal order of being. When we do so, our fictions have a strange tendency to become reality.

Borges once imagined what the universe would be like if it were to take the form of a library. This library, he wrote, would consist of interlocking hexagons, the walls of each containing a certain number of bookshelves and books, each containing a certain number of pages and lines. The inhabitants of the library would wander the hexagons for years, decades, and generations, trying to discover the meaning of their existence in the pages of these books. While no single book in the library would be identical to any other, Borges writes, what is certainly the case is that every possible combination of the letters of the alphabet would be found in its volumes. Among volumes and volumes of nonsense, in other words, would sit the complete works of Shakespeare, as well as another version of them that would be missing one letter, another missing two, and so on.

The remarkable insight of this tale lies in how the narrator, himself lost in this universe of meaning, describes the inevitability of our desire for a dimension other than that of our mundane temporality. By a seemingly flawless deduction, the library's scholars theorize the existence of a book of books, a book that contains a perfect explanation of the otherwise unorganized and overwhelming morass of information. Since every book exists, they reason, this one too must exist, and in its pages someone has read or at some point will read the timeless meaning and purpose of the library, and that person has become or will become enlightened. The narrator himself is uncertain about this hypothesis, but he expresses a hope that some day, someone, even if he himself will never know of it, will discover the ultimate meaning of the library, and that in this discovery the library itself and the existence of so many searchers will be justified.

When, in yet another story, Borges's character Ireneo Funes falls off his horse and hits his head, he loses his ability to forget. From that moment on, Funes remembers not only every object he has ever seen, but every aspect of every object, and every moment of his seeing of that object. The perfection of Funes's memory, though, soon becomes an overwhelming impediment. He is unable to assimilate new experiences; as Borges writes, he can perfectly recount the memory of an entire day, but it takes him an entire day to do so. He is surprised at his own reflection in the mirror every morning, since his perfect memory records every minute difference that time imposes on him; and he becomes increasingly annoyed at he generality of language, that we should use the same drab word, "dog," to refer to the animal facing one direction at 3:14 and another direction at 3:15.

The point of his thought experiment is that, as Funes himself puts it toward the end of the story, no matter how many languages he learns it is hard to say that he is capable of thought. In attaining a perfect memory, Borges's character also inevitably loses the ability to think, because thinking necessarily involves generalization, abstraction, and the forgetting of minor differences. In truth, Funes could never be surprised at his own reflection or annoyed at generality of our use of language; for in order to feel either surprised or annoyed, he would have had to perceive the connection between those different moments in time and hence forget the difference between them, if only long enough to see that they were two distinct moments.

At the most basic level, then, the sensual perception we depend on for any thought, any knowledge whatsoever, is itself produced through a complex process of remembering and forgetting, of immersion and abstraction that is inimical to the notion of knowledge as perfectible. In many ways the story is the ideal thought experiment for considering what a strictly timeless knowledge would be like if there could be such a thing, and how the very idea of such knowledge crumbles when confronted with the conditions under which we in fact encounter the world. And yet . . . and yet, as Borges would write, the ineluctable fact of our finitude does nothing to dampen the desire for eternity. Indeed, it may be that each is only conceivable against the absent ground of the other.

Reflection

MUSIC AND ETERNITY

Walter Frisch

Music is often considered one of the most temporal of the arts. Music unfolds in time and can only be experienced sequentially. Musical notation is an attempt to fix or capture that temporality, to render it graphically visual and reproducible. Yet, although we can take in a page of music at a single glance, we still need to hear or perform the piece in real time. Moreover, notated musics are a relatively small portion of all the musics that have existed or exist in the world today.

Although temporality and eternity are clearly not the same concepts, it is perhaps not surprising that creators of music have associated the two, since temporality extended can become eternity. Musicians have long sought to reflect ideas of eternity, both within the structural processes of music and in relation to the kinds of extramusical ideas (such as texts and images) often associated with music.

The topic of music and eternity obviously calls for a crosscultural, global approach, since these concepts are at play in other traditions. (A recent CD of Indian music is called *Eternity—The Soul of India*.) However, in this brief essay, I will only deal with the western art tradition, with which I am most familiar. I will focus on five pieces created across two and a half centuries—by J. S. Bach, Richard Wagner, Gustav Mahler, Olivier Messiaen, and Philip Glass—that

I believe exemplify (but by no means exhaust) the ways western composers have sought to convey the idea of eternity in music.

In western Europe, musical manifestations of eternity were initially linked with sacred music. It became customary in the Middle Ages—although the practice dates from ancient Jewish traditions—to end a Catholic rite or a chant with a phrase that became known as the Doxology.

In Latin, the familiar text is "Gloria Patri, et Filio, et Spiritui Sancto. Sicut erat in principio, et nunc, et semper, et in sæcula sæculorum. Amen." In English this is most often rendered as "Glory be to the Father, Son, and Holy Ghost. As it was in the beginning, is now, and ever shall be, world without end. Amen." (The translations of *semper* as "ever shall be" and *in saecula saeculorum* as "world without end" date from the Anglican Book of Common Prayer of the sixteenth century.)

The Doxology has been sung in churches for well over a thousand years. But musical settings that seek specifically to capture its image of eternity are more recent. Perhaps the finest and most exciting is that of J. S. Bach, at the end of his *Magnificat* of 1723 (BWV 243).

J. S. Bach, *Magnificat*

At the words *in saecula saeculorum* Bach's five-part choir breaks into a polyphonic imitative style; successive voices take up the phrase and sustain a long note on the *lo* of *saeculorum*. Imitation of this kind (of which Bach was a master, including in his many fugues) is a splendid musical emblem of eternity, because the successive entrance of voices could in principle go on forever. In the final moments of the *Magnificat*, however, all five vocal parts, accompanied by a full orchestra, including trumpets and drums, converge on the resonant final "Amen."

In the Romantic period, especially in Germany, musical eternity became more secularized, tied to sensual and spiritual (though not necessarily Christian) longing. Nowhere is this more apparent than in the epitome of German Romantic music, Wagner's opera *Tristan und Isolde* (1859). Although Wagner based his work on a medieval romance, his infatuation with the work of Schopenhauer led him to filter the original love story of Tristan and Isolde through the philsopher's concept of life as a cycle of frustration and unfulfillment that can only be released in death. In the famous love duet of act 2 of *Tristan*, the lovers sing of being united in the eternal bliss of death:

So starben wir,
um ungetrennt,
ewig einig
ohne End',
ohn' Erwachen,
ohn' Erbangen,
namenlos
in Lieb' umfangen,
ganz uns selbst gegeben,
der Liebe nur zu leben!

[Thus might we die, so that together, eternally one, without end, without waking, without fearing, namelessly enveloped in love, given up to each other, to live only for love!]

The word *ewig* ("eternal" or "forever") appears no fewer than a dozen times in the libretto for this scene.

In *Tristan* Wagner created a powerful musical language for the expression of infinite longing. Bach's imitative polyphony would not work here. Rather, Wagner writes in a style that delays firm resolution and keeps moving forward, because it does not settle down into stable harmonies or on firm cadences. Wagner's main device is the sequence, in which a melodic-harmonic phrase gets repeated successively at different pitch levels and (like Bach's imitative polyphony) could in theory go on indefinitely. The musical phrase for the first two lines of the libretto is repeated a third higher for the second two lines.

Richard Wagner, *Tristan und Isolde*, act 2, scene 2

The fifth and sixth lines are also treated sequentially.

Composing some fifty years after Wagner composed *Tristan*, Gustav Mahler was strongly influenced not only by Schopenhauer's longing but also by Nietzsche's idea of the "eternal return." Mahler's great symphonic song cycle *Das Lied von der Erde* (*The Song of the Earth*; 1909) is a setting of ancient Chinese poems as rendered in German by Hans Bethge. The final song, "Der Abschied" ("The Farewell"), ends with a long paean to the eternal renewal of nature:

Die liebe Erde allüberall
Blüht auf im Lenz und grünt Aufs neu!
Allüberall und ewig
Blauen licht die Fernen!
Ewig... ewig...

[The dear earth everywhere blooms in spring and grows green afresh! Everywhere and eternally, distant places have blue skies! Eternally... eternally.]

The way in which Mahler draws out and repeats the final word, *ewig*, was unprecedented in music at the time.

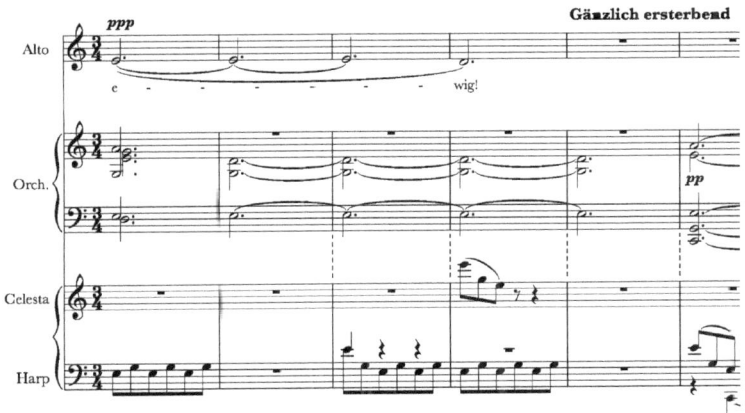

Gustav Mahler, *Das Lied von der Erde*, "Der Abschied"

The voice's two-note theme descends only to the note *above* the final tonic C, thus hovering without resolution. In the orchestra a harmonically ambiguous figure floats between the harps and a bell-like keyboard instrument, the celesta. Mahler closes with one of the most open-ended sounds available to him, a triad with a sixth added on top. There is no question this is Mahler's attempt to convey a musical idea of eternity.

For the twentieth-century French master Olivier Messiaen (1908–1992), the concept of eternity was intimately bound up with both his Catholic religious faith and his love of nature. One of his most powerful compositions bears the notion of eternity both in its title, *Quatuor pour la fin du temps* (*Quartet for the End of Time*, 1941) and in its fifth movement, "Louange à l'Éternité de Jesus" ("Praise to the Eternity of Jesus"). Messiaen wrote this quartet for clarinet, violin, cello, and piano while a prisoner of war in a Germany during World War II.

The composition was inspired by a passage from the Book of Revelation that included this text: "And the angel which I saw stand upon the sea and upon the earth lifted up his hand to heaven, and sware by him that liveth for ever and ever . . . that there should be time no longer" (Revelation 10:5–6). The tempo marking for the fifth movement of the quartet, probably unique in music, is in a certain sense unachievable: *infiniment lent,* "infinitely slowly." In his preface to the score, Messiaen writes that the theme "magnifies with love and reverence the eternity of the Word, powerful and gentle, 'whose time never runs out.'"

Recent American music often characterized as "minimalist" comes closest among modern art music styles to adumbrating eternity. This music unfolds by a process of gradual, incremental change and is as such very different from the more goal-oriented music of high modernism. Philip Glass's *Two Pages* of 1968, written for piano or electric keyboard, is a good example. Glass directs that a simple melodic figure of five notes is to be repeated rapidly thirty-four times.

Philip Glass, *Two Pages*

The figure is then modified by the addition of four more notes identical to the first four. This expanded nine-note figure is directed to be repeated eighteen times. This figure is in turn is expanded by three notes that are the first three notes of the original, and this new figure is repeated fourteen times. The musical process thus literally shapes the composition; it is the "form." As with some of the other music I have discussed, this process could in principle continue infinitely, into eternity. In fact, it does not: the piece lasts eighteen minutes.

Glass has remarked that for him music is like "an underground river," one that is constantly flowing and that he as a composer can choose to listen to and write down.[1] This fascinating image or metaphor captures vividly the relationship of his music to an idea of eternal flow. In some ways, Glass's comment reminds us of an early Romantic notion of music as articulated by E. T. A. Hoffmann (who adapted it from the physicist Johann Ritter): that music is always around us in nature and that composers hear these sounds as music, "first as individual chords, then as melodies with harmonic accompaniment."[2]

For many centuries, then, western composers have been preoccupied with ideas of eternity—in the realms of the secular or sacred, the physical or metaphysical, or the natural or supernatural. Bach's kinetic polyphony, Wagner's yearning sequences, Mahler's delicate dissonance, Messiaen's slow themes, and Glass's processes of incremental change: these are just a few examples of how a great composer can adapt the musical language of his time to convey, within the finite space of a composition, a musical image of eternity.

---

[1] Glass makes this comment during an interview in the documentary *Glass: A Portrait of Philip in Twelve Parts*, dir. Scott Hicks (2007; DVD 2009).

[2] E. T. A. Hoffmann, "Johannes Kreisler's Certificate of Apprenticeship," trans. Max Knight, *19th-Century Music* 5 (1982): 192.

Reflection

THE KADDISH

Abraham P. Socher

> And so, if following Plato, we wish to give things their right names, let us say that God is eternal, but the world is everlasting.
> —BOETHIUS, The Consolation of Philosophy (523 A.D.)

> We're all gonna be here forever
> So Momma don't you make such a stir
> Just put down that camera
> And come on and join up
> The last of the family reserve.
> —LYLE LOVETT, "Family Reserve," *Joshua Judges Ruth* (1992)

Once, when I was eleven or twelve, my mother said, "When I die, I expect to go to heaven and I expect to see my daddy there." I don't know exactly what occasioned her statement of metaphysical confidence at that particular moment, standing in the dining room. My grandfather had been dead for at least a decade, but she still missed him, regretted the sad circumstances of his final days, and never shied from bold declarations. "Well," said my father, "how old will you be and how old will he be?"

Such questions are almost as venerable as the hopes for a world to come that they challenge, but, of course, I didn't know that at the time, and my father's response startled me. I had neither been taught nor pondered any particular doctrine of eternal life, but it hadn't occurred to me that the idea might be not just unlikely but difficult to make any sense of at all. My father's question came back to me when I recited the Mourner's Kaddish for him a few years ago. I did not find the prayer consoling, but then it was not written to console and famously does not mention death or promise an afterlife. The Kaddish began sometime in the early rabbinic period as what historians of liturgy call a doxology, a statement of belief that, in this case was recited, at least at first, primarily after a sermon.[1]

The sonorous Aramaic chant of the Kaddish is familiar, but the words are worthy of close attention, in part because they do suggest a theological notion of eternity, though, strikingly, not one that is humanly attainable. The prayer begins: "Magnified and sanctified may His great name in the world which He created by His will. May He establish His kingdom in your lifetime and in your days and in the lifetime of the all the house of Israel, swiftly and soon. And say: Amen."[2]

What is devoutly wished for here is an achievement *within* time, "in your lifetime . . . and in the lifetime of the whole household of Israel." It is at this point that the entire congregation responds with its key declaration: "May his great name be praised forever and all time." The phrase Jonathan Sacks translates here as "forever and all time" and others have translated as "forever and ever" is *lealam uleolomei almaia*. Both are attempts to convey the sense of the prayer's threefold intensification of the Aramaic form of the Hebrew word *olam*, which in the Bible was often used to mean

---

[1] See Babylonian Talmud, *Sotah* 49a, and, for classic discussions of the historical origins of the Kaddish, see David Del Sola Pool, *The Kaddish* (Leipzig, 1909, reprinted in New York, 1964), and Joseph Heinemann, "Prayers of Beth Midrash Origin," *Journal of Semitic Studies* 5 (1960), 264–280. See, now, David Shyovitz, "You Have Saved Me From the Judgement of Gehenna": The Origins of the Mourner's Kaddish in Medieval Ashkenaz," *AJS Review* 39:1 (April 2105), pp. 49–73, which incorporates a useful literature review, as well as making several original points.

[2] *Koren Siddur*, with introduction, translation, and commentary by Jonathan Sacks (Jerusalem. 2009), 60.

permanent or indefinite or everlasting existence, though by the time of the Kaddish's composition it had also come to denote the world or universe.³ The content of the congregation's response, then, is not so much one of direct praise but rather the hope that such human praise will continue forever and ever. The classical Hebrew analogue to this Aramaic phrase is *barukh shem kevod malchuto le-olam va-ed*, "blessed be the name of His glorious kingship forever and ever," which was originally employed in response to the high priest's invocation of God's ineffable name.⁴

The reader's immediate reply to the congregation's expression of hope that God's name be praised forever reads not only as an immediate attempt to fulfill it, but also as a metaphysical clarification or caveat:

Blessed and praised, glorified and exalted, raised and honored, uplifted and lauded be the name of the Holy One blessed be He, [who is] beyond any blessings and hymns, praises and consolations which may be uttered in the world. And say Amen.

There are, it seems, two distinctions being made here. The first is between "the name of the Holy One Blessed be He"—itself repeatedly alluded to but strikingly never actually uttered in the prayer—and the divine bearer of that name. The second, more implicit, distinction is between the kind of eternity that can be predicated of a name or its praise, even *the* great name, and that which can be predicated of the bearer of that name. The first kind of eternity is, even if it continues forever, within time; it consists in

---

3 See, e.g., Ecclesiastes 1:4, Deuteronomy 15:17, and Genesis 9:16, respectively. A related meaning is ancient, as in, e.g., Isaiah 63:9.

4 This line is adapted from Psalm 72:19. From very early on in the Second Temple period, it was employed as a liturgical response to the classic statement of faith of the opening line of the *Shema*, "Hear O' Israel, the Lord is God, the Lord is One" (Deuteronomy 6:4). See Ismar Elbogen, *Jewish Liturgy: A Comprehensive History*, trans. Raymond Scheindlin (Philadelphia, 1993), 21–24.

an endless series of utterances "in the world."[5] The assertion that the bearer of the divine name is beyond all such praise suggests a different, perhaps more Platonic, sort of eternity.

The *Or Zarua*, a thirteenth-century work by the Ashkenazi pietist Isaac ben Moshe of Vienna, is apparently the first to describe the custom of reciting the *Kaddish Yatom*, the "Orphan's Kaddish," which was recited by an orphan at the end of the Sabbath. This was a key moment in the historical process of turning the Kaddish into a prayer of mourning.[6] That the Kaddish should have become such a prayer despite its failure to mention death or the afterlife or to offer any words of consolation, as do, for instance, medieval Catholic prayers for the dead, is surprising—an issue to which I shall return—but not entirely so.

For there is a well-established rabbinic practice of affirming *tzidduk hadin*, the justice of the (divine) decree, in the face of tragedy. Hence the still-current practice of saying *barukh dayan ha-emet* on hearing of someone's passing. I have heard this taken, and even offered, as an expression of consolation, but what it actually means is "Blessed is the true Judge," that is, the Judge is blessed even, or particularly, at a moment when the bereaved might be tempted to curse him, or deny that his judgment is true or even that the there is a divine judge at all. In fact, the most powerful expression of unbelief in the rabbinic tradition is *leit din ve-leit dayan*, "there is no judgment and there is no judge."[7] The Kaddish's determination that God be praised for all of eternity is, in the context of mourning, an affirmation that there is, and will always be, a divine judge.

---

5 The use of the Aramaic word *alma* here is, no doubt, a deliberate literary choice, meant to echo *lealam uleolomei almaia*, though here the primary meaning of the term is "world."

6 Two other thirteenth-century Ashkenazi texts also mention the practice, the *Mahzor Vitry* and the commentary to the Siddur of Eleazar b Judah of Worms. See Israel Ta-Shma, *Early Franco-German Ritual and Custom* (Jerusalem, 1994), 299–310 (in Hebrew), and Shyovitz, op. cit.

7 Most famously attributed to rabbinic Judaism's arch-heretic Elisha ben Abuya, Babylonian Talmud Kiddushin 39b.

At least one reason for choosing the Kaddish to express faith in God's judgment is a strange story about the great second-century figure Rabbi Akiva, which appears in an odd early medieval compilation of rabbinic legends called the *Alphabet of Rabbi Akiva* and is retold in the *Or Zarua* and elsewhere. In this story, Rabbi Akiva is walking in a cemetery and meets a naked man who is carrying a heavy load of wood and "running like a horse." Rabbi Akiva, thinking that he is speaking to a live human being, asks if there is any way he can relieve him of his duties. But it turns out that the man is, in fact, dead and suffering the punishments of hell (Gehenna) for his sins as a tax collector. How can he be relieved of his torments? Only if his abandoned son comes before a congregation, says *barkhu et adonai ha-mevorakh* ("let us bless God Who is blessed") and is answered with *yehe shmei rabba mevorakh* ("may his great name be blessed"). With some difficulty, Rabbi Akiva manages this. This, says the *Or Zarua*, is the precedent for an orphan to recite Kaddish.

Several scholars have noted the medieval Christian context and flavor of the idea that prayer can relieve the purgatorial suffering of our loved ones, and Leon Wieseltier has meditated on this theme in his extraordinary philosophical memoir *Kaddish*.[8] This is an important historical point, but I want to return here to the Kaddish's sense of eternity and what strikes me as a deep philosophical irony in choosing this, of all prayers, as the vehicle for easing our loved ones' afterlife. To return to the distinctions I identified in the Kaddish: God is not to be identified with his name, and it is only his name, a word in mortal human mouths and documents, that can be praised forever and ever. It will, in short, be endlessly praised if we never stop praising it. But God itself is, the Kaddish asserts, untouched by such praise: "He is beyond

---

8 Leon Wieseltier, *Kaddish* (New York: Vintage, 1998), 36–49.

all blessings and hymns, praises and consolations which may be uttered in the world."

That is, the prayer itself seems to with Boethius, among others, in distinguishing between on the one hand the everlastingness that can characterize the world and that which is in it and on the other hand eternity as a kind of timelessness, which is beyond it. The mystery and power of "the great Name," at least here, would seem to be that it is a part of this world while referring to that which is beyond it.

One could map these rabbinic thoughts about God, eternity, and time onto this or that philosophical theory. And the evident importance given to—and perhaps philosophical puzzlement over—the this-worldly name of an unworldly and timeless being in the text of the Kaddish *is* philosophically suggestive. But I think that, here and elsewhere, it would do violence to rabbinic thinking to over systematize it.

I will end, rather, with the irony that nothing in the text of the Kaddish itself suggests that anything in the world, including us, or any temporal successors we may have after our death, can escape into a timeless afterlife. Indeed, such timelessness would appear to be, at least here, an exclusive property of God.

Glossary

ANCIENT GREEK AND LATIN

The most important terminological distinctions are those concerned with capturing the conceptual difference between everlastingness in time and eternity proper, that is, the timeless (and for some thinkers even durationless) mode of being. Readers new to this material should begin by noting the Greek and Latin terms traditionally associated with each of these conceptions and by observing that there is a general correspondence between these Greek and Latin terms. Everlastingness, then, came to be referred to by the Greek noun ἀϊδιότης (*aidiotês*) and adjective ἀΐδιος (*aidios*) and the Latin *sempiternitas* (noun) and *sempiternus* (adjective), and eternity by the Greek noun αἰών (*aiôn*) and adjective αἰώνιος (*aiônios*) and the Latin *aeternitas* (noun) and *aeternus* (adjective). Another Latin word, *aevum* (noun), also corresponds to αἰών, with both originally having the sense of "life" or "life span," but *aevum*, unlike *aeternitas*, is rarely if ever employed to pick out eternity in contradistinction to everlastingness. (The fixed sense intermediate between *sempiternitas* and *aeternitas* that *aevum* will acquire in medieval Latin philosophy is not yet found in ancient texts.) This terminological mapping is certainly correct, insofar as it is possible to find late antiquity authors employing these terms in this way, for example, Olympiodorus, *In Aristotelis Meteora Commentaria* 15–23, and Boethius, *De sancta trinitate* 4 (see chapter 1, note 1). Yet one should bear in mind that this distinction was in some sense a work in progress and that these terms cannot be assumed to have these meanings in earlier authors. When Cicero, for example, speaks of the whole *aeternitas* of past time (*De natura deorum* 2.13 [36]), he is obviously not using *aeternitas* in any timeless sense. In fact, these terms can be employed in ways that do not conform to the straightforward terminological mapping described above. Proclus, for example, sometimes follows this convention of using αἰών and ἀιδιότης for eternity and everlastingness, respectively (e.g., *Theologia Platonica* 3.55.11–14), but he is also comfortable speaking of two senses of ἀιδιότης: one eternal and one in time (*Elementatio theologica* 55). Likewise, Plotinus explicitly raises the question whether these terms are synonymous (*Enneads* 3.7.3.3), and although he decides that they are not, for Plotinus they refer to two different aspects of timeless eternity (*Enneads* 3.7.5.15–18).

Medieval Hebrew

In Biblical Hebrew, eternity is usually designated by the word עולם (*olam*). Thus, Genesis 13:15 in the King James Version: "For all the land that which thou seest, to thee I will give it, and to thy seed for ever" renders עד עולם as "for ever." Similarly, the phrase in Psalms 106:48 מן העולם ועד העולם is translated by the Vulgate *ab aeterno et usque in aeternum*. In later Mishnaic and medieval Hebrew *olam* came to denote *a world*. Yet medieval poets played with both the old and new meanings of the term in phrases such as אדון עולם, which means both "Lord of the World" and "Lord of Eternity." The eternal connotation of *olam* is still alive in modern Hebrew in phrases such as מעולם (never) or לעולם (for ever).

Medieval philosophical Hebrew uses both קדום (*qadum*) and קדמון (*qadmon*) for eternity *a parte ante*. נצח (*netzah*) and נצחי were used in medieval Hebrew for eternity *a parte post* or for sempiternity. Moses Narboni (1300–1362[?]), the Hebrew Averroist commentator on Maimonides's *Guide for the Perplexed*, employs נצח as a translation of the Arabic *dahr*. (See Warren Zev Harvey, "Albo's Discussion of Time," *Jewish Quarterly Review* 70 [1980]: 221.) In modern Hebrew נצח acquired the additional sense of atemporal existence and became the standard translation for the English "eternity" and its modern European equivalents.

Medieval Arabic

Arabic has an embarrassment of riches when it comes to expressing the notion of eternity. Particularly noteworthy are the following: *abad, azal, dahr, lam yazal, sarmad.*

Abad and *azal* (often frequently seen in the adjectival forms *abadī* and *azalī*) both tend to refer to temporal everlastingness, especially in the past up to the present moment. This meaning is found in al-Kindī, for instance, who in his *Book of Definitions* defines *al-azalī* as "what was never nonbeing and has no need for anything else in its subsistence." It is worth noting that *abadan* is a standard way to say "always," that is, "at all times."

Dahr is often closer to "everlasting" and is sometimes associated with the eternity of the universe. The term *dahrī*, "eternalist," was also used to describe a group (on whom see the contribution of Hans Hinrich Biesterfeldt in this volume) who take a materialist approach to explaining the universe.

Lam yazal is actually a verb with a negative modifier; it literally means "has not ceased." This term is therefore often used to refer to the notion of eternity *ex parte ante*.

Sarmad, like *azal*, can have the connotation of everlastingness (it is used in the Qur'ān with this sense at 28:71–72) or of timeless eternity and divine transcendence above time and change. Avicenna famously sets out the following distinction: "the relation of some unchanging [thābit] things to others, and the simultaneity that belongs to them from this perspective, is a notion above the everlasting [dahr]. It seems more worthy to be called eternity [sarmad]" (Avicenna, *Healing: Physics* 2.13.7, McGinnis and Reisman trans. [See chapter 2, note 33 in this book], p. , modified). Avicenna's idea that *dahr* refers to the relation between the unchanging and the changing was taken up by the later Safavid thinker Mīr Dāmād in his teaching on "perpetual creation" (*ḥudūth dahrī*).

Medieval Latin

The term *aevum* already appears in ancient Latin writers, such as Lucretius and Horace, as well as in Jerome's Vulgate. In medieval philosophy it acquires the more specific sense of the duration of a created yet incorruptible being, that is, an angel. *Nunc aevi* (the present of everlasting duration) has been distinguished from *nunc aeternitatis seu stans* (the present of eternity or the static present), and *nunc temporis seu fluens* (the present of time or the fleeting present). The Scholastics also distinguished between *aeternitas divina* (God's eternity) and *aeternitas participata* (the eternity participated in and shared by creatures).

Bibliography

LITERATURE BEFORE 1900

Abū Bakr Muḥammad ibn Zakariyyā' al-Rāzī. *Kitāb al-Shukūk ʿalā Jālīnūs*. Edited by Mehdi Mohaghegh. Tehran: Society for the Appreciation of Cultural Works and Dignitaries, 1993.

Abū Bakr Muḥammad ibn Zakariyyā' al-Rāzī. *Rasā'il falsafiyya (Opera philosophica)*. Edited by Paul Kraus. Cairo: Paul Barbey, 1939.

Adamson, Peter, and Peter E. Pormann. "More Than Heat and Light: Miskawayh's Epistle on Soul and Intellect." *Muslim World* 102 (2012): 478–524.

Adamson, Peter, and Peter E. Pormann, trans. *The Philosophical Works of al-Kindī*. Karachi: Oxford University Press, 2012.

Ammonius. *In Aristotelis De interpretatione commentarius*. Edited by A. Busse. In *Commentaria in Aristotelem Graeca*. Vol. 4, pt. 5. Berlin: Reimer, 1897.

Antiochus of Athens (ap. Rhetorus). "On the Seven Planets in an Epitome from Antiochus." In *Catalogus Codicum Astrologorum Graecorum VII*, edited by F. Boll. Brussels: Henri Lamertin, 1908, 127–128.

Antiochus of Athens (ap. Rhetorus). "Thesauroi." In *Catalogus Codicum Astrologorum Graecorum I*, edited by F. Boll and F. Cumont. Brussels: Henri Lamertin, 1898, 140–166.

Aristotle. *The Complete Works of Aristotle*. Edited by Jonthan Barnes. 2 vols. Princeton: Princeton University Press, 1984.

Atkinson, M. *Plotinus: Ennead V.1 On the Three Principal Hypostases*. Oxford: Oxford University Press, 1983.

Augustine. *Confessions*. Translated by H. Chadwick. Oxford: Oxford University Press, 2009.

Averroes. *Tahāfut al-Tahāfut*. Edited by Maurice Bouyges. Beirut: Dar el-Machreq, 1930, 6.

Avicenna. *The Metaphysics of the Healing*. Edited and translated by Michael E. Marmura Provo: Brigham Young University Press, 2005.

Badawī, ʿAbdurraḥmān, ed. *Dirāsāt wa-nuṣūṣ fī l-falsafa wa-l-ʿulūm ʿinda l-ʿArab*. Beirut: al-Muʾassasa al-ʿArabīya, 1981, 57–97.

Badawī, ʿAbdurraḥmān, ed. *Plotinus apud Arabes*. Cairo: Dirāsa Islamiyya, 1947.

Bardenhewer, Otto, ed. *Die pseudo-aristotelische Schrift Ueber das reine Gute*. Freiburg im Breisgau: Herder, 1882.

Baumgarten, Alexander. *Metaphysis*. Translated by Courtney D. Fugate and John Hymers. London: Bloomsbury, 2013.

Bayle, Pierre. *Dictionnaire Historique et Critique*. 5th ed. Amsterdam: P. Brunel, 1740.

Beierwaltes, W. *Plotin. Über Ewigkeit und Zeit (Enneade III 7)*. Frankfurt: Klostermann, 1967.

Berkeley, George. *Three Dialogues between Hylas and Pilonous*. Edited by Jonathan Dancy. Oxford: Oxford University Press, 1998.

Boethius. *Consolation of Philosophy*. Translated by J. C. Relihan. Indianapolis: Hackett, 2001.

Boethius. *De consolatione philosophiae. Opuscula theological*. Edited by C. Moreschini. Munich and Leipzig: K.G. Saur, 2005.

Boethius. *Theological Tractates: Consolation of Philosophy*. Translated by "I.T." (1609), rev. by H. F. Stewart. Loeb Classical Library. Cambridge, MA: Harvard University Press, 1968.

Bolzano, Bernard. *Theory of Science*. Translated by Jan Berg. Dordrecht: D. Reidel, 1973.

Bolzano, Bernard. *Wissenschaftslehre*. Sulzbach: Seidel, 1837.

Bradley, F. H. *Appearance and Reality* (1893). 9th ed. Oxford: Oxford University Press, 1930.

Bradley, F. H. *Collected Works of F. H. Bradley*. 12 vols. Bristol: Thoemmes Press, 1999–.

Bradley, F. H. *The Principles of Logic* (1883). 2nd ed. Oxford: Oxford University Press, 1922.

Clarke, Samuel. *A Demonstration of the Being and Attributes of God (and Other Works)*. Edited by Enzio Vailati. Cambridge: Cambridge University Press, 1998.
Clarke, Samuel. *The Works* (1738). 4 vols. New York: Garland, 1978.
Cohen de Herrera, Abraham. *Puerta del Cielo*. Edited by Kenneth Krabbenhoft. Madrid: Fundacion Universitaria Española, 1987.
Cohen de Herrera, Abraham. *Gate of Heaven*. Translated by Kenneth Krabbenhoft. Leiden: Brill, 2002.
Conway, Anne. *The Principles of the Most Ancient and Modern Philosophy*. Edited by Alison P. Coudert and Taylor Corse. Cambridge: Cambridge University Press, 1996.
Cornford, F. M. *Plato's Cosmology*. London: Routledge, 1937.
Curd, P., and R. D. McKirahan, Jr. *A Presocratics Reader: Selected Fragments and Testimonia*. Indianapolis: Hackett, 1996.
Dante Alighieri. *The Divine Comedy of Dante Alighieri*. Translated by Allen Mandelbaum. Berkeley: University of California Press, 1980–82.
Dante Alighieri. *La Commedia seconde l'antica vulgata*. Edited by Giorgio Petrocchi. 2nd edition. Milan: Mondadori, 1994
Descartes, René. *Descartes' Conversation with Burman*. Translated by John Cottingham. Oxford: Clarendon Press, 1976.
Descartes, René. *Oeuvres*. 11 vols. Ed. Charles Adam and Paul Tannery. New CNRS ed. Paris: J. Vrin, 1974–86.
Descartes, René. *The Philosophical Writings of Descartes*. 3 vols. Translated by John Cottingham, Robert Stoothof, and Dugald Murdoch. Cambridge: Cambridge University Press, 1985.
Endress, Gerhard. *Proclus arabus: Zwanzig Abschnitte aus der "Institutio theologica" in arabischer Übersetzung*. Beirut: Imprimerie Catholique, 1973.
Fakhr al-Dīn al-Rāzī. *al-Maṭālib al-'āliya*. Edited by A. Ḥ. al-Saqqā. 9 vols. Beirut: Dār al-Kitāb al-'Arabī, 1987.
Friedman, Dov Baer. *Maggid Devarav le-Ya'akov*. Edited by Rivka Schatz-Uffenheimer. Jerusalem: Magnes Press, 1976.
Friedman, Dov Baer. *Or Torah*. Brooklyn, 2011.
Galen. *De Placitis Hippocratis et Platonis* [On the opinions of Plato and Hippocrates]. Edited by Philip De Lacy. Berlin: Akademie, 1978–84.
Gassendi, Pierre. *Opera Omnia in sex tomos divisa*. 6 vols. Lyon: Laurent Anisson and Jean-Baptiste Devenet, 1658. Reprinted in facsimile and with an introduction by Tullio Gergory. Stuttgart: Friedrich Frohmann, 1964.
al-Ghazālī. *Freedom and Fulfillment: An Annotated Translation of al-Ghazālī': "al-Munqidh min al-Ḍalāl*." Translated by Richard Joseph McCarthy, S.J. Boston: Twayne, 1980.

al-Ghazālī. *The Incoherence of the Philosophers*. Edited and translated by Michael E. Marmura. Provo: Brigham Young University Press, 1997.

Goethe, Johann Wolfgang von, *Faust: Der Tragödie zweiter Teil* (1832). Project Gutenberg. Accessed November 23, 2015.

Hegel, Georg Wilhelm Friedrich. *Lectures on the Philosophy of World History*. Translated by Hugh Nesbit. Cambridge: Cambridge University Press, 1981.

Hegel, Georg Wilhelm Friedrich. *Phenomenology of Spirit*. Translated by A. V. Miller. Oxford: Clarendon Press, 1977.

Hegel, Georg Wilhelm Friedrich. *Philosophy of Nature*. Edited and translated by M. J. Petry. 3 vols. London: Allen and Unwin, 1970.

Hegel, Georg Wilhelm Friedrich. *Sämtliche Werke*. Edited by Hermann Glockner. Stuttgart: Friedrich Fromann, 1964.

Hegel, Georg Wilhelm Friedrich. *Science of Logic*. Translated by A. V. Miller. London: Allen and Unwin, 1969.

Heller, Meshulam Faivush. *Likkutim Yekarim*. Jerusalem, 1974.

Hobbes, Thomas. *The English Works of Thomas Hobbes of Malmesbury*. 11 vols. Edited by William Molesworth. London: J. Bohn, 1839–45.

Hobbes, Thomas. *Leviathan*. Edited by Edwin Curley. Indianapolis: Hackett, 1994.

Hoffmann, E. T. A. "Johannes Kreisler's Certificate of Apprenticeship." Translated by Max Knight. *19th-Century Music* 5 (1982).

al-Jāḥiẓ, ʿAmr b. Baḥr. *In Praise of Books*. Translated by James Montgomery. Edinburgh: Edinburgh University Press, 2013.

al-Jāḥiẓ, ʿAmr b. Baḥr. *K. al-Ḥayawān*. 7 vols. Edited by ʿAbdassalām Muḥammad Hārūn. Cairo: Maktabat Muṣṭafā al-Bābī al-Ḥalabī, 1938–45.

Kant, Immanuel. *Critique of the Power of Judgment*. Translated by Paul Guyer and Eric Matthews. Cambridge: Cambridge University Press, 2000.

Kant, Immanuel. *Critique of Practical Reason*. In *Immanuel Kant: Practical Philosophy*. Edited and Translated by Mary Gregor. Cambridge: Cambridge University Press, 1997, pp. 133–271.

Kant, Immanuel. *Critique of Pure Reason*. Edited and translated by Allan Wood and Paul Guyer. Cambridge: Cambridge University Press, 1999.

Kant, Immanuel. Kants *gesammelte Schriften*. Berlin: Königlich Preußische Akademie der Wissenschaften. 23 vols. 1910–.

Kant, Immanuel. *Groundwork for the Metaphysics of Morals*. 2nd ed. Translated by Mary Gregor and Jens Timmerman. Cambridge: Cambridge University Press, 2012.

Kant, Immanuel. *Religion within the Boundaries of Mere Reason*. Translated and edited by Allen Wood and George Di Giovanni. Cambridge, Cambridge University Press, 1998.

al-Khwārazmī. *Mafātīḥ al-ʿulūm*. Edited by Gerlof van Vloten. Leiden: Brill, 1895.

Kierkegaard, Søren. *The Concept of Anxiety*. Translated and edited by Reidar Thomte in collaboration with Albert B. Anderson. Princeton, NJ: Princeton University Press, 1980.

Kierkegaard, Søren. *Johannes Climacus: Philosophical Fragments*. Translated by Howard V. Hong and Edna H. Hong. Princeton, NJ: Princeton University Press, 1985.

Kierkegaard, Søren. *Kierkegaard's Concluding Unscientific Postscript*. Translated by David F. Swenson. Princeton, NJ: Princeton University Press, 1968.

Kierkegaard, Søren. *Søren Kierkegaards Samlede Vaerker*. 14 Volumes. Edited by A. B. Drachmann, J. L. Heiberg, and H. O. Lange. Copenhagen: Gyldendal, 1962.

Kirk, G. S., J. E. Raven, and M. Schofield. *The Presocratic Philosophers*. Cambridge: Cambridge University Press, 1983.

Koetschet, Pauline. "Galien, al-Rāzī, et l'éternité du monde. Les fragment du traité Sur la Démonstration IV dans les Doutes sur Galien," *Arabic Sciences and Philosophy* 25 (2015): 167–198.

Kumarila Bhatta. *Çlokavārtika*. Translated by Ganganath Jha. Calcutta: Asiatic Society, 1909.

Leibniz, Gottfried Wilhelm. *The Labyrinth of the Continuum: Writings on the Continuum Problem, 1672–1686*. Translated and edited by Richard T. W. Arthur. New Haven, CT: Yale University Press, 2001.

Leibniz, Gottfried Wilhelm. *The Leibniz-Clarke Correspondence*. Edited by H. G. Alexander. Manchester: Manchester University Press, 1970.

Leibniz, Gottfried Wilhelm. *Nachgelassene Schriften physikalischen, mechanischen und technischen Inhalts*. Edited by E. Gerland. Hildsheim: Olms 1995 [1906].

Leibniz, Gottfried Wilhelm. *New Essays on Human Knowledge*. Translated and edited by Peter Remnant and Jonathan Bennett. Cambridge: Cambridge University Press, 1981.

Leibniz, Gottfried Wilhelm. *Philosophical Essays*. Edited and translated by R. Ariew and D. Garber. Indianapolis: Hackett, 1989.

Leibniz, Gottfried Wilhelm. *Philosophical Papers and Letters*. 2nd ed. Edited and translated by Leroy E. Loemker. Dordrecht: Reidel, 1969.

Leibniz, Gottfried Wilhelm. *Die philosophischen Schriften*. Berlin: C. I. Gerhardt, 1875–90.

Leibniz, Gottfried Wilhelm. *Sämtliche Schriften und Briefe*. Berlin: Akademie-Verlag, 1923–.

Leibniz, Gottfried Wilhelm. *De summa rerum. Metaphysical Papers 1675–1676.* Edited and translated by G. H. R. Parkinson. New Haven, CT: Yale University Press, 1992.

Lessing, Gotthold Ephraim, *Anti-Goeze: Eine Duplik* (1778) in *Werke*, edited by H. Göpfert (München: Hanser Verlag, 1979), vol. 8: 32–33.

Levi Isaac of Berdichev. *Kedushat Levi*. Edited by M. Derbarmadiger. Monsey, NY: 1995.

Locke, John. *An Essay Concerning Human Understanding*. Edited by Peter H. Nidditch. Oxford: Clarendon Press, 1975.

Lotze, Hermann. *Lotze's System of Philosophy, Part I: Logic*. Translated by Bernard Bosanquet. Oxford: Clarendon Press, 1884.

Maimonides. *Dalālat al-Ḥā'irīn*. Edited by Solomon Munk and Issachar Joel Jerusalem: 1929.

Maimonides. *The Guide of the Perplexed*. Translated by Shlomo Pines. Chicago: University of Chicago Press, 1963.

Maimonides. *The Guide for the Perplexed*. Translated by Michael Friedländer. New York: Dover, 1956.

Malebranche, Nicolas. *Dialogues on Metaphysics and on Religion*. Translated by Nicholas Jolley and David Scott. Cambridge: Cambridge University Press, 1997.

Malebranche, Nicolas. *The Search after Truth*. Translated by Thomas M. Lennon and Paul J. Olscamp. Cambridge: Cambridge University Press, 1997.

McGinnis, Jon, and David C. Reisman. *Classical Arabic Philosophy: An Anthology of Sources*. Indianapolis: Hackett, 2007.

McGuire, J. E. "Newton on Place, Time and God: An Unpublished Source." *British Journal for the History of Science 11* (1978): 114–129.

McGuire, J. E., and S. K. Strange. "An Annotated Translation of Plotinus *Ennead* iii 7: On Eternity and Time." *Ancient Philosophy* 8 (1988): 251–271.

Mendelssohn, Moses. *Morning Hours: Lectures on God's Existence*. Translated by Daniel O. Dahlstorm and Corey Dyck. Dordrecht: Springer, 2011.

Michael Psellus. *Philosophica Minora*. Vol. 1. *Opuscula Logica, Physica, Allegorica, Alia*. Edited by J. M. Duffy. Leipzig: Teubner, 1992.

Nietzsche, Friedrich. *The Gay Science*. Translated by Walter Kaufmann. New York: Vintage, 1974.

Nietzsche, Friedrich. *Thus Spake Zarathustra*. Translated by Adrian del Caro. Cambridge: Cambridge University Press, 2006.

Nietzsche, Friedrich. *Will to Power*. Translated by Walter Kaufmann. New York: Vintage, 1967.

Numenius. *Fragments*. Edited by É. des Places. Paris: Les Belles Lettres, 1973.
Olympiodorus. *Eis ton Paulon <Heliodorou>. Heliodori, ut dicitur, in Paulum Alexandrinum Commentarium*. Edited by Emilie Boer. Leipzig: B. G. Teubner, 1962.
Olympiodorus. *In Aristotelis Meteora Commentaria*. Edited by G. Stüve. In *Commentaria in Aristotelem Graeca*. Vol. 12, pt. 2. Berlin: Reimer, 1900.
Orffyreus. *Perpetuum mobile triumphans*. Cassel: 1719.
Paulus Alexandrinus. *Elementa Apotelesmatica*. Edited by Emilie Boer. Leipzig: B. G. Teubner, 1958.
Paulus Alexandrinus and Olympicdorus. *Late Classical Astrology: Paulus Alexandrinus and Olympiodorus with the Scholia from Later Commentators*. Translated by Dorian Gieseler Greenbaum. Reston, VA: ARHAT, 2001.
Pellat, Charles, trans. *Le "Kitāb at-Tarbīʿ wa-t-tadwīr" de Ǧāḥiẓ*. Damascus: Institut Français de Damas, 1955
Philoponus, John. *De aeternitate mundi contra Proclum*. Edited by H. Rabe. Leipzig: Teubner, 1899.
Philoponus, John. *In Aristotelis De anima libros commentaria*. Edited by M. Hayduck. In *Commentaria in Aristotelem Graeca*. Vol. 15. Berlin: Reimer, 1897.
Philoponus, John. *Against Proclus's "On the Eternity of the World 1–5*. Translated by M. Share. Ithaca: Cornell University Press, 2005.
Plato. *Meno and Phaedo*. Edited by David Sedley and Alex Long. Cambridge: Cambridge University Press, 2010.
Plotinus. *Opera. Porphyrii vita Plotini. Enneades I-VI*. 3 vols. Edited by P. Henry and H.-R. Schwyzer. Oxford: Clarendon Press, 1964–1982.
Plotinus. *Enneads*. 7 vols. Edited and translated by A. H. Armstrong. Cambridge, MA: Harvard University Press, 1966–88.
Porphyry. A. Smith, ed. *Porphyrius Fragmenta*. Stuttgart: Teubner, 1993.
Porphyry. *To Gaurus on How Embryos Are Ensouled*. Translated by J. Wilberding. London: Duckworth, 2011.
Proclus. *Elements of Theology*. Edited with commentary by E. R. Dodds. 2nd ed. Oxford: Clarendon Press, 1963.
Proclus. *In Platonis Parmenidem commentaria*. 3 vols. Edited by C. Steel. Oxford: Clarendon Press, 2007–9.
Proclus. *In Platonis Timaeum commentaria*. 3 vols. Edited by E. Diehl. Leipzig: Teubner, 1903–6.
Proclus. *On the Eternity of the World*. Translated by Helen S. Lang and A. D. Macro. Berkeley: University of California Press, 2001.
Proclus. *Theologia Platonica (Théologie Platonicienne)*. 5 vols. Edited by H. D. Saffrey and L. G. Westerink. Paris: Les Belles Lettres, 2003.

Rhetorius. "'Selected Chapters.'" In *Catalogus Codicum Astrologorum Graecorum VIII, pars IV*, edited by F. Cumont. Brussels: Henri Lamertin, 1921, 115–224.

Richard of St. Victor. *De Trinitate*. Ed. G. Salet. Paris: Cerf, 1999.

Rowson, Everett K. *A Muslim Philosopher on the Soul and Its Fate: Al-ʿĀmirī's "Kitāb al-Amad ʿalā l-abad."* New Haven, CT: American Oriental Society, 1988.

Saʿādia Gaon. *The Book of Beliefs and Opinion*. Translated by Samuel Rosenblatt. New Haven, CT: Yale University Press, 1948.

Saʿādia Gaon. *Kitāb al-Amānāt wa 'l-Iʿtiqādāt*. Edited by Samuel Landauer. Leiden: Brill, 1880.

Schelling, F. W. J. *Ages of the World*. In Slavoj Žižek/F. W. J. Schelling *The Abyss of Freedom/Ages of the World*, pp. 105–182. Translated by Judith Norman. Ann Arbor: University of Michigan Press, 1997.

Schelling, F. W. J. *Die Weltalter: Fragmente in den Urauffassungen von 1811 und 1813*. Edited by M. Schröter. Munich: Biederstein, 1946.

Schelling, F. W. J. *Philosophical Investigations into the Essence of Human Freedom*. Translated by Jeff Love and Johannes Schmidt. Albany: State University of New York Press, 2006.

Schelling, F. W. J. *Schellings Werke*. Edited by M. Schröter. 12 vols. Munich: Beck, 1927–59.

Schlegel, Friedrich. *Friedrich Schlegel's "Lucinde" and the Fragments*. Translated and edited by P. Firchow. Minneapolis: University of Minnesota Press, 1970.

Schlegel, Friedrich. *Kritische Friedrich Schlegel Ausgabe*. Ed. E. Behler, J.-J. Anstett, and H. Eichner. Paderborn: Schöningh Verlag, 1958–.

Schmid, C.C.E. *Versuch einer Moralphilosophie*. Jena: Cröker, 1790.

Schopenhauer, Arthur. *Sämtliche Werke*. Edited by Arthur Hübscher. 4th ed. 7 vols. Wiesbaden: Brockhaus, 1988.

Schopenhauer, Arthur. *The World as Will and Representation. Volume 1*. Translated and Edited by Judith Norman, Alistair Welchman, and Christopher Janaway. Cambridge: Cambridge University Press, 2010.

Shapira, Tsevi Elimelekh. *Benei Yissakhar*. Benei Berak, Israel: 2005.

Siddur Koren. With introduction, translation, and commentary by Jonathan Sacks. Jerusalem: Koren, 2009.

Shne'ur Zalman of Liady. *Tefillot mi-Kol ha-Shanah*. Brooklyn, 2008.

Simplicius. *In Aristotelis Categorias commentarium*. Edited by K. Kalbfleisch. Vol. 8 in Commentaria in Aristotelem Graeca. Berlin: Reimer, 1907.

Simplicius. *In Aristotelis Physica*. Edited by H. Diels. Vol. 10 in Commentaria in Aristotelem Graeca. Berlin: Reimer, 1882.

Simplicius. *In Aristotelis Physica*. Edited by H. Diels. Vol. 11 in Commentaria in Aristotelem Graeca. Berlin: Reimer, 1895.

Spinoza, Benedict. *The Collected Works of Spinoza*. Edited and translated by Edwin Curley. Vol. 1. Princeton, NJ: Princeton University Press, 1985.
Spinoza, Benedict. *Opera*. Edited by Carl Gebhardt. 4 vols. Heidelberg: Carl Winter, 1925.
Spinoza, Benedict. *Theological Political Treatise*. Translated by Michael Silverthorne and Jonathan Israel. Cambridge: Cambridge University Press, 2007.
Spinoza, Benedict. *The Vatican Manuscript of Spinoza's "Ethics."* Edited by Leen Spruit and Pina Totaro. Leiden: Brill, 2011.
Suárez, Francisco. *Disputationes metaphysicae*. 2 vols. Reinheim: Olms, 1965.
Suárez, Francisco. *On the Essence of Finite Being as Such, On the Existence of that Essence and Their Distinction (Metaphysical Disputation XXXI)*. Translated by Norman J. Wells. Milwaukee: Marquette University Press, 1983.
Taormina, D. P. *Iamblique. Critique de Plotin et de Porphyre. Quatre Études*. Paris: J. Vrin, 1999.
Tarán, L. *Parmenides. A Text with Translation, Commentary and Critical Essays*. Princeton, NJ: Princeton University Press, 1965.
Tasso, Torquato. *Prose*. Milan: Riccardo Ricciardi, 1959.
Thomas Aquinas. *Commentary on the Book of Causes*. Washington, DC: Catholic University Press, 1996.
Thomas Aquinas. *Summa Contra Gentiles*. Translated by James F. Anderson. Notre Dame, IN: University of Notre Dame Press, 1975.
Twersky, Menahem Nahum. *Me'ir Einayyim*. Benei Berak, Israel 1997.
Twersky, Menahem Nahum. *Yismah Lev*. Benei Berak, Israel: 1997.
Vettius Valens, *Anthologiarum libri novem*. Edited by David Pingree. Leipzig: B. G. Teubner, 1986.
Wolff, Christian. *Mathematisches Lexicon*. Leipzig, 1716.

LITERATURE SINCE 1900

Adamson, Peter. *Al-Kindī*. New York: Oxford University Press, 2007.
Adamson, Peter. *The Arabic Plotinus: A Philosophical Study of the "Theology of Aristotle."* London: Duckworth, 2002.
Adamson, Peter. "Correcting Plotinus: Soul's Relationship to Body in Avicenna's Commentary on the Theology of Aristotle." In *Philosophy, Science and Exegesis in Greek, Arabic and Latin Commentaries*, edited by Peter Adamson, Han Baltussen, and M. W. F. Stone. London: Institute of Classical Studies, 2004, vol. 2, 59–75.
Adamson, Peter. "From the Necessary Existent to God." In *Interpreting Avicenna: Critical Essays*, edited by Peter Adamson. Cambridge: Cambridge University Press, 2013, 170–189.

Adamson, Peter. "Galen and Abū Bakr al-Rāzī on Time." In *Medieval Arabic Thought: Essays in Honour of Fritz Zimmermann*, edited by Rotraud Hansberger and Charles Burnett. London: Warburg Institute, 2012, 1–14.

Adamson, Peter. "Galen on Void." In *Philosophical Themes in Galen*, edited by Peter Adamson, Rotraud Hansberger, and James Wilberding. London: Institute of Classical Studies, 2014, 197–211.

Adamson, Peter. "Miskawayh's Psychology." In *Classical Arabic Philosophy: Sources and Reception*, edited by Peter Adamson. London: Warburg Institute, 2007, 39–54.

Aertsen, Jan A. "The Eternity of the World: The Believing and the Philosophical Thomas. Some Comments." In *The Eternity of the World in the Thought of Thomas Aquinas and His Contemporaries*, edited by Josef Wissink. Leiden: Brill, 1990, 9–19.

Agamben, Giorgio. *Homo Sacer: Sovereign Power and Bare Life*. Translated by Daniel Heller-Roazen. Stanford: Stanford University Press, 1998.

Allison, Henry. *Kant's Transcendental Idealism*. Rev. and enl. ed. New Haven, CT: Yale University Press, 2004.

Altmann, Alexander, and Samuel Miklos Stern. *Isaac Israeli: A Neoplatonic Philosopher of the Early Tenth Century*. Oxford: Oxford University Press, 1958.

Aquila, Henry. *Matter in Mind: A Study of Kant's Transcendental Deduction*. Bloomington: Indiana University Press, 1989.

Armstrong, A. H. "Eternity, Life and Movement in Plotinus' Account of Nous." In *Le Néoplatonisme*. Paris: CNRS, 1971, 67–74.

Badiou, Alain. *Being and Event*. Translated by Oliver Feltham. London: Continuum, 2006.

Badiou, Alain. *Deleuze: The Clamor of Be*ing. Translated by Louise Burchill. Minneapolis: University of Minnesota Press, 1999.

Badiou, Alain. *Logics of Worlds: Being and Event, 2*. Translated by Alberto Toscano. London: Continuum, 2009.

Badiou, Alain. *Saint Paul: The Foundation of Universalism*. Translated by Ray Brassier. Stanford: Stanford University Press, 2003.

Baltes, M. *Die Weltentstehung des platonischen Timaios nach antiken Interpreten*. Vol. 1. Leiden: Brill, 1976.

Baltes, M. *Γέγονεν* (Platon, Tim. 28b7). Ist die Welt real entstanden oder nicht?" In *ΔIANOHMATA Kleine Schriften zu Platon und zum Platonismus*, edited by M. Baltes. Stuttgart: Teubner, 1999, 301–325.

den Bok, N. *Communicating the Most High: A Systematic Study of Person and Trinity in the Theology of Richard of St. Victor*. Turnhout: Brepols, 1996.

Benveniste, E. "Expression indo-européenne de l'Éternité." *Bulletin de la société linguistique de Paris* 38 (1937): 103–112.

Bianchi, Luca. "Abiding Then: Eternity of God and Eternity of the World from Hobbes to the Encyclopédie." In *The Medieval Concept of Time: The Scholastic Debate and Its Reception in Early Modern Philosophy*, edited by Pasquale Porro. Leiden: Brill, 2001, 543–560.

Bonansea, Bernardino. "The Impossibility of Creation from Eternity According to St. Bonaventure." *Proceedings of the American Catholic Philosophical Association* 48 (1974): 121–135.

Bonansea, Bernardino. "The Question of an Eternal World in the Teaching of St. Bonaventure." *Franciscan Studies* 34 (1974): 7–33.

Borges, Jorge Luis. *Obras completas*. 4 vols. Buenos Aires: Emecé, 1996.

Bosenquet, Bernard. "Idealism and the Reality of Time." *Mind* 23 (1914): 91–95.

Bourne, Craig. *A Future for Presentism*. Oxford: Oxford University Press, 2006.

Bricker, Phillip. "The Fabric of Space: Intrinsic vs. Extrinsic Distance Relations." *Midwest Studies in Philosophy* 13 (1993): 271–294.

Broad, C. D. *Scientific Thought*. Paterson, NJ: Littlefield and Adams, 1959.

Burrell, David B. "God's Eternity." *Faith and Philosophy* 1 (1984): 389–406.

Callender, Craig. *The Oxford Handbook of the Philosophy of Time* Oxford: Oxford University Press, 2011.

Callender, Craig, and Nick Huggett. *Physics Meets Philosophy at the Planck Scale: Contemporary Theories in Quantum Gravity*. Cambridge: Cambridge University Press, 2001.

Carlisle, Clare. "Kierkegaard's Repetition: The Possibility of Motion." *British Journal for the History of Philosophy* 13 (2005): 521–541.

Cherniss, H. *Aristotle's Criticism of Plato and the Academy*. Vol. 1. Baltimore: Johns Hopkins University Press, 1944.

Cherniss, H. *The Riddle of the Early Academy*. Berkeley: University of California Press, 1945.

Chiaradonna, Riccardo. "Le traité de Galien Sur la démonstration et sa postérité tardo-antique." In *Physics and Philosophy of Nature in Greek Neoplatonism*, edited by Riccardo Chiaradonna and Franco Trabattoni. Leiden: Brill, 2009, 43–77.

Coulter, D. M. *Per Visibilia ad Invisibilia: Theological Method in Richard of St. Victor (d. 1173)*. Turnhout: Brepols, 2006.

Crone, Patricia. "Dahrīs." In *Encyclopedia of Islam*, 3. referenceworks.brillonline.com/entries/encyclopaedia-of-islam-3.

Crone, Patricia. "The Dahrīs According to al-Jāḥiẓ." *Mélanges de l'Université Saint-Joseph* 63 (2010–11): 63–82.

Crone, Patricia. "Ungodly Cosmologies." In *The Oxford Handbook of Islamic Theology*, edited by Sabine Schmidtke. Oxford Handbooks Online, March 2014.

Cross, Richard A. "Duns Scotus on Eternity and Timelessness." *Faith and Philosophy* 14 (1997): 3–25.

Cunning, David. "Descartes' Modal Metaphysics." In *The Stanford Encyclopedia of Philosophy*, spring 2014 ed., edited by Edward N. Zalta. http://plato.stanford.edu/archives/spr2014/entries/descartes-modal/.

Dales, Richard C. *Medieval Discussions of the Eternity of the World*. Leiden: Brill, 1990.

D'Ancona, Cristina. "La Teologia neoplatonica di 'Aristotele' e gli inizi della filosofia arabo-musulmana." In *Entre Orient et Occident: La philosophie et la science gréco-romaines*, edited by Ulrich Rudolph and Richard Goulet. Vandoeuvres, France: Fondation Hardt, 2011, 135–195.

Darwish, Mahmoud. *Fī ḥaḍrat al-ġiyāb*. Beirut: Riad El-Rayyes Books, 2006.

———. *In the Presence of Absence*. trans. Sinan Antoon. New York: Archipelago Books, 2011.

———. *Kazahr al-lawz aw ab'ad*. Beirut: Riad El-Rayyes Books, 2005.

———. *La ta'tathir 'amma fa'alt*. Beirut: Riad El-Rayyes Books, 2004.

———. *Ǧidārīyah*. Beirut: Riad El-Rayyes Books, 2001.

Davidson, Herbert A. "John Philoponus as a Source of Medieval Islamic and Jewish Proofs of Creation." *Journal of the American Oriental Society* 89 (1969): 357–391.

Davidson, Herbert A. "Maimonides' Secret Position on Creation." In *Studies in Medieval Jewish History and Literature*, edited by Isadore Twersky. Cambridge, MA: Harvard University Press, 1979, 16–40.

Davidson, Herbert A. *Proofs for Eternity, Creation and the Existence of God in Medieval Islamic and Jewish Philosophy*. New York: Oxford University Press, 1987.

Davies, Daniel. *Method and Metaphysics in Maimonides' "Guide for the Perplexed."* New York: Oxford University Press, 2011.

Davies, Brian. *The Thought of Thomas Aquinas*. Oxford: Oxford University Press, 1992.

de Blois, François C. "Zindīḳ." In *Encyclopaedia of Islam*. New ed., vol. 11, 510–513. Leiden: Brill, 2002.

de Callataÿ, Godefroid. *Annus Platonicus: A Study of World Cycles in Greek, Latin and Arabic Sources*. Paris: Peeters, 1996.

de Callataÿ, Godefroid. "Platón astrólogo: La teoría del Gran Año y sus primeras deformaciones." In *Homo Mathematicus, Actas del Congreso Internacional sobre Astrólogos Griegos y Romanos celebrado en Benalmádena*, edited by A. Pérez Jiménez and R. Caballero. Malaga: Charta Antiqua, 2002, 317–324.

de Callataÿ, Godefroid. "World Cycles and Geological Changes According to the Ikhwān al-Ṣafā." In *In the Age of al-Fārābī: Arabic Philosophy in the Fourth/Tenth Century. Proceedings of the Conference held at the Institute of Classical Studies and the Warburg Institute (London, 19–21 June 2006)*, edited by P. Adamson. Warburg Institute Colloquia, 12. London: Warburg Institute, 2008, 179–193.

Deleuze, Gilles. *Cinema 1: The Movement Image*. Translated by Hugh Tomlinson and Barbara Habberjam. Minneapolis: University of Minnesota Press, 1986.

Deleuze, Gilles. *Difference and Repetition*. Translated by Paul Patton. New York: Columbia University Press, 1994.

Deleuze, Gilles. *Nietzsche and Philosophy*. Translated by Hugh Tomlinson. London: Athlone Press, 1983.

Della Rocca, Michael. "Spinoza's Substance Monism." In *Spinoza. Metaphysical Themes*, edited by Olli Koistinen and John Biro. Oxford: Oxford University Press, 2003, 11–37.

Denzinger, H. *Compendium of Creeds, Definitions and Declarations on Matters of Faith and Morals*. San Francisco: Ignatius Press, 2012.

Dillon, J. *The Heirs of Plato*. Oxford: Oxford University Press, 2003.

Donagan, Alan. *Spinoza*. Chicago: University of Chicago Press, 1988.

Dorato, Mauro. "On Becoming, Cosmic Time, and Rotating Universes." In *Time, Reality and Experience*, edited by Craig Callender. Cambridge: Cambridge University Press, 2002, 253–276.

Dörrie, H. "Hypostasis: Wort und Bedeutungsgeschichte." *Nachrichten der Akademie der Wissenschaften in Göttingen* 3 (1955): 35–92.

Dummett, Michael. *The Interpretation of Frege's Philosophy*. Cambridge, MA: Harvard University Press 1981.

Dupré, Louis. "Of Time and Eternity in Kierkegaard's Concept of Anxiety." *Faith and Philosophy* 2 (1984): 160–176.

Dyer, L.-A. *Translating Eternity in the Twelfth-Century Renaissance*. Ph.D. diss., University of Notre-Dame, 2011.

Earman, John. *Bangs, Crunches, Whimpers, and Shrieks: Singularities and Acausalities in Relativistic Spacetime*. Oxford: Oxford University Press, 1995.

Efros, Israel. *Studies in Medieval Jewish Philosophy*. New York: Columbia University Press, 1974.

Eire, Carlos. *A Very Brief History of Eternity*. Princeton, NJ: Princeton University Press, 2010.

Elbogen, Ismar. *Jewish Liturgy: A Comprehensive History*. Translated by Raymond Scheindlin. Philadelphia: Jewish Publication Society, 1993.

Ellenberger, F. *Histoire de la géologie, I: Des anciens à la première moitié du XVIIe siècle*. Paris: Lavoisier, 1988.

Endress, Gerhard. "Ṭīna." In *Encyclopaedia of Islam*. New ed. Leiden: Brill, 2000, X 530.

Frank, Manfred. *The Philosophical Foundations of Early German Romanticism*. Translated by Elizabeth Millán-Zaibert. Albany: State University of New York Press, 2004.

Fränkel, H. *Wege und Formen frühgriechischen Denkens*. Munich: Beck, 1970.

Frankfurt, Harry. "Descartes on the Creation of the Eternal Truths." *Philosophical Review* 86 (1977): 36–57.

Frege, Gottlob. "The Thought: A Logical Inquiry." *Mind* 65 (1956): 289–311.
Freudenthal, Gideon. "Perpetuum mobile: The Leibniz-Papin Controversy." *Studies in History and Philosophy of Science* 33 (2002): 567–637.
Futch, Michael J. *Leibniz's Metaphysics of Time and Space*. Berlin: Springer, 2008.
Galperine, M.-C. "Le temps intégral selon Damascius." *Les études philosophiques* 3 (1980): 325–341.
Garrett, Don. "Spinoza's Necessitarianism." In *God and Nature: Spinoza's Metaphysics*, edited by Yirmiyahu Yovel. Leiden: Brill, 1991, 97–118.
Gay, Peter. *The Naked Heart: The Bourgeois Experience—Victoria to Freud*. New York: Norton, 1996.
Geach, Peter. *Truth, Love and Immortality: An Introduction to McTaggart's Philosophy*. Los Angeles: University of California Press, 1979.
Glasner, Ruth. *Averroes' Physics: A Turning Point in Natural Philosophy*. Oxford: Oxford University Press, 2009.
Godden, David, and Nicholas Griffin. "Psychologism and the Development of Russell's Account of Propositions." *History and Philosophy of Logic* 30 (2009): 171–186.
Gödel, Kurt. *Collected Works*. Vol. 2. *Publications 1938–1974*. Oxford: Oxford University Press, 1990.
Gödel, Kurt. "A Remark about the Relationship between Relativity Theory and Idealistic Philosophy." In *Albert Einstein: Philosopher-Scientist* (1949), edited by Paul Schlipp. La Salle, IL: Open Court, 1970.
Goldziher, Ignaz, and Amélie-Marie Goichon. "Dahriyya." In *Encyclopaedia of Islam*. New ed. Leiden: Brill, 1993, vol. 2, 95–97.
Gorham, Geoffrey. "Descartes on God's Relation to Time." *Religious Studies* 44 (2008): 413–431.
Gorham, Geoffrey. "Newton on God's Relation to Space and Time: The Cartesian Framework." *Archiv für Geschichte der Philosophie* 93 (2011): 281–320.
Green, Arthur. *The Language of Truth: Teachings from the Sefat Emet by Rabbi Judah Leib Alter of Ger*. Philadelphia: Jewish Publication Society, 1998.
Green, Arthur. *Speaking Torah: Spiritual Teachings from Around the Maggid's Table*. With Ebn Leader, Ariel Evan Mayse, and Or N. Rose. Woodstock, VT: Jewish Lights, 2013.
Greenbaum, Dorian Gieseler, *The Daimon in Hellenistic Astrology: Origins and Influence*. Ph.D. diss., Warburg Institute, University of London, 2009.
Greenbaum, Dorian Gieseler. *The Daimon in Hellenistic Astrology: Origins and Influence*. Leiden/Boston: Brill, 2015.
Grunbaum, Adolf. *Philosophical Problems of Space and Time*. New York: Knopf, 1963.
Gueroult, Martial. *Spinoza: Dieu (Ethique 1)*. Paris: Aubier, 1968.

Hallett, H. F. *Aeternitas: A Spinozistic Study*. Oxford: Clarendon Press, 1930.
Häring, N. M. "A Commentary on the Pseudo-Athanasian Creed by Gilbert of Poitiers." *Medieval Studies* 27 (1965): 23–53.
Harvey, Warren Zev. "Albo's Discussion of Time." *Jewish Quarterly Review* 70 (1980): 210–238.
Heidegger, Martin. *Being and Time*. Translated by John Macquarrie and Edward Robinson. New York: Harper and Row, 1962.
Heidegger, Martin. *The Concept of Time*. Oxford: Blackwell, 1992
Heidegger, Martin. *Martin Heidegger: Basic Writings*. Edited by David Farrell Krell. San Francisco Harper, 1993.
Heilen, Stephan. "'Some Metrical Fragments from Nechepsos and Petosiris."', In *La poésie astrologique dans l'Antiquité*, edited by I. Boehm and W. Hübner. Paris: De Boccard, 2011, 23–93.
Heilen, Stephan. 'Hadriani Genitura' – *Die astrologischen Fragmente des Antigonos von Nikaia*. Berlin/Boston: De Gruyter, 2015, 2 vols.
Heinemann, Joseph. "Prayers of Beth Midrash Origin." *Journal of Semitic Studies* 5 (1960): 264–280.
Hoffmann, P. "Jamblique exegete du pythagoricien Archytas: Trois originalités d'une doctrine du temps." *Les études philosophiques* 3 (1980): 307–323.
Hoffmann, P. "Paratasis: De la description aspectuelle des verbes grecs à une définition du temps dans le néoplatonisme tardif." *Revue des études grecques* 96 (1983): 1–26.
Horwitz, Rivka. "Mendelssohn's Commentary on the Tetragrammaton: The Eternal." *Jewish Studies* 37 (1997): 185–214.
Houlgate, Stephen. "Schelling's Critique of Hegel's Science of Logic." *Review of Metaphysics* 53 (1999): 99–128.
Hughes, Christopher. *A Complex Theory of a Simple God*. Ithaca, NY: Cornell University Press, 1989.
Husserl, Edmund. *Logical Investigations* Vols. 1 and 2. Translated by J. N. Findlay. London: Routledge, 2001.
Idel, Moshe. *Absorbing Perfections: Kabbalah and Interpretation*. New Haven, CT: Yale University Press, 2002.
Idel, Moshe. "Infinities of Torah in Kabbalah." In *Midrash and Literature*, edited by G. H. Hartman and S. Budick. New Haven, CT: Yale University Press, 1986, 141–157.
Jaspers, Karl K. *Nietzsche: An Introduction to the Understanding of His Philosophical Activity*. Translated by C. F. Wallraff and F. J. Schmitz. Tucson: University of Arizona Press, 1966.
Johansen, T. K. *Plato's Natural Philosophy*. Cambridge: Cambridge University Press, 2004.
Karofsky, Amy. "Suarez' Doctrine of Eternal Truth." *Journal of the History of Philosophy* 39 (2001): 23–47.

Kenny, Anthony. *The God of the Philosophers*. Oxford: Clarendon Press, 1979.
Kiefer, Claus. "Time in Quantum Gravity." In *The Oxford Handbook of the Philosophy of Time*, edited by Craig Callender. Oxford: Oxford University Press, 2011, 663–678.
Kirsanov, V. "Leibniz in Paris." In *The Global and the Local: The History of Science and the Cultural Integration of Europe*, edited by M. Kokowski. Krakow: 2008, 353–361.
Klatzkin, Jacob. *Theseaurus Philosophicus Linguae Hebraicae et Veteris et Recentioris*. 4 vols. Berlin: Eschkol Verlag, 1930.
Klemm, F. 1965. "Vom Perpetuum mobile zum Energieprinzip." In F. Klemm and H. Schimank, *Julius Robert Mayer zum 150. Geburtstag*. Munich: R. Oldenbourg, 5–23.
Klossowski, Pierre. *Nietzsche and the Vicious Circle*. Translated by Dan Smith. Chicago: University of Chicago Press, 1998.
Komorowska, Joanna, *Vettius Valens of Antioch: An Intellectual Monography*. Kraków: Ksiegarnia Akademicka, 2004.
Kosch, Michelle. *Freedom and Reason in Kant, Schelling and Kierkegaard*. Oxford: Clarendon Press, 2006.
Kovach, Francis J. "The Question of the Eternity of the World in St. Bonaventure and St. Thomas: A Critical Analysis." *Southwestern Journal of Philosophy* 5 (1974): 141–172.
Kneale, Martha. "Eternity and Sempiternity." In *Spinoza: A Collection of Critical Essays*, edited by Marjorie Green. New York: Anchor Books, 1973, 227–240.
Kretzmann, Norman. *The Metaphysics of Creation: Aquinas's Natural Theology in Summa Contra Gentiles II*. Oxford: Oxford University Press, 1999.
Kusch, Martin. *Psychologism: A Case Study in the Sociology of Philosophical Knowledge*. London: Routledge, 1995.
Lackeit, C. Aion. *Zeit und Ewigkeit in Sprache und Religion der Griechen*. Königsberg: Hartungsche Buchdruckerei, 1916.
Laerke, Mogens. *Leibniz lecteur de Spinoza: La genese d'une opposition complexe*. Paris: Honoré champion, 2008.
Lapointe, Sandra. *Bolzano's Theoretical Philosophy: An Introduction*. New York: Palgrave Macmillan, 2011.
Leftow, B. *Time and Eternity*. Ithaca, NY: Cornell University Press, 1991.
Le Poidevin, Robin. *Travels in Four Dimensions: The Enigmas of Space and Time*. Oxford: Oxford University Press, 2003
Lévi-Strauss, Claude. *Structural Anthropology*. Translated by Claire Jacobson and Brooke Grundfest Schoepf. New York: Anchor Books, 1967.
Lewis, David K. *On the Plurality of Worlds*. Oxford: Blackwell, 1986.

Lloyd, A. C. "The Principle That the Cause Is Greater Than Its Effect." *Phronesis* 21 (1976): 146–156.
Lolordo, Antonia. *Pierre Gassendi and the Birth of Early Modern Philosophy.* Cambridge: Cambridge University Press, 2007.
Lovejoy, Arthur. "The Obsolescence of the Eternal." *Philosophical Review* 18 (1909): 479–502.
Lucchetta, Giulio A. *La natura e la sfera: La scienza antica e le sue metafore nella critica di Rāzī.* Bari: Milella, 1987.
Markosian, Ned. "Simples." *Australasian Journal of Philosophy* 76 (1998): 213–226.
Mayer, Toby. "Avicenna's Burhān al-Siddiqīn." *Journal of Islamic Studies* 12 (2001): 18–39.
McDaniel, Kris. "Brutal Simples." *Oxford Studies in Metaphysics* 3 (2007): 233–266.
McDaniel, Kris. "John M. E. McTaggart." In *The Stanford Encyclopedia of Philosophy*, winter 2013, edited by Edward N. Zalta. http://plato.stanford.edu/archives/win2013/entries/mctaggart/.
McDowell, John. *Mind and World.* Cambridge, MA: Harvard University Press, 1996.
McTaggart, J. M. E. *The Nature of Existence.* Vol. 2. Cambridge: Cambridge University Press, 1927.
McTaggart, J. M. E. "The Unreality of Time." *Mind* 17 (1908): 456–473.
Meinong, Alexius. "The Theory of Objects." (1904). Translated by Isaac Levi. In *Realism and the Background of Phenomenology*, edited by Roderick Chisholm. Glencoe, IL: Free Press 1960.
Melamed, Yitzhak Y. "The Building Blocks of Spinoza's Metaphysics Substance, Attributes, and Modes." In *The Oxford Handbook of Spinoza*, edited by Michael Della Rocca. Oxford: Oxford University Press. Forthcoming.
Melamed, Yitzhak Y. "A Glimpse into Spinoza's Metaphysical Laboratory: The Development of Spinoza's Concepts of Substance and Attribute." In *The Young Spinoza*, edited by Yitzhak Y. Melamed. Oxford: Oxford University Press, 2015, 272–286.
Melamed, Yitzhak Y. "Spinoza's Deification of Existence." *Oxford Studies in Early Modern Philosophy* 6 (2012): 75–104.
Melamed, Yitzhak Y. *Spinoza's Metaphysics: Substance and Thought.* New York: Oxford University Press, 2013.
Melamed, Yitzhak Y. "What Is Time?" In *The Routledge Handbook of Eighteenth Century Philosophy*, edited by Aaron Garrett. London: Routledge, 2014, 232–244.
Mercer, Christia. *Exploring the Philosophy of Anne Conway.* Unpublished book manuscript.

Mercer, Christia. *Leibniz's Metaphysics: Its Origins and Development.* Cambridge: Cambridge University Press, 2001.
Meyer, Ulrich. "Time and Modality." In *The Oxford Handbook of the Philosophy of Time*, edited by Craig Callender. Oxford: Oxford University Press, 2011, 91–121.
Miller, James. *Measures of Wisdom: The Cosmic Dance in Classical and Christian Antiquity.* Toronto: University of Toronto Press, 1986.
Moore, G. E. "The Nature of Judgment." *Mind* 8 (1899): 176–193.
Moore, G. E. *Some Main Problems of Philosophy.* London: Humanities Press, 1953.
Morison, Benjamin. "Logic." In *The Cambridge Companion to Galen*, edited by R. J. Hankinson. Cambridge: Cambridge University Press, 2008, 66–115.
Newton-Smith, W. H. "The Beginning of Time." In *The Philosophy of Time*, edited by R. Le Poidevin and M. MacBeath, 168–182. Oxford: Oxford University Press, 1993.
Nizet, Jean. "La temporalité chez Soren Kierkegaard." *Revue philosophique de Louvain* 71 (1973): 225–246.
Norman, Judith. "The Work of Art in German Romanticism." In *Internationales Jahrbuch des Deutschen Idealismus* 6. Edited by Karl Ameriks and Jürgen Stolzenberg. 2009.
O'Brien, D. "L'être et l'éternité." In *Études sur Parménide*, vol. 2, edited by P. Aubenque. Paris: J. Vrin, 1987, 135–162.
O'Brien, D. "Temps et éternité dans la philosophie grecque." In *Mythes et representations du temps*, edited by D. Tiffenau. Paris: CNRS, 1985, 59–85.
O'Brien, D. "Temps et intemporalité chez Parménide." *Les études philosophiques* 3 (1980): 257–272.
O'Meara, D. J. *Plotinus: An Introduction to the "Enneads."* Oxford: Oxford University Press, 1993.
Onians, R. B. *The Origins of European Thought about the Body, the Mind, the Soul, the World, Time and Fate.* Cambridge: Cambridge University Press, 1951.
Owen, G. E. L. "Eleatic Questions." *Classical Quarterly* 10 (1960): 84–102.
Passmore, John. *A Hundred Years of Philosophy.* London: Duckworth, 1957.
Pessin, Sarah. "Saadia Gaon." In *The Stanford Encyclopedia of Philosophy*, http://plato.stanford.edu/entries/saadya.
Petit, A. "L'éternel retour, un paradox plotinien." In *Études sur Plotin*, edited by M. Fattal. Paris: Editions L'Harmattan, 2000, 75–86.
Pines, Shlomo. "The Philosophic Purport of Halachic Works and the Purport of the *Guide of the Perplexed*." In *Maimonides and Philosophy*, edited by Shlomo Pines and Yirmiyahu Yovel. Dordrecht: Kluwer, 1986.
Pingree, D. *The Thousands of Abū Ma'shar.* London: 1968.

Pippin, Robert. *Hegel's Idealism: Satisfactions of Self-Consciousness*. Cambridge: Cambridge University Press, 1989.
Plass, P. C. "Timeless Time in Neoplatonism." *Modern Schoolman* 55 (1977): 1–19.
Plath, Sylvia. *The Bell Jar*. London: Faber, 1963.
Prauss, Gerold. *Kant und das Problem der Dinge an sich*. Bonn: Bouvier, 1974.
Rashed, Marwan. "Abū Bakr al-Rāzī et la kalām." *Melanges de l'Institut Dominicain d'Etudes Orientales du Caire* 24 (2000): 39–54.
Rashed, Roshdi, and Marwan Rashed, eds. *Sciences and Philosophy in 9th Century Baghdad: Thābit Ibn Qurrā (826-901)*. Berlin: de Gruyter, 2009.
Richter-Bernburg, Lutz. "Abū Bakr al-Rāzī and al-Fārābī on Medicine and Authority." In *In the Age of al-Fārābī: Arabic Philosophy in the Fourth/Tenth Century*, edited by Peter Adamson. London: Warburg Institute, 2008, 119–30.
Ritter, Hellmut. *The Ocean of the Soul: Man, the World and God in the Stories of Farīd al-Dīn 'Aṭṭār*. Leiden: Brill, 2003. German original: *Das Meer der Seele*. Leiden: Brill, 1978.
Rudolph, Ulrich. *Al-Māturīdī and the Development of Sunnī Theology in Samarqand*. Leiden: Brill, 2015. German original: *Al-Māturīdī und die sunnitische Theologie in Samarkand*. Leiden: Brill, 1997.
Sambursky, S., and S. Pines. *The Concept of Time in Late Neoplatonism*. Jerusalem: Israel Academy of Sciences and Humanities, 1971.
Savitt, Steven. "The Replacement of Time." *Australasian Journal of Philosophy* 72 (1994): 463–474.
Schmid, C. C. E. *Versuch einer Moralphilosophie*. Jena: Cröker, 1790.
Schmitt, Carl. *Political Theology: Four Chapters on the Concept of Sovereignty*, translated by G. Schwab. Cambridge, MA: MIT Press, 1985.
Schmaltz, Tad M. *Radical Cartesianism: The French Reception of Descartes*. Cambridge: Cambridge University Press, 2002.
Schofield, M. "Did Parmenides Discover Eternity?" *Archiv für Geschichte der Philosophie* 52 (1970): 113–135.
Scholem, Gershom. "The Meaning of the Torah in Jewish Mysticism." In Scholem, *On the Kabbalah and Its Symbolism*, translated by Ralph Manheim. New York: Schocken Books, 1996, 32–86.
Scholem, Gershom. "Revelation and Tradition as Religious Categories in Judaism." In Scholem, *The Messianic Idea in Judaism and Other Essays on Jewish Spirituality*. New York: Schocken Books, 1971, 282–303.
Sedley, D. *Creationism and Its Critics in Antiquity*. Berkeley: University of California Press, 2007.
Seeskin, Kenneth. *Maimonides on the Origin of the World*. Cambridge: Cambridge University Press, 2005.

Sellars, Wilfrid. "Empiricism and the Philosophy of Mind." In *Science, Perception and Reality*. London: Routledge & Kegan Paul, 1963, 127–196.
Shaki, Mansour, and Daniel Gimaret. "Dahrī." In *Encyclopaedia Iranica*. Costa Mesa, CA: Mazda, 1993, vol. 6, 587–590.
Shanley, Brian J. "Eternity and Duration in Aquinas." *Thomist* 61 (1997): 525–548.
Sharples, R. W. *Cicero "On Fate" and Boethius "The Consolation of Philosophy" IV.5–7, V*. Warminster: Aris and Philipps, 1991.
Shyovitz, David. "'You Have Saved Me from the Judgment of Gehenna': The Origins of the Mourner's Kaddish in Medieval Ashkenaz." *AJS Review* 39 (2105): 49–73.
Sluga, Hans. *Gottlob Frege*. London: Routledge, 1980.
Smeenk, Chris, and Christian Wüthrich. "Time Travel and Time Machines." In *The Oxford Handbook of the Philosophy of Time*, edited by Craig Callender. Oxford: Oxford University Press, 2011, 577–631.
Smith, A. "Eternity and Time." In *The Cambridge Companion to Plotinus*, edited by L. Gerson. Cambridge: Cambridge University Press, 1996, 196–216.
Smith, Daniel. "On the Becoming of Concepts." In Smith, *Essays on Deleuze*. Edinburgh: Edinburgh University Press, 2012, 122–145.
Sorabji, Richard. *The Philosophy of the Commentators, 200–600 ad: A Sourcebook in Three Volumes*. London: Duckworth, 2004.
Sorabji, Richard. *Time, Creation and the Continuum*. London: Duckworth, 1983.
Steel, C. "The Neoplatonic Doctrine of Time and Eternity and Its Influence on Medieval Philosophy." In *The Medieval Concept of Time: Studies on the Scholastic Debate and Its Reception in Early Modern Philosophy*, edited by P. Pasquale. Leiden: Brill, 2001, 3–31.
Steinberg, Diane. "Spinoza's Theory of the Eternity of the Mind." *Canadian Journal of Philosophy* 11 (1981): 55–65.
Stone, Alison. *Petrified Intelligence: Nature in Hegel's Philosophy*. Albany: State University of New York Press, 2005.
Strange, S. K. "Plotinus on the Nature of Eternity and Time." In *Aristotle in Late Antiquity*, edited by L. Schrenk. Washington, DC: Catholic University of America Press, 1994, 22–53.
Stump, E., and N. Kretzmann. "Eternity." *Journal of Philosophy* 58 (1981): 429–458.
Stump, E., and N. Kretzmann. "Eternity." In *Routledge Encyclopedia of Philosophy*, edited by Edward Craig. London: Routledge, 1998, vol. 3, 422–427.
Ta-Shma, Israel. *Early Franco-German Ritual and Custom*. Jerusalem: Magnes Press, 1994.
Tarán, L. "Perpetual Duration and Atemporal Eternity in Parmenides and Plato." *Monist* 62 (1979): 43–53.

Tucker, Gordon. "Taking in the Torah of the Timeless Present." In *Jewish Mysticism and the Spiritual Life: Classical Texts, Contemporary Reflections*, edited by Lawrence Fine, Eitan P. Fishbane, and Or N. Rose. Woodstock, VT: Jewish Lights, 2011, 67–71.

Turner, C. H. "A Critical Text of the "Quicumque Vult"," *Journal of Theological Studies* 11, 43 (1910), 401–411.

Vailati, Enzio. *Leibniz and Clarke: A Study of Their Correspondence.* Oxford: Oxford University Press, 1997.

van Ess, Josef. *Der Eine und das Andere: Beobachtungen an islamischen häresiographischen Texten.* 2 vols. Berlin: de Gruyter, 2011.

van Ess, Josef. *Theologie und Gesellschaft im 2. und 3. Jahrhundert Hidschra: Eine Geschichte des religiösen Denkens im frühen Islam.* 6 vols. Berlin: de Gruyter, 1991–97.

Van Inwagen, Peter. "How to Think about Free Will." *Journal of Ethics* 12 (2008): 327–341.

van Veldhuijsen, Peter. "The Question on the Possibility of an Eternally Created World: Bonaventura and Thomas Aquinas." In *The Eternity of the World in the Thought of Thomas Aquinas and His Contemporaries*, edited by Jozef Wissink. Leiden: Brill, 1990, 20–38.

Vollert, Cyril, Lottie H. Kendzierski and Paul M. Byrne. *St. Thomas Aquinas, Siger of Brabant, St. Bonaventure: On the Eternity of the World.* Milwaukee: Marquette University Press, 1964.

Wagner, M. *The Enigmatic Reality of Time.* Leiden: Brill, 2008.

Wakelnig, Elvira. "The Other Arabic Version of Proclus' De aeternitate mundi: The Surviving First Eight Arguments." *Oriens* 40 (2012): 51–95.

Walker, Joel Thomas. "Against the Eternity of the Stars: Disputation and Christian Philosophy in Late Sasanian Mesopotamia." In *La Persia e Bisanzio*, edited by Gherardo Gnoli and Antonio Panaino, 518–535. Rome: Accademia Nazionale dei Lincei, 2001.

Wandinger, Nikolaus. "Der Begriff der 'Aeternitas' bei Thomas von Aquin." *Zeitschrift für Katholische Theologie* 116 (1994): 301–320.

Welchman, Alistair, and Judith Norman. "Creating the Past: Schelling's Ages of the World." *Journal of the Philosophy of History* 4 (2010): 23–43.

Westphal, Merold. "Kierkegaard and Hegel." In *The Cambridge Companion to Kierkegaard*, edited by Alastair Hannay and Gordon Daniel Marino. Cambridge: Cambridge University Press, 1997, 101–124.

Whittaker, J. "Ammonius on the Delphic E." *Classical Quarterly* 19 (1969): 185–192.

Whittaker, J. "The 'Eternity' of the Platonic Forms." *Phronesis* 13 (1968): 131–144.

Whittaker, J. *God, Time, Being: Studies in the Transcendental Tradition in Greek Philosophy.* Oslo: Universitetsforlaget, 1971.
Wieseltier, Leon. *Kaddish.* New York: Vintage, 1998.
Wiesheipl, James. "The Date and Context of Aquinas' De aeternitate mundi." In *Graceful Reason*, edited by Lloyd P. Gerson. Toronto: Pontifical Institute, 1983, 239–271.
Wilberding, James. "Intelligible Kinds and Natural Kinds in Plotinus." *Études Platoniciennes* 8 (2011): 53–73.
Wippel, John F. "Did Thomas Aquinas Defend the Possibility of an Eternally Created World?" *Journal of the History of Philosophy* 19 (1981), 21–37.
Wippel, John F. "Thomas Aquinas on God's Freedom to Create or Not." In Wippel, *Metaphysical Themes in Aquinas II.* Washington, DC: Catholic University Press, 2007), 218–239.
Wisnovsky, Robert. *Avicenna's Metaphysics in Context.* London: Duckworth, 2003.
Witten, Edward. "Reflections on the Fate of Spacetime." In *Physics Meets Philosophy at the Planck Scale: Contemporary Theories in Quantum Gravity*, edited by Craig Callender and Nick Huggett. Cambridge: Cambridge University Press, 2001, 125–137.
Wohlman, Avital. *Thomas d'Aquin et Maïmonide*: Un dialogue exemplaire. Paris: Cerf, 1988.
Wolfson, Elliot R. *Language, Eros, Being: Kabbalistic Hermeneutics and Poetic Imagination.* New York: Fordham University Press, 2005.
Wolfson, Harry Austryn. *The Philosophy of Spinoza.* 2 vols. New York: Schocken, 1969.
Yourgrau, Palle. *A World without Time: the Forgotten Legacy of Gödel and Einstein.* New York: Basic Books, 2005.
Zimmermann, Albert. "Mundus est aeternus? Zur Auslegung dieser These bei Bonaventura und Thomas von Aquin." In *Die Auseinandersetzungen an der Pariser Universität im XIII. Jahrhundert*, edited by Albert Zimmermann. Berlin: de Gruyter, 1976, 317–330.

Index

act, eternal, 4, 9, 11, 23, 41–43, 46, 49, 53, 79, 82, 91, 138–139, 172, 181, 190–191, 194, 200, 202, 204, 206, 208–210, 215, 219, 222–223, 225
Adamson, Peter, 68n5, 80n4, 82n11, 86n25, 90n35, 95n52, 98n61
aevum, 297, 300
Agamben, Giorgio, 9, 182, 219–220, 225
agency, 11, 82, 87, 101, 202–204, 223, 264
Akiva, Rabbi, 294
Ammonius, 17n11, 31n43, 97n58
angels, 75n1, 171–172, 232, 279, 300
Antiochus of Athens, 59
a priori, 101, 182–186, 188, 196, 206, 212, 222, 250–251
archê, 79, 95
Aristotle, 2–3, 5, 12, 21–31, 34, 42, 47, 52, 56, 63, 67, 76, 78–79, 82, 85–86, 88–95, 97, 109–110, 122, 206n65
  On the Heavens, 22, 110
  Metaphysics, 22, 47n102, 79, 91
  Physics, 91
  Topics, 110
aseity, 127
astrology, 56–63, 65, 67–68, 123
Augustine, 4, 45–48, 50–51, 65
Averroes. *See* Ibn Rushd
Avicenna. *See* Ibn Sīnā

Bach, Johann Sebastian, 283–286, 289
Badiou, Alain, 9, 179–180, 182, 220–221, 225
Baumgarten, Alexander, 12, 165
Bayle, Pierre, 142
Bergson, Henry, 9, 219
Berkeley, George, 134
Bible, 166, 229, 291
Boethius, 4, 14, 16, 51–54, 105, 107, 114–116, 130, 140–141, 156, 157n87, 168, 295, 297

Bolzano, Bernard, 246–247, 251–253
Bonaventure, 113
Borges, Jorge Luis, 2, 277–282
Bradley, F.H., 256–263, 266
Broad, C.D., 263–264

cause, 5, 21–25, 40, 53–54, 57, 77–83, 88, 94, 96–101, 105, 109, 111, 121, 133, 135, 147–149, 155, 157, 165, 174, 185, 188, 191
   efficient, 135, 144, 155, 166
   eternal, 21, 25, 78, 100, 107
   external, 7, 202
   free, 151
   intrinsic, 99
   uncaused, 97–99, 114
change, 16, 18, 22–28, 45, 49, 61, 68, 72, 82, 91, 99–101, 104, 110–111, 125–126, 139, 172, 185, 221, 234, 253–254, 260–264, 275–276, 299
changelessness, 19–20, 22, 40, 125, 221. 253
Cicero, 298
Clarke, Samuel, 131–132, 135n16, 165–166, 176
Cohen de Herrera, Abraham, 130, 133, 155n78
completeness, 19, 31, 39, 52, 188
contingency, 110, 112, 147, 205n64, 253, 271–272, 274–276
Conway, Anne, 140–141
creation, 4, 8, 24, 27, 37, 45–46, 49, 71, 77–80, 82, 99, 105–106, 118–119, 121, 135, 145, 169, 171–172, 181, 199–202, 204, 206–207, 224–225, 236–237, 279–300
   eternal, 231, 300

Dahriyya, 117–123
Daimon, 57–60, 62
damnation, eternal, 226
Dante Alighieri, 168–172
Darwish, Mahmoud, 239–244
Davies, Daniel, 109
death, 170, 180, 239–242, 244, 260–261, 285, 291, 293, 295
Deleuze, Gilles, 214n71, 218–220
Della Rocca, Michael, 151n62
Demiurge, 23–25, 27, 34, 76, 95
Descartes, René, 6, 130–134, 137–139, 142–148, 174
   Meditations, 145

Principles of Philosophy, 138, 142–143
determinism, 66, 191, 202–203
divisibility, 83, 133, 257
duration, 2, 4–6, 14, 20, 22–23, 27–31, 36, 38–48, 50, 52–53, 86, 93–4, 105, 110, 115, 127, 129–132, 137–142, 152–154, 156–162, 165–166, 168–169, 199, 212–213, 215, 257, 297, 300

emanation, 4, 111. *See also* procession and reversion.
ET-simultaneity, 255–256
existence, 1–2, 6–7, 11, 20, 23–24, 29, 34–36, 39–40, 45, 60, 65, 71–72, 74, 77, 79, 97–100, 103, 105–110, 121–122, 127, 12, 131–135, 138–139, 142–143, 149, 152–166, 172, 180, 184, 186, 188, 195, 205–206, 208, 214–215, 217, 221, 224, 231, 236, 243, 255, 261, 264, 266, 268, 271, 280, 292, 299
   necessary, 7, 97, 98–100, 133, 152, 154–155, 163, 186, 276
   self-necessitated, 7, 133, 149, 155–156, 159–161, 164–166
extension, 5, 18–20, 30, 32–33, 47–48, 53, 92, 106, 132, 139

faith, 7, 76–77, 216, 221, 287, 294
al-Fārābī, 95–96, 110, 122
First Cause, 80–83, 96, 100, 133
fortune, 57–62
freedom, 7–9, 53, 77–78, 104, 187, 189–191, 195, 199–201, 204–209, 222, 224–225
Frege, Gottlob, 246–247, 251, 253

Galen, 5, 88–91, 95, 102, 123
Garrett, Don, 155n77
al-Ghazālī, 6, 75, 77–78, 94, 98–99, 101–107, 111, 114, 122
Glass, Philip, 288–289
God, 6, 8, 22, 48–49, 73, 76–77, 103–104, 108–110, 119, 146–148, 150, 292–295
   eternity as, 32
   eternity of, 4, 5–6, 10–11, 45, 50–53, 78, 80, 82–84, 87, 93, 95, 99–100, 114–116, 124, 128, 131–134, 136, 139–141, 146, 160–163, 172, 204, 224, 231, 295

INDEX 325

existence of, 7, 77, 98, 129, 134, 138, 140, 149, 158, 165–166, 180, 231
freedom of, 78, 104, 106, 143–144, 147–148, 151, 204–206, 224
having essence involving existence, 138, 153
immensity, 139
incorporeality, 82, 109
knowledge, 53–54, 104–105, 134, 138
life of, 11, 15, 32, 132, 190
omnipotence, 76, 78, 114
omniscience, 4, 78
and time, 5–6, 14–15, 46, 48–49, 79–81, 105, 132, 235, 255
as above both time and eternity, 41, 50 80–82
Gödel, Kurt, 10, 249, 265–272
Goethe, Johann Wolfgang von, 226, 228
Great Year, 28, 31, 39–40, 60, 64–69

Hallett, H. F., 158n88, 163n103
Hasidism, 231–238
Hegel, Georg Wilhelm Friedrich, 179, 181, 195–201, 211–215, 223–224
 Phenomenology, 195–197
 Science of Logic, 197
Heidegger, Martin, 183, 197, 215, 239–242
Hobbes, Thomas, 129, 134–137
Husserl, Edmond, 179, 215, 246–247, 250–253

Ibn Rushd, 6, 96, 98n60, 103, 107n68
Ibn Sīnā, 5, 75n1, 88, 94–102, 104–107, 111, 114, 122, 155n77, 299–300
immortality, 29, 62–63, 103, 183
infinite/infinity, 85, 92, 120, 125, 133, 136, 138, 142, 228, 231–234, 289
 actual, 75n1, 103
 existence, 153n67
 mind, 147
 mode, 154, 158–161
 paradoxes of, 136–137
 potential, 103
 of space, 85, 135, 185
 of time/duration, 83, 94, 103, 105, 129–130, 132, 136–137, 140–142, 156, 159, 165, 198–199, 266

substance, 150–151
Intellect, 23, 33–37, 40–42, 50–51, 80–81, 83, 145, 150
 pre-, 35
 active, 110–111
Isaac ben Moshe of Vienna, 293–294

al-Jāḥiẓ, ʿAmr b. Baḥr, 118–123

Kabbalah, 130, 232n4
Kaddish, 290–295
kalam, 5, 97n59, 121, 123
Kant, Immanuel, 7–9, 12, 179–195, 198–201, 203–206, 209–212, 215, 219, 222–224, 230
 antinomies, 187, 198–199
 Critique of Practical Reason, 187, 192n34
 Critique of Pure Reason, 182, 186
 Groundwork, 192
 Religion within the Boundaries of Mere Reason, 191, 215n72
Kierkegaard, Søren, 8, 180–181, 195, 201, 209–219, 221–224
 Philosophical Fragments, 211
 Concept of Anxiety, 211–215, 216
al-Kindī, 5, 75, 77–87, 96–97, 99, 103, 105, 114, 299
Kneale, Martha, 153nn68–69, 155n76
Kretzmann, Norman, 43n88, 73n117, 113n87, 129, 255–256
Kumarila Bhatta, 71

Laws of Nature, 7, 143–149, 187–189, 267–269
Leibniz, Gottfried Wilhelm, 6–7, 132, 137n19, 157n87, 163n104, 164–165, 173–178
Lessing, Gotthold Ephraim, 229–230
Levi Isaac of Berdichev, Rabbi, 231–232, 237–238
Lévinas, Emmanuel, 9
'life' (without modification): 15–16, 25n28, 32–34, 36, 38–41, 47–48, 51, 52
 life, eternal, 22, 290
literature, perfect, 229
Locke, John, 6, 129, 132, 141
logic, 143–145, 247, 250–254, 264

Lot of Daimon, 59, 62
Lot of Fortune, 59, 62
Lotze, Hermann, 246, 251
Lovett, Lyle, 290

Maggid of Mezritch (Rabbi Dov Baer Fridman), 232–238
Mahler, Gustav, 283, 286–287, 289
Maimonides, Moses, 4, 6, 75–76, 107–112, 166–167
Malebranche, Nicolas, 6, 132, 134n14, 139–140
mathematics, 2, 143, 220, 272
McDaniel, Kris, 275n46
McTaggart, J. M. E., 10, 248–249, 256–267, 267
Meinong, Alexius, 247n3
Melamed, Yitzhak Y., 150n56, 155n78, 157n85, 158n90, 164n106, 166n118
Menahem Nahum of Chernobyl, Rabbi, 233–235
Mendelssohn, Moses, 2, 166
Mersenne, Marin, 144–145
Messiaen, Olivier, 283, 297–289
Mimamsakas, 71, 74
mind, 58–59, 134, 138, 235. *See also* Intellect
  eternity, 153, 162–163
  God's, 134, 147
  human, 134, 138, 143, 147, 154, 162–163
Miskawayh, 95–99
moment, 8, 52, 72, 83, 91–93, 102–103, 108–110, 133, 139, 181, 199, 206, 213–218, 264, 277, 279, 282, 285, 299
monism, 12–13
Moon, 27–28, 57–64
More, Henry, 6, 132
motion, 4–5, 22, 25, 27–28, 33, 56, 58–59, 65, 89, 91, 93, 97, 100, 105, 107, 111, 115, 130, 141, 171, 173–178, 269
Muʿtazila, 118

Narboni, Moses, 298
necessity, 35, 45, 77, 98–100, 133, 148, 150–152, 156, 159, 186, 188, 205, 208, 253, 272, 276
  and eternity, 11, 79, 82, 97–100, 112, 154–165, 249
  holy, 63
  not requiring a cause, 98–99
  self-, 7, 98–99, 133, 155–165
Neo-Platonism. *See* Platonism, late
Newton, Isaac, 6, 131–132, 176, 268
Nietzsche, Friedrich, 217–219, 225, 286
Novalis, 226–228
nunc-stans, 135–137

Olympiodorus, 14n2, 297
One, the, 33–36, 40, 43, 50
Oresme, Nicole, 69

Parmenides, 2–3, 16–21, 29, 32, 40, 49–50, 101n64
Paulus Alexandrinus, 62
perpetuum mobile, 173–178
Philo Judaeus, 15, 58
Philoponus, John, 4, 6, 24n22, 42n84, 46–49, 53, 77, 83–86, 103, 123, 136
physics, 2, 10, 66, 173–178, 248–249, 255, 265–275
Plato, 3–4, 21–34, 36–37, 39–41, 43–46, 48, 50, 52, 56, 58–60, 64–67, 72, 76, 79–80, 85, 95, 102, 109, 111, 115, 217, 227, 272, 290, 293
  Meno, 206, 211
  Parmenides, 29, 36
  Republic, 24n23, 65
  Symposium, 29, 227–228
  Timaeus, 4–5, 23–31, 36–37, 39–40, 48, 58, 60, 64–65, 76, 79, 85, 95, 140
  form-principles, 37–39, 50
  the Forms, 24–25, 28–29, 34–37, 39–42, 50–51, 72
Platonism, 4–6, 16–17, 24, 31–54, 58–60, 67, 72, 116, 245, 272, 293
  Late, 4–6, 16–17, 31–54, 116, 129n2
  Middle, 58
Plotinus, 4–5, 15–16, 23n21, 25, 31–51, 78–83, 95, 97, 116, 130, 298
Porphyry, 38n76, 41n83, 42n84
predetermination, 53
present, 11, 18–19, 33, 45, 50–53, 83, 100, 132–133, 135–136, 139, 141, 169, 181, 184, 198, 201, 206–207, 212–216, 220, 223–224, 259, 261–266, 275, 299–300

presentism, 259, 261, 264, 266
Prime Mover, 22–23, 28, 42, 76
procession and reversion, 34–36, 40–43. *See also* emanation
Proclus, 5, 14n2, 16–18, 39n79, 48n106, 51n114, 78–81, 83, 91n41, 99, 123, 298
providence, 59, 63, 119
purpose, 104, 210

Qurʾān, 78, 82, 84, 117, 295

al-Rāzī, Abū Bakr, 5, 75n1, 88–96, 101–102
relativity
　general, 10, 248, 265–267, 269–270, 273
　special, 10, 248–249, 266
return, eternal, 8, 66, 182, 217–219, 225, 286
Richard of St. Victor, 126–128
Romanticism, 226–230, 285

Saʿādia Gaon, 4, 6, 77, 84–83, 96, 99, 122
Schelling, Friedrich Wilhelm Joseph, 201–209
　Ages of the World, 201–202, 204–205, 207–208
　Freedom Essay, 8, 180–181, 201, 204, 207–212, 215, 222–224
Schlegel, Friedrich, 227–230
Schopenhauer, Arthur, 8, 180–181, 189n23, 209–211, 223, 285–286
sempiternity, 14, 153, 184, 184, 298. *See also* infinite time/duration
Simplicius, 17n12, 18n13, 42n84, 77
Socrates, 36–39, 49–50, 65, 206, 211–214, 227
Soul, 33–34, 37, 41, 44, 57–59, 63, 80–81, 83, 93–95, 101, 103, 141, 183, 236
　World-, 27–28
space, 7, 66, 85, 168, 170, 184–186, 198, 210, 232, 246, 249, 257–260, 265
　-time, 10, 171, 245, 248–249, 266–276
Spinoza, Benedict de, 6–7, 132–134, 137n19, 147–167
　Ethics, 134n12, 137n19, 147–148, 150–165
　Cogitata Metaphysica, 129, 152n64, 157–158, 160
Stoics, 65–66, 90, 96
Stump, Eleonore, 53n117, 129, 255–256

Suárez, Francisco, 6, 130r 6, 132, 144n40
Sun, 27–28, 31, 56–64
substance, 7, 26, 120–121, 125, 127, 149–152, 157, 159, 161–166, 184–186, 248, 278

tense, 11, 18–21, 28–29, 31, 105, 231–235, 264
　-logic, 264
Tetragrammaton, 2, 166, 231, 235, 292, 295
Theology of Aristotle, 75n1, 79, 83, 95
Theophrastus, 92
Thomas Aquinas, 6, 75–76, 80n6, 107–116, 144
time
　empty, 5, 93, 105
　as "moving image of eternity", 4, 15, 30
　passage of, 62, 135, 213, 266–267, 269, 272, 278
　reality/unreality of, 10, 248–249, 256–267, 270, 272–273, 276, 277–279
　travel, 267, 269n35
timelessness, 2, 4, 6, 16, 18–19, 27, 30, 105, 105, 114–116, 166, 168, 199, 219, 252, 295
Torah, 109, 232–234
Trinity, 83, 124–128
truths, eternal, 11, 74, 134, 142–149, 153
Tsevi Elimelech Shapira of Dinov, Rabbi, 236–237

Vettius Valens, 56–63
vis viva, 173–178

Wagner, Richard, 283, 285–286, 289
Welchman, Alistair, 204n61
Wilberding, James G., 37n73, 101n64, 168
Wippel, John, 110n78, 113–114
world
　corruption of, 89–98
　eternity of, 5–6, 76–79, 82–83, 85–86, 88–116, 120–122, 136–137, 173, 183, 198
　creation of, 5, 45, 77, 84, 87, 94, 98, 102–106, 108, 110, 119–123, 135, 139, 199–200, 231, 279. *See also* creation
　as *perpetuum mobile*, 176
　possible, 246, 253–254, 265–272, 274
Wolff, Christian, 12, 175